Lecture Notes in Mechanical Engineering

Series Editors

Fakher Chaari, National School of Engineers, University of Sfax, Sfax, Tunisia

Francesco Gherardini , Dipartimento di Ingegneria "Enzo Ferrari", Università di Modena e Reggio Emilia, Modena, Italy

Vitalii Ivanov, Department of Manufacturing Engineering, Machines and Tools, Sumy State University, Sumy, Ukraine

Mohamed Haddar, National School of Engineers of Sfax (ENIS), Sfax, Tunisia

Editorial Board

Francisco Cavas-Martínez , Departamento de Estructuras, Construcción y Expresión Gráfica Universidad Politécnica de Cartagena, Cartagena, Murcia, Spain

Francesca di Mare, Institute of Energy Technology, Ruhr-Universität Bochum, Bochum, Nordrhein-Westfalen, Germany

Young W. Kwon, Department of Manufacturing Engineering and Aerospace Engineering, Graduate School of Engineering and Applied Science, Monterey, CA, USA

Justyna Trojanowska, Poznan University of Technology, Poznan, Poland

Jinyang Xu, School of Mechanical Engineering, Shanghai Jiao Tong University, Shanghai, China

Lecture Notes in Mechanical Engineering (LNME) publishes the latest developments in Mechanical Engineering—quickly, informally and with high quality. Original research reported in proceedings and post-proceedings represents the core of LNME. Volumes published in LNME embrace all aspects, subfields and new challenges of mechanical engineering.

To submit a proposal or request further information, please contact the Springer Editor of your location:

Europe, USA, Africa: Leontina Di Cecco at Leontina.dicecco@springer.com
China: Ella Zhang at ella.zhang@springer.com
India: Priya Vyas at priya.vyas@springer.com
Rest of Asia, Australia, New Zealand: Swati Meherishi at swati.meherishi@springer.com

Topics in the series include:

- Engineering Design
- Machinery and Machine Elements
- Mechanical Structures and Stress Analysis
- Automotive Engineering
- Engine Technology
- Aerospace Technology and Astronautics
- Nanotechnology and Microengineering
- Control, Robotics, Mechatronics
- MEMS
- Theoretical and Applied Mechanics
- Dynamical Systems, Control
- Fluid Mechanics
- Engineering Thermodynamics, Heat and Mass Transfer
- Manufacturing
- Precision Engineering, Instrumentation, Measurement
- Materials Engineering
- Tribology and Surface Technology

Indexed by SCOPUS, EI Compendex, and INSPEC.

All books published in the series are evaluated by Web of Science for the Conference Proceedings Citation Index (CPCI).

To submit a proposal for a monograph, please check our Springer Tracts in Mechanical Engineering at https://link.springer.com/bookseries/11693.

Maurizio Barberio · Micaela Colella ·
Angelo Figliola · Alessandra Battisti
Editors

Architecture and Design for Industry 4.0

Theory and Practice

Volume 2

Springer

Editors
Maurizio Barberio
Dipartimento di Meccanica
Matematica e Management
Politecnico di Bari
Bari, Italy

Angelo Figliola
Dipartimento di Pianificazione
Design e Tecnologia dell'Architettura
Sapienza Università di Roma
Rome, Italy

Micaela Colella
Dipartimento di Architettura
Costruzione e Design
Politecnico di Bari
Bari, Italy

Alessandra Battisti
Dipartimento di Pianificazione
Design e Tecnologia dell'Architettura
Sapienza Università di Roma
Rome, Italy

ISSN 2195-4356 ISSN 2195-4364 (electronic)
Lecture Notes in Mechanical Engineering
ISBN 978-3-031-36921-6 ISBN 978-3-031-36922-3 (eBook)
https://doi.org/10.1007/978-3-031-36922-3

© The Editor(s) (if applicable) and The Author(s), under exclusive license to Springer Nature Switzerland AG 2024

This work is subject to copyright. All rights are solely and exclusively licensed by the Publisher, whether the whole or part of the material is concerned, specifically the rights of translation, reprinting, reuse of illustrations, recitation, broadcasting, reproduction on microfilms or in any other physical way, and transmission or information storage and retrieval, electronic adaptation, computer software, or by similar or dissimilar methodology now known or hereafter developed.
The use of general descriptive names, registered names, trademarks, service marks, etc. in this publication does not imply, even in the absence of a specific statement, that such names are exempt from the relevant protective laws and regulations and therefore free for general use.
The publisher, the authors, and the editors are safe to assume that the advice and information in this book are believed to be true and accurate at the date of publication. Neither the publisher nor the authors or the editors give a warranty, expressed or implied, with respect to the material contained herein or for any errors or omissions that may have been made. The publisher remains neutral with regard to jurisdictional claims in published maps and institutional affiliations.

This Springer imprint is published by the registered company Springer Nature Switzerland AG
The registered company address is: Gewerbestrasse 11, 6330 Cham, Switzerland

Preface

Ten years after the introduction of the concept of Industry 4.0 at the Hannover fair in 2011, the enabling technologies of the Fourth Industrial Revolution are gradually being implemented in various industrial sectors. Among these, the AEC sector has also begun to accept the challenges dictated by the 4.0 paradigm, continuing that path of hybridization with other disciplinary fields and industrial sectors (aerospace, naval, automotive, etc.) that characterized the first digital era. The 4.0 challenge leads industry operators to introduce a series of new paradigms that affect the entire supply chain, from first-level training courses to the creation of innovative companies, passing through the design and construction of digital and performative architectures.

The affirmation of computational design, the increasingly transversal and multi-disciplinary flow of knowledge, and the democratization of machines open new possibilities in a historical moment where the combination of technological innovation and design can play an essential role in environmental sustainability and reduced consumption of resources. The development of CAD/CAM and robotics digital manufacturing technologies has helped to reduce the gap produced by the increase in computational power in the generation of the form compared to the materialization of the same. Through this process, architecture regains its own tectonic identity, and the architect can regain a material sensitivity which risks being dissolved in virtual space. The democratization of digital manufacturing tools and the ubiquity of computational design require a new material dimension and a new figure capable of controlling the entire process. In the post-digital era, where the essence of design lies in the control and information of the process that holistically involves all the aspects mentioned above, rather than in formal research, it is necessary to understand technologies and analyze the advantages that they can bring in terms of environmental sustainability and product innovation.

This book intends to systematize from a theoretical and practical point of view, the best contributions, and the best experiences in the professional and entrepreneurial, academic, and research fields of architecture and design based on this new design paradigm. The main purpose of the proposed systematization is to create a widespread awareness necessary to initiate technology transfer processes involving the public sector, universities and research centres, and the private sector consisting of innovative companies. The issues addressed in the book are central to the development of a total 4.0 awareness for architects, engineers and designers, and digital entrepreneurs: advanced and computational digital design, virtualization of the project and production and construction processes, use of cyber-physical systems, advanced and customized prefabrication, additive manufacturing, automated manufacturing and construction, artificial intelligence, as well as the story of significant experiences of public and private self-entrepreneurship.

Bari, Italy	Maurizio Barberio
Bari, Italy	Micaela Colella
Rome, Italy	Angelo Figliola
Rome, Italy	Alessandra Battisti

Contents

Theory

The Big Vision: From Industry 4.0 to 5.0 for a New AEC Sector 3
Micaela Colella, Maurizio Barberio, and Angelo Figliola

Achieving SDGs in Industry 4.0. Between Performance-Oriented Digital Design and Circular Economy 19
Alessandra Battisti and Livia Calcagni

Industry 4.0 for AEC Sector: Impacts on Productivity and Sustainability 33
Ilaria Mancuso, Antonio Messeni Petruzzelli, and Umberto Panniello

Programming Design Environments to Foster Human-Machine Experiences 51
Giovanni Betti, Saqib Aziz, and Christoph Gengnagel

Designing with the Chain 67
Stefano Converso and Lorenzo Pirone

Adauctus Architectus Novus **on the Definition of a New Professional Figure** 89
Giuseppe Fallacara, Francesco Terlizzi, and Aurora Scattaglia

The Future of Architecture is Between Oxman and Terragni 101
Mario Coppola

Open-Source for a Sustainable Development of Architectural Design in the Fourth Industrial Revolution 113
Giuseppe Gallo and Giovanni Francesco Tuzzolino

Educating the Reflective Digital Practitioner 133
Ioanna Symeonidou

Teaching Digital Design and Fabrication to AEC's Artisans 151
Maurizio Barberio

The Corona Decade: The Transition to the Age of Hyper-Connectivity and the Fourth Industrial Revolution 169
Alexandros Kallegias, Ian Costabile, and Jessica C. Robins

Quasi-Decentralized Cyber-Physical Fabrication Systems—A Practical Overview 185
Ilija Vukorep and Anatolii Kotov

Latent Design Spaces: Interconnected Deep Learning Models for Expanding the Architectural Search Space 201
Daniel Bolojan, Shermeen Yousif, and Emmanouil Vermisso

From Technology to Strategy: Robotic Fabrication and Human Robot Collaboration for Increasing AEC Capacities 225
Dagmar Reinhardt and M. Hank Haeusler

Overview on Urban Climate and Microclimate Modeling Tools and Their Role to Achieve the Sustainable Development Goals 247
Matteo Trane, Matteo Giovanardi, Anja Pejovic, and Riccardo Pollo

Industry 4.0 and Bioregional Development. Opportunities for the Production of a Sustainable Built Environment 269
Luciana Mastrolonardo and Matteo Clementi

Towards Construction 4.0: Computational Circular Design and Additive Manufacturing for Architecture Through Robotic Fabrication with Sustainable Materials and Open-Source Tools 291
Philipp Eversmann and Andrea Rossi

RFId for Construction Sector. Technological Innovation in Circular Economy Perspective 315
Matteo Giovanardi

Digital Tools for Building with Challenging Resources 331
Christopher Robeller

Digital Deconstruction and Data-Driven Design from Post-Demolition Sites to Increase the Reliability of Reclaimed Materials 345
Matthew Gordon and Roberto Vargas Calvo

Impact and Challenges of Design and Sustainability in the Industry 4.0 Era: Co-Designing the Next Generation of Urban Beekeeping 359
Marina Ricci, Annalisa Di Roma, Alessandra Scarcelli, and Michele Fiorentino

Resolve Once—Output Many (ROOM): Digital Design and Fabrication at the Service of Social Equity 373
Blair Gardiner and Sofia Colabella

From Analogue to Digital: Evolution of Building Machines Towards Reforming Production and Customization of Housing 387
Carlo Carbone and Basem Eid Mohamed

Virtual, Augmented and Mixed Reality as Communication and Verification Tools in a Digitized Design and File-To-Factory Process for Temporary Housing in CFS 411
Monica Rossi-Schwarzenbeck and Giovangiuseppe Vannelli

Digital Processes for Wood Innovation Design 431
Fabio Bianconi, Marco Filippucci, and Giulia Pelliccia

Technologies

Visual Programming for Robot Control: Technology Transfer Between AEC and Industry 453
Johannes Braumann, Karl Singline, and Martin Schwab

Design, Robotic Fabrication and Augmented Construction of Low-Carbon Concrete Slabs Through Field-Based Reaction–Diffusion 471
Roberto Naboni, Alessandro Zomparelli, Anja Kunic, and Luca Breseghello

Digitally Designed Stone Sculpting for Robotic Fabrication 485
Shayani Fernando, Jose Luis García del Castillo y López, Matt Jezyk, and Michael Stradley

MycoCode: Development of an Extrudable Paste for 3D Printing Mycelium-Bound Composites 503
Fatima Ibrahim, Giorgio Castellano, Olga Beatrice Carcassi, and Ingrid Maria Paoletti

3D-Printing of Viscous Materials in Construction: New Design Paradigm, from Small Components to Entire Structures 521
Valentino Sangiorgio, Fabio Parisi, Angelo Vito Graziano, Giosmary Tina, and Nicola Parisi

A Study on Biochar-Cementitious Composites Toward Carbon–Neutral Architecture 539
Nikol Kirova and Areti Markopoulou

DigitalBamboo_Algorithmic Design with Bamboo and Other Vegetable Rods 579
Stefan Pollak and Rossella Siani

Virtual Reality Application for the 17th International Architecture Exhibition Organized by La Biennale di Venezia 593
Giuseppe Fallacara, Ilaria Cavaliere, and Dario Costantino

Towards a Digital Shift in Museum Visiting Experience. Drafting the Research Agenda Between Academic Research and Practice of Museum Management 609
Giuseppe Resta and Fabiana Dicuonzo

Practice

The Humanistic Basis of Digital Self-productions in Every-Day Architecture Practice 651
Marco Verde

Digital Twins: Accelerating Digital Transformation in the Real Estate Industry 673
Mattia Santi

The Right Algorithm for the Right Shape 699
Inês Caetano, António Leitão, and Francisco Bastos

Volatile Data: Strategies to Leverage Datasets into Design Applications 733
Edoardo Tibuzzi and Georgios Adamopoulos

Simulating Energy Renovation Towards Climate Neutrality. Digital Workflows and Tools for Life Cycle Assessment of Collective Housing in Portugal and Sweden 747
Rafael Campamà Pizarro, Adrian Krężlik, and Ricardo Bernardo

Configurator: A Platform for Multifamily Residential Design and Customisation 769
Henry David Louth, Cesar Fragachan, Vishu Bhooshan, and Shajay Bhooshan

From Debris to the Data Set (DEDA) *a Digital Application for the Upcycling of Waste Wood Material in Post Disaster Areas* 807
Roberto Ruggiero, Roberto Cognoli, and Pio Lorenzo Cocco

From DfMA to DfR: Exploring a Digital and Physical Technological Stack to Enable Digital Timber for SMEs 837
Alicia Nahmad Vazquez and Soroush Garivani

Spatial Curved Laminated Timber Structures 859
Vishu Bhooshan, Alicia Nahmad, Philip Singer, Taizhong Chen, Ling Mao, Henry David Louth, and Shajay Bhooshan

Unlocking Spaces for Everyone 887
Mattia Donato, Vincenzo Sessa, Steven Daniels, Paul Tarand, Mingzhe He, and Alessandro Margnelli

Lotus Aeroad—Pushing the Scale of Tensegrity Structures 925
Matthew Church and Stephen Melville

Data-Driven Performance-Based Generative Design and Digital Fabrication for Industry 4.0: Precedent Work, Current Progress, and Future Prospects 943
Ding Wen Bao and Xin Yan

Parameterization and Mechanical Behavior of Multi-block Columns 963
D. Foti, M. Diaferio, V. Vacca, M. F. Sabbà, and A. La Scala

Technologies

Visual Programming for Robot Control: Technology Transfer Between AEC and Industry

Johannes Braumann , Karl Singline , and Martin Schwab

Abstract For a long time, the construction sector has been considered a field with a low degree of digitization and automation with architects and designers looking for inspiration in other industries. Today, the construction sector is steadily innovating and automating, prompted by the lack of skilled labor. While robots are gradually starting to be used in situ for construction, robotic arms—also referred to as industrial robots—have already created new ways for the creative industries to develop innovative machinic processes at 1:1 scale. As the field of architecture eagerly moved towards robotics with an open mindset and little existing infrastructure or established protocols, architects and designers were quick to adapt the key themes of Industry 4.0 for their purposes. A core enabling factor has been the field's expertise in advanced, geometry-focused visual programming tools, which have since been adapted for robotic fabrication in order to enable individualized fabrication processes and mass customization. This chapter explores this development through several case studies and provides an outlook how visual programming and robotics may lead to a more sustainable, local, decentralized, and innovative post-industrial manufacturing in the creative industries and beyond.

Keywords Visual programming · Technology transfer · Creative industries · Robotic fabrication · Mass customization

United Nations' Sustainable Development Goals 9. Industry, Innovation and Infrastructure · 11. Sustainable Cities and Communities · 12. Responsible Consumption and Production

---

J. Braumann (✉)
Creative Robotics and Association for Robots in Architecture, 4020 Linz, Austria
e-mail: johannes@robotsinarchitecture.org

K. Singline · M. Schwab
Creative Robotics, University of Arts and Industrial Design Linz, 4020 Linz, Austria

© The Author(s), under exclusive license to Springer Nature Switzerland AG 2024
M. Barberio et al. (eds.), *Architecture and Design for Industry 4.0*, Lecture Notes in Mechanical Engineering, https://doi.org/10.1007/978-3-031-36922-3_26

1 Introduction

For a long time, the construction sector has been considered a field with a low degree of digitization and automation [1], with architects and designers looking for inspiration in other industries. Today, the construction sector is steadily innovating and automating, prompted by the lack of skilled labor. While robots are gradually starting to be used in situ for construction, robotic arms—also referred to as industrial robots—have already created new ways for the creative industries to develop innovative machinic processes at 1:1 scale. Startups from the field of architecture are developing new additive fabrication technologies [2] or radically rethinking existing processes like shotcrete [3] while established construction companies innovate their fabrication workflows (Fig. 1). This innovation can partly be attributed to universities, where many architecture and design students are today exposed to those technologies during their studies.

This leap in competence, from a few key research institutions to a much wider range of users, coincided with the definition of Industry 4.0 at Hannover Fair 2011, as marked by the first conference on Robotic Fabrication in Architecture, Art, and Design, ROB|ARCH a year later in 2012.

As the field of architecture eagerly moved towards robotics with an open mindset and little existing infrastructure or established protocols, architects and designers were quick to adapt the key themes of Industry 4.0 for their purposes.

Fig. 1 Large-scale robotic fabrication process defined in a visual programming environment at ZÜBLIN Timber, Germany

This chapter traces the connection between Industry 4.0 and robotic fabrication in architecture, with a specific focus on the role of visual, flow-based programming as a driver for algorithmic thinking, resulting in innovative robotic fabrication workflows that are today also increasingly adopted by other industries.

2 Industry 4.0

2.1 Individualization

While research into robotic fabrication in architecture has been ongoing since as early as the 1980s [4] the results of that research did not reach the larger community of the creative industry but was primarily rooted in academic research and engineering applications. Now in the fourth industrial revolution, innovative architectural applications using robotic arms are no longer just trailing the perfectly orchestrated mass-fabrication in industry but matching or even exceeding the state of the art of industry regarding individualization and mass-customization.

While the "smart factory" as the overlaying idea of Industry 4.0 does not directly apply to many architectural applications, it contains a range of sub-topics that strongly resonate with the creative industries. Custom manufacturing enables individualization and small lot sizes, while digital twins facilitate process simulation and rapid (design) iterations.

An enabling factor for that development can be found in the repurposing of digital tools originating from the creative industries for automation and robotics, which allow architects and designers to apply their deep knowledge and understanding of working with small lot sizes to the—for them—still new field of robotics, particularly robotic arms, or industrial robots.

The development and repurposing of tools has become important because while individualization is of course actively used in industry for a variety of applications [5], these developments are mostly built on custom, purpose-built software, rather than accessible programming environments. Though there are commercial software packages for a variety of tasks like milling, 3D printing, 3D scanning, etc., they only cover the most common, commercially relevant robotic tasks.

2.2 Digital Twin

Currently, a main selling point of industrial manufacturing software is the digital twin, which promises to speed up the development of new production processes as well as the quality of their output by enabling an accurate simulation of entire production lines.

When looking closer at these processes, it becomes apparent that while their scale and orchestration makes them highly complex, the individual actions making up those processes are often comparably simple, moving elements from A to B, performing spot welding, or dispensing glue along a curve.

The limitations of digital twins become apparent once non-standard processes are involved that go beyond the state of the art. Even industrially applied processes like fused filament fabrication still cannot be efficiently simulated due to the complex interactions of heat and material [6]. Thus, these fabrication processes rely on their software making viable assumptions about the process parameters, along with the process expertise of the machine operator that is often the results of years or decades of working with a given manufacturing method (Fig. 2).

2.3 Collaboration

However, when working with completely new processes, no viable assumptions or simulation frameworks yet exist. In industry, interdisciplinary teams solve these challenges in a collaborative approach with each discipline contributing their expertise, from geometry to material to robotics. This approach can also be seen in the early robotics projects in the field of architecture, where mathematicians collaborated with architects to realize complex brick stacking patterns [7].

Beyond large-scale industry, it often is not feasible for smaller enterprises, especially in the creative industries, to assemble interdisciplinary teams, instead having to rely on the local process experts with a deep understanding of a given material, but less programming and robotics expertise.

Geometry-focused, visual programming like McNeel Grasshopper and Autodesk Dynamo today provides a pathway for these user groups to apply their process knowledge to a robotic process through an accessible, responsive interfaces that fosters experimentation and by design facilitates iteration and individualization.

3 Visual Programming

Today, visual or flow-based programming is often associated with "low-code" [8] strategies that make complex processes accessible to non-experts, such as allowing designers to create complex, parametric geometries [9] or non-coders to automate processes and workflows [10]. However, visual programming is today also frequently used in industry, from complex factory automation to the definition of robotic processes.

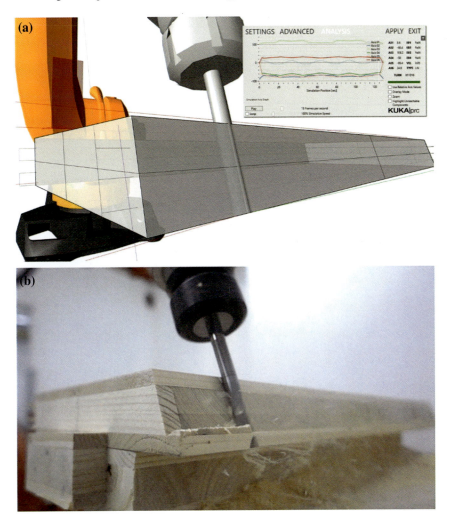

Fig. 2 Digital, idealized representation of a robotic milling process (Fig. 2a), actual complexity of timber as a cross-laminated, anisotropic material (Fig. 2b)

3.1 Visual Programming in Industry

The underlying programming paradigm of flow-based programming was first developed by Morrison [11] in the late 60 s for IBM, building upon the concept of co-routines that was sketched out even earlier [12].

The concept of flow-based programming is that the process of programming does not happen in a textual, but in a graphical way. This often results in a flowchart-like appearance, where modules, containing certain functionality or processes, are connected via lines or arrows, thus defining their parametric relationship (Fig. 3, left).

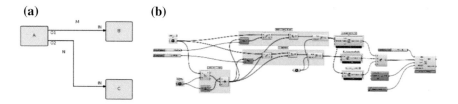

Fig. 3 Flow-based programming by Morrison dating back to the 60 s (Fig. 3a). CAD-oriented visual programming through McNeel Grasshopper (Fig. 3b)

Within the area of robotics, a basic visual programming language for PLCs is the Function Block Diagram (FBD) which was standardized as part of IEC 1131–3 [13] and allows the user to e.g. create individual functions via ST (Structured Text) and then make them more easily useable via the graphical representation as a FBD.

More recently there has been a large number of new, visual programming environments presented within the scope of industrial robots, such as drag&bot [14], Fox|Core by Faude, and the Desk software used to program Franka Emika robots, all of which are running within the browser. Within the greater area of robotic research there are also educational environments like Scratch [15] and Blockly, whose block-based visual programming has been expanded to include industrial robots [16–18].

These environments are highly flexible and can be used to program a wide variety of tasks, even allowing the easy integration of a wide array of external sensors and data sources. However, their primary use-cases can be found in the area of pick-and-place, bin picking and assembly, as they lack CAD integration and more complex geometric functionality that would be needed to e.g., derive toolpaths from an imported surface geometry.

3.2 *Visual Programming in Architecture, Art, and Design*

Geometry-focused approaches towards visual programming are instead often found in the creative industry. Users in the field of architecture and design utilize tools such as McNeel Grasshopper and Autodesk Dynamo to define parametric objects, artists create real-time visualization and digital art through VVVV and Max/MSP and entire video games are developed through Unreal Blueprints and Unity Visual Scripting. In the area of 3D-modelling, visual programming has been pioneered for the definition of photorealistic materials, defining how different textures are blended and linked via a node-based system, but has since been expanded to geometric operations, e.g. via Blender Animation Nodes, XPresso, Houdini and others.

A fundamental difference in these systems is whether they are real-time based or not. Programs like VVVV, but also FBDs in PLCs, are generally running in real-time, i.e., the graph is constantly being refreshed at a high rate, ideally at 60 Hz for live-visuals and potentially much more frequently for PLC-programs that might be triggered every millisecond or less, for 1000 Hz and more.

CAD-oriented programming environments like Grasshopper (Fig. 3, right) on the other hand are optimized for longer-running, more complex processes and by default not constantly updating: operations are represented as function blocks with inputs on the left and outputs on the right side, creating a directed, acyclic graph. Only when an input changes, the downstream components refresh automatically to reflect the change in input values, creating a highly reactive system that lends itself to a very intuitive interaction with data and geometry. Commonly, such function blocks include geometric operations like e.g., creating a point out of three numeric values representing its XYZ coordinates, but through plugins like KUKA|prc, HAL, RoboDK, and Robots, the range of function blocks can be expanded to include robotic fabrication, thus for example defining the robot's programmed position through a coordinate system.

3.3 Towards Robotics

Since 2007 Grasshopper has established itself as the core platform for applications utilizing industrial robots within the field of architecture and design. There are several advantages of using such a platform for robotic processes: Geometry and toolpath strategies can be generated at the same time, so that it is possible to intelligently have the geometry respect certain fabrication-related parameters, while the fabrication-related data can be automatically assigned to the relevant geometric features. This automatically translates both into the potential to be used for the automated batch-generation for mass-customization, as well as providing immediate feedback on the feasibility of the developed fabrication process, thus speeding up the learning process for new users. Furthermore, as the user is defining their own fabrication strategy, the potential scope of robotic fabrication is expanded beyond the functionality offered by CAM software such as milling, wire-erosion and plasma cutting.

3.4 Limitations

However, going beyond the state of the art also leads to an increased complexity for the user, as the development of such strategies requires both in-depth robotic knowledge, as well as material expertise and proficiency with the visual programming environment. Unlike many CAM environments, visual programming frameworks generally offer little official documentation and training, instead relying on third party and community-led initiatives.

A related general criticism of flow-based visual programming in Grasshopper by Davis et al. [19] is the lack of modularization to facilitate efficient code re-use as well the infrequent use of clearly named parameters.

Integrated development environments (IDE) for text-based programming like Microsoft Visual Studio support the user in creating a clear and easily readable syntax, often offering refactored code automatically. This process is based on consistent coding conventions for programming languages, which do not yet exist for visual programming.

Finally, especially within the scope of robotic fabrication, a limitation of CAD-oriented flow-based programming environments is that while they are very efficient in defining data flows, they are less optimized for process flows, i.e., conditional clauses, parallel processes [20]. This can be especially an issue for flexible fabrication processes that incorporate sensors and feedback loops (see Sect. 4.2).

4 Robotic Workflows and Dataflows

4.1 Defining a Robotic Process

A core appeal of using robotic arms lies in the abstraction of complexity. Rather than having to design and fabricate a bespoke, complex machine, robotic arms instead allow users to deploy an extremely reliable, well-tested and readily available manipulator, that then must be equipped with a suitable tool to perform a given task. The robot therefore forms the basis of a larger, robotic setup (Fig. 4). While they may not reach the level of performance of a purpose-built machine [21], their capabilities generally exceed the tolerances required at construction sites.

Their programming consists mostly of a sequence of movement commands and IO operations, structured by conditionals and logical expressions. A movement command defines where to move in relation to the current position of the robot,

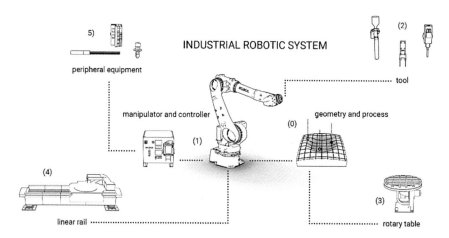

Fig. 4 Overview industrial robotic system—schematic layout for robot-based fabrication

the speed of the movement, and how to interpolate between the current and subsequent position. The robot's position can either be expressed in axis values, with a numerical value corresponding to each of its degrees of freedom, or in a Cartesian format, expressing a coordinate system through an XYZ coordinate and three Euler angles ABC in the case of KUKA robots. Common interpolation methods are linear interpolation, that moves in a straight line, circular interpolation that creates an arc through an auxiliary point, and point-to-point interpolation that interpolates at the axis level, creating a highly efficient trajectory.

4.2 Robotic Fabrication Through Visual Programming

In a visual programming environment, these movements can be represented by individual nodes. In most robotics-focused, accessible programming environments, the axis values or Cartesian position of the robot's tool is extracted from the current position of the—simulated or real—robotic arm. Therefore, the robot is moved, the position saved, and then the process repeated until the final program exists—similar to teaching by demonstration [22] done on a physical robot, but through the flow-based programming with an easier control over the structure of a program.

Within a geometry-focused visual programming environment like Grasshopper, the toolpath logic can be extracted from the underlying geometry: A NURBS curve is projected onto a free-formed surface and then segmented into a series of point objects. Based on the parametrization of the surface closest to each point, a coordinate system is defined. Along with a numerical value for the movement speed, each coordinate system results a linear movement, that is then traced by the robot, moving along the given surface while keeping the tool axis perpendicular to it.

To achieve a reliable simulation, the entire robotic setup must be known: As there are barely any standardized tools for industrial robots, calibration of each tool is required so that the robot is aware of the position and orientation of the tool center point in relation to the robot's flange. Similarly, local coordinate systems are defined in relation to the robot's world coordinate system—commonly at its base—to define the program origin and orientation. This modularity can also be well represented in a visual programming environment, coupling the robot node with different tools or external equipment to define the so-called robot cell. The typical setup for robot-based manufacturing shown in Fig. 4 consists of the product and associated geometry and process parameters (0), the industrial robot unit consisting of manipulator and controller (1), the end effector—with tools like gripper, extruder or spindle (2), a static or dynamic base for the workpiece—e.g. a rotary table for increased reachability (3), a static or dynamic robot foundation—e.g. a linear rail for expanding the working volume (4) and peripheral elements like industrial control systems, sensors, actuators and safety equipment (5).

4.3 Dataflows for Robotic Fabrication

Having both the accurate robotic setup and the parametric design within a single environment allows architects and designers to constrain the toolpath generation to parametric geometry, automatically creating an individual robot control data file for each design variation and thus fostering individualization.

Thus, a robotic process in a geometry-focused visual programming environment commonly consists of three parts (Fig. 5): The generation of the parametric geometry, the extraction of toolpaths, most commonly as sets of coordinate systems, and finally the robotic simulation and code generation. This sequence forms a directed graph where the user can easily interact with a complex, parametric system that covers both design and fabrication. Once a process is free of collisions and other problems, the resulting file can be copied to the robot, or streamed to the robot in real-time.

As geometry-focused visual programming environments create acyclic, directed graphs, they are best suited for bringing parametric designs to fabrication. However, fabrication processes that incorporate sensor feedback or user interaction are challenging to represent within such a system, as it becomes necessary to differentiate between data flow and process flow. This can be achieved implementing behavior models like state machines. Recent projects have implemented Unity Visual Scripting for that purpose as it differentiates between flow-graphs—creating similar graphs like Grasshopper—and state graphs—controlling the process flow from one state to another, e.g., informed by user interaction [20].

Due to the flexibility of the paradigm of visual programming, it has become the predominant way of programming applications that go beyond the state of the art in robotic fabrication in the creative industry, at academia and startups as well as within the established industry.

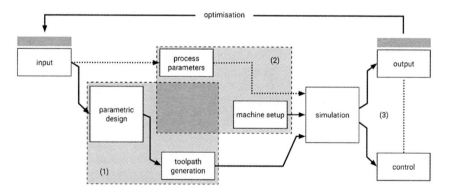

Fig. 5 Schematic workflow diagram—visual programming setup for robot-based fabrication: (1) parametric design and toolpath generation (2) digital representation of physical machine setup and process parameters (3) simulation, code generation and manual interpretation

5 Case Studies

This section provides an overview of innovative robotic applications within the field of architecture, where visual programming has become a core enabling factor for the realization of innovative processes. From large-scale, multinational construction companies to architectural offices, innovative construction startups and the wide dissemination at maker spaces.

5.1 Architecture and Design Office: Matter Make, Malta

Located on the island country of Malta, holistic design, and fabrication studio Matter Make identified the complexity of operating within the disconnected region regarding the lack of easy access to building materials (Fig. 6). This absence of access to a singular raw material led them to reconsider traditional architectural and interior design techniques and incorporate a workflow built upon a flexible visual programming as a core component of their working environment focused on complete processes based on the products available for import.

This approach prompted Matter Make to add value through means of design creativity that was supported by the novel workflow and flexibility of an industrial robotic arm which allowed a range of materials and processes. This meant projects could differ vastly between materials such as copper to plywood to high density foam without requiring additional software or equipment.

Within their visual programming environment, the same digital model is created and evolved from initial conception through to finalized design form can feature layered data of varying complexities which is used for a multitude of tasks such as quantifying a bill of materials, while simultaneously generating toolpaths required to produce their designs robotically.

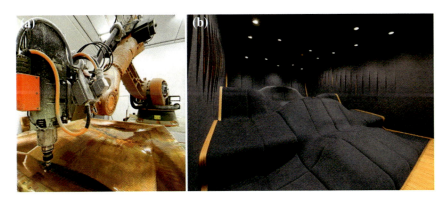

Fig. 6 Robotic incremental sheet forming of copper (Fig. 6a), free-formed home theatre by Matter Make (Fig. 6b)

Fig. 7 Large-scale robotically fabricated timber formwork for free-formed concrete columns

5.2 Large-Scale Construction Company: ZÜBLIN Timber, Germany

ZÜBLIN Timber is a leading timber construction company based in Germany that has worked on high-end projects such as the Metropol Parasol in Sevilla and Stuttgart 21 in Germany (Fig. 7). A pioneer in robotic fabrication in architecture, ZÜBLIN Timber set up their first robotic timber fabrication setup already in 1995 and has since upgraded it to be able to cover 160m^2 with a large robot and two linear axes.

ZÜBLIN Timber uses visual programming in-house for both general production preparation and the definition of robotic processes. This facilitates the integration of advanced algorithms coming from architects and fabrication consultation offices and greatly streamlines the workflow from design to production: Where parametric geometry is often reduced to plan drawings that are then manually processed at the fabricator, ZÜBLIN Timber can keep the entire process within a single environment.

5.3 Construction Startup: REPRECT, Austria

Austrian based start-up, Reprect, is a research and development group with a focus on innovating precast concrete through adaptation of novel technologies. In 2020 Reprect explored the challenges and advantages of that the introduction of an industrial six-axis robotic arm could be for digital fabrication across multiple aspects of precast concrete manufacturing techniques. This was completed over several steps: research into existing workflows for traditional precast concrete fabrication, identifying key areas where automation could play a significant role and frameworks that would be required to support such an integration.

A significant challenge was the countless variables introduced when handling custom building components, rarely were two precast elements identical. To ensure the workflow was feasible, it had to accommodate precast panels of various dimensions, penetrations, and modifications: horizontal and vertical drilling, surface milling and contouring. The visual programming environment Grasshopper was used for its ability to process traditional CAD data that was combined with domain-specific meta-data in the form of a JSON file. Based on that, toolpaths are calculated to provide visual feedback in the form of a full robotic simulation almost immediately. This research resulted in a software solution was designed to allow for the growth of projects without a requirement of significant retooling.

5.4 Education

Educational institutions take up a central role in the popularization of new technologies. Within the field of construction robotics, there are now dedicated Master programs at several universities such as RWTH Aachen, ICD Stuttgart and ETH Zurich where visual programming is used as a core tool to get students exposed to robotics.

A different approach is to specifically use robotics as a cross-sectional, interdisciplinary tool that can be applied in a wide variety of creative disciplines. At the research department Creative Robotics in Linz, Austria, students from a variety of programs such as architecture, industrial design, fashion, and interactive media take robotic courses where they are encouraged to apply robotic technologies within their own field of expertise.

Visual programming environments like Grasshopper and Unity Visual Scripting are therefore taught to students with widely varying degrees of experience in working with CAD software and digital software tools. Through that, robotics and visual programming put together become an interface that encourages interdisciplinary collaboration and experimentation, that has already resulted in several startups, such as Print-a-Drink for robotically fabricated cocktails and YOKAI Studios for innovative textile processes (Fig. 8, left). Research-led teaching also contributed to experimental projects like a mobile 3D-printing platform that was exhibited at the Ars Electronica Festival (Fig. 8, right).

That process is facilitated by the close collaboration of Creative Robotics with the Grand Garage, Europe's largest maker space, embedding academic research in a semi-public environment that encourages interdisciplinary thinking.

5.5 Case Study Summary

In summary, the collated projects described above demonstrate the discrete capabilities of incorporating visual programming and automation techniques already utilized in industry 4.0 within a contrasting range of architectural and related fields, from education to small bespoke design and large-scale fabrication for construction.

Fig. 8 Education resulting in student-led startups, like YOKAI Studios in the field of fashion and textiles (Fig. 8a), research-led teaching for the "Wandering Factory" at Ars Electronica Festival (Fig. 8b)

Although each case study is operating at a distinct scale and output, all actively demonstrate capabilities for non-standardized and mass customization construction methods facilitated by robotics that was previously deemed unavailable due to economic and efficacy constraints [23].

6 Outlook

The field of architecture has taken a pioneering role in the rapid adoption of robotic technology, questioning established standards and through that realizing advanced applications that have also caught the eyes of industry. While the ultimate goal of an automated construction site still lies in the future, current projects with industrial robots have moved beyond academic research, towards creating the foundation for innovative startups in the fields of architecture and design.

An enabling factor for this process has been the critical mass of users within architecture and design, often in academia, who have laid a foundation of knowledge for new users to build upon. Achieving this critical mass can at least partly be attributed to the development of domain-specific visual programming tools that allowed architects and designers to build upon their existing knowledge of parametric and generative design but expand it to cover also robotic fabrication. This can be seen as a testament to the efficiency of peer-to-peer teaching for knowledge transfer between different fields.

Other fields in the creative industries, such as fashion and textiles or crafts, are also starting to increasingly adopt digitization and concepts of Industry 4.0. However, their fundamental knowledge as well as their goals and the scale and valorization of their output differ significantly from architecture and design. It will therefore be necessary to adapt and modify the workflows and dataflows developed by architects and designs, towards again creating bespoke, domain-specific tools that have the potential of greatly accelerating the adoption and integration of new technologies.

The underlying technologies for simulation and control can be maintained, requiring mostly the interaction metaphors and concepts, including the degree of abstraction, to be adapted to new user groups. Building on the current approach of linking functional elements that often represent only basic geometrical methods, reduced workflows can be developed that group functionalities (Fig. 9) and automatically assign data to the correct inputs and outputs, guided by intelligent assistant systems. Having a clearly defined data structure opens possibilities for an accessible, automated optimization of robotic process parameters that can consider the individual degrees of freedom provided by each production process.

This reduction in complexity and scope enables new users to focus on the domain-specific input and associated process parameters, while providing them with a pathway to extend their level of control over a process by interacting with the underlying building blocks.

Ultimately, robotics combined with powerful and accessible tools for robot programming offers startups the potential to create highly innovative, disruptive applications, but also gives the much wider field of smaller architectural offices the possibility to take control of the fabrication process with the potential to realize high-end construction processes that strongly incorporate individualization, thus bridging the gap to much larger architecture and construction firms that employ specialized, interdisciplinary teams to fabricate their designs.

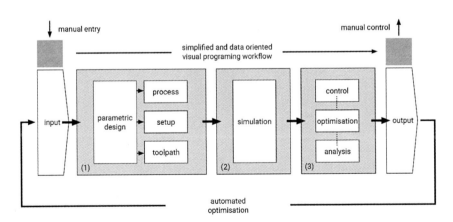

Fig. 9 Outlook for data-based workflow approaches: design-based geometry and process parameter generation (1) process related simulation (2) computer-aided analysis and automated optimization for design and process adjustment (3)

When looking at the impact of digital fabrication combined with visual programming on a higher level, the increasing accessibility of customization and multi-level-optimization leads to a potential reduction of energy consumption and use of resources whereas the low system costs enable and support a return to local and decentralized production.

By combining high flexibility with scalability, we expect individualized robotic fabrication informed by visual programming to support the creation of sustainable and innovative post-industrial manufacturing, in architecture and beyond.

References

1. IFR: World Robotics 2018—Industrial Robots. (2018)
2. Branch Technology. https://www.branch.technology. Last Accessed 11 June 2021
3. Aeditive—Revolutionäre Effizienz im Betonbau. https://www.aeditive.de/en/. Last Accessed 11 June 2021
4. Bock, T.: Innovationen im bauwesen: Roboter auf japanischen Baustellen. Bauingenieur. **63**, 121–124 (1988)
5. Da Silveira, G., Borenstein, D., Fogliatto, F.S.: Mass customization: Literature review and research directions. Int. J. Prod. Econ. **72**, 1–13 (2001). https://doi.org/10.1016/S0925-527 3(00)00079-7
6. Rashid, A.A., Koç, M.: Fused filament fabrication process: a review of numerical simulation techniques. Polymers (Basel). **13**, 3534 (2021). https://doi.org/10.3390/polym13203534
7. Bärtschi, R., Knauss, M., Bonwetsch, T., Gramazio, F., Kohler, M.: Wiggled brick bond. In: Ceccato, C., Hesselgren, L., Pauly, M., Pottmann, H., and Wallner, J. (eds.) Advances in architectural geometry 2010, pp. 137–147. Springer, Vienna (2010). https://doi.org/10.1007/978-3-7091-0309-8_10
8. Fryling, M.: Low code app development. J. Comput. Sci. Coll. **34**, 119 (2019)
9. Mamou-Mani, A.: Structural innovation through digital means: Wooden waves, galaxia, conifera, sandwaves, polibot, silkworm. http://mamou-mani.com. Last Accessed 27 April 2022
10. Brell-Cokcan, S., Braumann, J.: Industrial robots for design education: robots as open interfaces beyond fabrication. In: Zhang, J., Sun, C. (eds.) Global design and local materialization, pp. 109–117. Springer, Berlin, Heidelberg (2013). https://doi.org/10.1007/978-3-642-38974-0_10
11. Morrison, J.P.: Flow-based Programming: A new approach to application development. J.P. Morrison Enterprises, (2010)
12. Conway, M.E.: Design of a separable transition-diagram compiler. Commun. ACM. **6**, 396–408 (1963). https://doi.org/10.1145/366663.366704
13. Maslar, M.: PLC standard programming languages: IEC 1131–3. In: Conference Record of 1996 Annual pulp and paper industry technical conference, pp. 26–31. (1996). https://doi.org/10.1109/PAPCON.1996.535979.
14. Naumann, M., Wegener, K., Schraft, R.D., Lachello, L.: Robot cell integration by means of application-P'n'P. In: In: Proceedings of ISR 2006. (2006)
15. Resnick, M., Maloney, J., Monroy-Hernández, A., Rusk, N., Eastmond, E., Brennan, K., Millner, A., Rosenbaum, E., Silver, J., Silverman, B., Kafai, Y.: Scratch: programming for all. Commun. ACM. **52**, 60–67 (2009). https://doi.org/10.1145/1592761.1592779
16. Mateo, C., Brunete, A., Gambao, E., Hernando, M.: Hammer: An android based application for end-user industrial robot programming. Presented at the (2014). https://doi.org/10.1109/MESA.2014.6935597

17. Trower, J., Gray, J.: Blockly language creation and applications: Visual programming for media computation and bluetooth robotics control. In: Proceedings of the 46th ACM Technical symposium on computer science education, pp. 5–5. (2015)
18. Weintrop, D., Afzal, A., Salac, J., Francis, P., Li, B., Shepherd, D., Franklin, D.: Evaluating CoBlox: A comparative study of robotics programming environments for adult novices. Presented at the (2018). https://doi.org/10.1145/3170427.3186599
19. Davis, D., Burry, J., Burry, M.C.: Understanding visual scripts: Improving collaboration through modular programming. Int J Archit Comput. **9**, (2011). https://doi.org/10.1260/1478-0771.9.4.361
20. Braumann, J., Gollob, E., Bastan, A.: Towards AR for large-scale robotics. In: 2022 IEEE conference on virtual reality and 3d user interfaces. Christchurch, New Zealand (2022)
21. Perez, R., Gutierrez Rubert, S.C., Zotovic, R.: A study on robot arm machining: advance and future challenges. In: Katalinic, B. (ed.) DAAAM proceedings, pp. 0931–0940. DAAAM International Vienna (2018). https://doi.org/10.2507/29th.daaam.proceedings.134.
22. Biggs, G., MacDonald, B.: A survey of robot programming systems, vol. 10
23. Apolinarska, A.A., Knauss, M., Gramazio, F., Kohler, M.: The Sequential Roof. In: Advancing wood architecture. Routledge (2016)

Design, Robotic Fabrication and Augmented Construction of Low-Carbon Concrete Slabs Through Field-Based Reaction–Diffusion

Roberto Naboni, Alessandro Zomparelli, Anja Kunic, and Luca Breseghello

Abstract Constructions have a tremendous impact on global warming and are responsible for 39% of annual carbon emissions. Designers will increasingly focus on developing design methods and solutions that mitigate the impact of buildings over the next few years. Accordingly, this chapter focuses on developing an accessible computational design method to investigate the design, engineering and construction of ribbed concrete slabs with low levels of embodied carbon by minimising the use of structural materials, maximising bending resistance and surface area for recarbonation through convoluted geometry. We discuss using a Reaction–Diffusion system for performance-driven generative structural design, informed by the outputs of Finite Element Analysis in the form of scalar and vector fields. To streamline the production of geometrically complex slabs, a field-based robotic milling approach is introduced to process styrofoam concrete formworks. Mixed Reality is used to assist construction operations and realise non-standard rebar reinforcements. The results consist of proof-of-concept ribbed-slab prototypes characterised by structural efficiency, high resolution, and low-machining time.

Keywords Computational design · Robotic milling · Mixed reality · Reaction–diffusion · Field-based design · Finite element analysis · Concrete slabs · Low-carbon

United Nations' Sustainable Development Goals 12. Responsible consumption and production

---

R. Naboni (✉) · A. Zomparelli · A. Kunic · L. Breseghello
CREATE, University of Southern Denmark, Odense, Denmark
e-mail: ron@iti.sdu.dk

1 Introduction

The climate emergency has radically shifted our architecture, design and construction priorities. The mitigation of climate change is a fundamental planetary goal for the survival of our ecosystem [14]. As a reflection of this, the design focus for architects and engineers is moving towards reducing embodied carbon associated with the life cycle of building materials and structures. Carbon dioxide emissions (CO_2) from the manufacturing process of buildings, including material extraction, transportation, fabrication, installation, operations and end-of-life, account for 49% of the total carbon emissions from new construction [18]. Among all building materials, concrete is still necessary to fulfil the need for construction over the following decades, already being the most used material in the world after water [19]. This extensive use of concrete significantly impacts GHG, being responsible for about 8% of global emissions [7]. Replacing structural concrete with more sustainable materials is currently not a realistic and scalable option. Designing novel structures requires embodied carbon reductions through advanced design optimisation in this context. Contemporary Construction 4.0 technologies offer vast opportunities for achieving such a goal [13]. In particular through the combined use of (i) data-driven generative design; (ii) computationally-driven design to fabrication workflows that enable strategic use of building materials; (iii) smart digital twins that streamline robotic construction and Mixed Reality (MR and enhance the level of functional integration across phases. This chapter discusses integrating such features in the context of advanced structural geometry, characterised by high material efficiency and low embodied carbon.

2 Carbon-Driven Design for Horizontal Structures. Precedents and State-Of-Art

Horizontal structures, such as slabs, beams and roof elements that carry perpendicular loads to their longitudinal direction, constitute about 43% of the total use of structural concrete in buildings [3]. Minimising such structures' carbon impact is critical in achieving sustainable construction, as recent studies have documented [6, 9, 10]. At SDU CREATE, we conducted a study on the cradle-to-cradle life-cycle assessment of 3D concrete printed beams [8], where we formulated a design strategy for achieving carbon-efficient design: maximising bending resistance, minimising the use of material (reduction of cement and steel reinforcement), maximising the surface area of concrete for recarbonation [5]. This design rationale provides a framework for exploring geometrically convoluted ribbed slabs, where material and structural efficiency is synergistically tackled. Historically, the design of such structures has been studied in connection with the use of Principle Moment Lines (PML). In 1951, Pierluigi Nervi designed the wool-factory building *Lanificio Gatti*, where a pre-compressed concrete slab was reinforced using ribbed reinforcements along the direction of primary and secondary PML [1]. A similar approach was later used to design the zoology hall at the University of Freiburg, a large-span building that was efficiently constructed with this principle [2]. More recently, the integration of Finite Element Analysis (FEA) and parametric modelling environments has allowed more accessible access to such structural patterns. However, existing workflows do not support a straightforward translation of structural analysis into structural shapes. In a recent study, we demonstrated a method to generate lightweight 3D concrete printed beam elements using PML to inform the planning of a continuous printing path. Tan and Muller (2015) developed an approach to overcome common software's inconsistent generation of stress and moment lines. In this chapter, we focus on an alternative approach, where PML and other structural analyses inform the emergent shape generation of ribbed slabs, leading to an interactive exploration of efficient structural patterns.

3 Approach and Methods

One of SDU CREATE's areas of expertise is formulating computational workflows for carbon-efficient design. The opportunities for data-driven structural morphogenesis based on FE have been discussed in multiple studies from SDU CREATE [4, 11, 12]. To render such investigations more accessible, we recently conceived a field-based approach for automating the design and fabrication of concrete slab structures (Fig. 1).

This approach analyses structural, morphological and functional properties with a set of 2D plots. These are collected as multilayer performance maps, representing scalar and vector fields. We introduce a Reaction–Diffusion (RD) system to read,

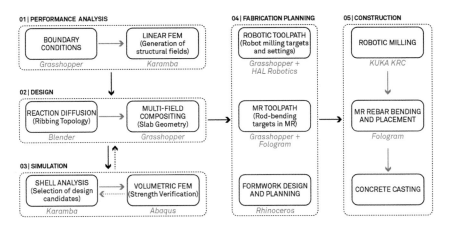

Fig. 1 Software pipeline for field-based design and fabrication of concrete ribbed-slabs

interpret and react to such fields and output the underlying patterns to generate three-dimensional ribbed slabs. This avoids inflexible parametric modelling operations, which limit the exploration of topologically diverse design options.

4 Simulation and Field-Based Design

4.1 Generation of Structural Fields

A computational design pipeline was developed to explore different loading and support scenarios for a 20 by 10 m rectangular slab. Starting from given input 2D shapes, the loading and support boundary conditions, and given material properties, a linear FEA was run using Karamba for Grasshopper [15]. The analysis includes primary and secondary PMLs, intensity, and sign information. Moment principals show the direction of the highest and lowest moment of inertia along the analysed surface. Through a custom exporter which translates mesh face properties into RGB colours, a series of 2D maps were stored for the following steps of the design workflow: (i) two maps of primary and secondary moment vectors directions, where the red (R and (G channels correspond to X and Y directions of the vectors; (ii) two grayscale scalar maps showing the absolute values of the moment intensities; (iii) two grayscale scalar maps outlining the sign of the principal moments, i.e. whether it is positive or negative (Fig. 2). The analyses were applied onto two different slab cases: (01) four-point supports placed at an offset from the perimeters; (02) two diagonal linear supports evenly placed along the two axes of the rectangular slab. Both elements are calculated with their self-load and a distributed load equivalent to 6 kN/m^2.

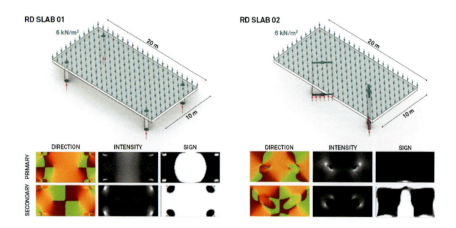

Fig. 2 Slabs' boundary conditions and field maps for the primary and secondary PML

5 Isostatics Structural Patterns with Reaction–Diffusion

In this phase, structural patterns for a ribbed slab are found with a Reaction–Diffusion (RD) system. In conventional structural design, topology and load paths for a ribbed slab are determined case by case using parametric modelling with relatively low flexibility. In our approach, an RD system was developed as a generative tool for quickly and interactively exploring isostatic structural patterns which adapt to the provided input fields. RD algorithms reproduce the chemical reaction of two or more substances spreading along a surface at a different rate [17]. When in contact, the two substances react by altering their concentration values. An anisotropic version of the Gray-Scott model was deployed to describe the variation over the time of the two substances according to parameters F and k, representing respectively the *feed* and *kill* values:

$$\frac{\partial A}{\partial t} = Diff_A \nabla^2 A - AB^2 + F(1 - A),$$

$$\frac{\partial B}{\partial t} = Diff_B \nabla^2 B - AB^2 + (F + k)B.$$

The variation of these parameters generates a highly varied range of patterns, including continuous, closed cellular, maze-like networks, and dotted configurations (Fig. 3). In this work, the growth of an RD is controlled using input fields from the preliminary structural analysis and general design to manufacturing considerations. In particular, the system needs to allow (1) the control of the directionality according to principal moments direction; (2) the variation of density according to moment intensity; (3) the dimensional control of individual rib elements according to construction and fabrication constraints. To do so, vector maps describing the principal moment curves and the scalar map of moment intensity were used to control the direction (laplacian) and diffusion rate of a pattern (F, k, and scale), respectively, obtaining anisotropic and variably dense patterns that were used as structural geometry (Fig. 4).

Using a Graphical User Interface (GUI), the students were to experiment and generate different patterns according to input images and tune the pattern scaling. The process was applied for primary and secondary principal moments to generate perpendicular ribs. Additionally, secondary design elements such as holes can be explored using additional grey-scale fields, controlling F and k, or used as an inlet or outlet for one of the two substances. With this exploration phase, recurrent patterns have been identified.

The anisotropic RD behaviour converges to some dominant configurations according to the parameters F and k. Independent longitudinal structural elements and a network of longitudinal structural elements (they could also be seen as one the negative of the other). For structural continuity, the network is considered a more convenient choice (Fig. 5).

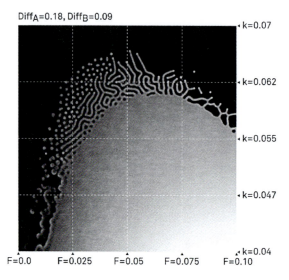

Fig. 3 Emerging RD patterns from variations of F and k parameters

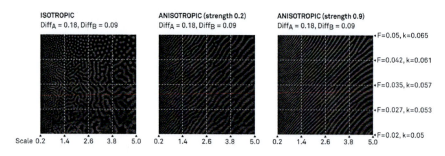

Fig. 4 Comparison between isotropic and anisotropic growth in the employed RD

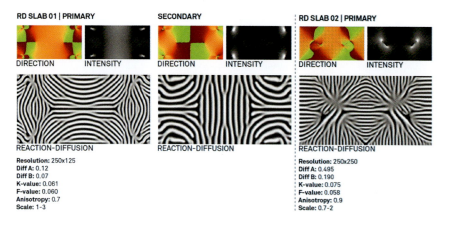

Fig. 5 Selected ribbing patterns for RD Slab 01 (two ways) and RD Slab 02

6 Multi-field Design Compositing

In this phase, the three-dimensional geometry of ribbed slabs was obtained by combining the previously collected information layers. Scalar 2D maps representing design elements, structural data, and RD patterns are combined into a single height map with a compositing method for generating the slab geometry. The RD pattern was used to create a series of ribbed elements in the slab's bottom surface to resist the tensional component induced by the bending moment. The ribs' cross-section shape was tuned mathematically through function curves according to the size of reinforcements and needed concrete cover. The ribs' sectional height (H) for the primary ribs was parameterised according to the moment intensity (I) to obtain a sufficient moment of inertia for any given point moment in the slab without the need for additional steel reinforcements. This is achieved with the following formulation H = RD*I*T. For the case of a two-way ribbed slab, the secondary ribs' sectional height was parametrised on a different layer. Afterwards, the information on the cross-section height for primary and secondary ribs was combined using the maximum value at every point of the slab: h = max(h_1, h_2). The map-based compositing for the supports provided a mathematical control for the transition among the various slab parts (Fig. 6).

7 Structural Feedback and Simulation

Volumetric FEA was run in Abaqus to verify the outcomes of the design optimisation process and compare them to non-optimised slabs with the same boundary conditions and weight. The analysis tested the slabs at their Ultimate Limit State (ULS) for a distributed load of 6 kN/m^2. The two slab prototypes proved a reduction of the weight-to-displacement ratio of 33 and 37.5%, respectively.

8 Robotic Fabrication and Augmented Construction

8.1 *Field-Based Robotic Milling of Complex Molding*

Robotic CNC milling was used to manufacture styrofoam moulds bypassing conventional CAD/CAM processors. Various milling strategies were developed parametrically and tested for 7-axis robotic milling to minimise the processing time. Two main strategies were investigated. In the first one, the heightmap obtained from the multi-layer compositing method was used to determine tool orientation, density/resolution of the machining steps and the nature of robot motion (e.g. continuous/punctual). On the other hand, the second strategy started from a mesh geometry generated from the compositing, which was further processed to achieve different surface finishing

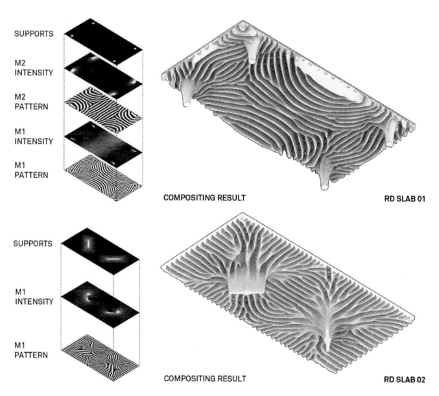

Fig. 6 The variable geometry of the slabs is controlled by the combined values of various maps

patterns. The tool was oriented vertically, while angle and milling resolution were the focus parameters in the contouring method. Initial toolpath explorations involved small-scale experiments performed on 20 by 20 cm specimens, which were evaluated in terms of production time, kinematics, and visual features. Five methods were studied: (1) *punch-milling* with variable tool inclination and punch depth informed by the composited maps (Fig. 7a); (2) *blading* through a path informed by the composited maps (Fig. 7b), and through the same path manipulated with sinusoidal function; (3) *parallel contouring* following 45 and 90 degrees angles with different milling resolution/overstep distance (3-8 mm) (Fig. 7c) and manipulated toolpath by application of sinusoidal functions along Z and XY plane orientations; (4) *geodesic offset contouring* to optimise the milling resolution in relation to the geometry and (5) *perpendicular contouring* to the structural ribs, performed as a continuous zig-zag toolpath. After these initial tests, the *geodesic offset contouring* and *perpendicular contouring* strategies were selected to fabricate the two slabs on a scale of 1:10. They have been considered a mid-way solution between the faster toolpaths, which compromised geometric complexity and formal precision such as blading, and highly detailed but time-consuming punch-milling tool-paths.

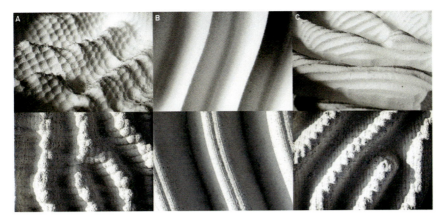

Fig. 7 Top—Robotically fabricated styrofoam formworks with different milling strategies; Bottom—Cast concrete panels

RD Slab 01—Geodesic offset contouring. A shell mesh geometry was used to generate geodesic contours with a 10 mm offset, originating from the support points and propagating towards the centre of the slab. The obtained three-dimensional curves are then translated into robotic target frames with a point distance of 5 mm (Fig. 8).

RD Slab 02—Contouring perpendicular to structural ribs. For the second slab, the medial axis of each structural rib was extracted from the Voronoi skeleton of the RD pattern. For each region, it is then created a continuous zig-zag pattern perpendicular to the medial axis following the ribs' slopes. The distance between the perpendicular milling paths was set to 10 mm, along which robot target frames were generated at a distance of 5 mm, keeping the same resolution settings as the first slab (Fig. 9).

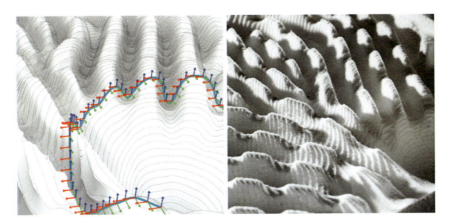

Fig. 8 Left: Robotic toolpath for geodesic offset contouring of RD Slab 01 and mesh visualisation of the milled EPS formwork; Right: Image of the milled EPS formwork

Fig. 9 Left: Robotic toolpath for the contouring perpendicular to structural ribs of RD Slab 02 and mesh visualisation of the milled EPS formwork; Right: Image of a milled EPS formwork

9 Mixed Reality Bending of Reinforcement Bars

Although the provision of reinforcement bars was not necessary according to the FEA, minimal longitudinal and transversal reinforcement was included to obtain ductile structural behaviour. This was achieved by placing (i) one 4 × 50x50mm BST-500 mesh and (ii) 6 mm B550 reinforcement bars. The reinforcement layout for the individual bars was derived from an analysis of tensional areas of the RD structural patterns. This resulted in a complex three-dimensional curve network. MR allowed the user to bend and check the individual reinforcement bars quickly, a crucial pre-assembly step to ensure that parts fit together as intended. Mixed Reality (MR) was adopted to efficiently achieve the accurate bending of 3D rebars, which was manually executed with the help of a manual rebar bending device (Fig. 10). Bending sequences were communicated to users wearing a Microsoft HoloLens 2, which was a guide for manually bending the bars into their positions.

10 Robotic Fabrication and Concrete Casting

The used robotic setup is a KUKA KR240 R3330 industrial robot installed on a six-metre long linear unit equipped with a 12 kW rotary spindle holding a foam rasp cutter of 20 mm diameter and 300 mm length (Fig. 11). The formwork consisted of two glued 1200 × 1200 × 275 mm EPS 150 blocks. It was laterally restrained to a steel welding table. Both formworks were fabricated with a spindle speed of 4500 RPM and a feed rate of 450 mm/s, resulting in an average fabrication time of around four hours. The milled formworks were lightly sanded to remove residual burrs and to prepare the surfaces for applying protective coatings. A single layer

Fig. 10 Three-dimensional rebar bending in MR

of a water-based paint primer was sprayed onto the milled surface. This provided a smooth surface for applying a wax release agent and waterproofing the formwork for the casting process.

Once the surface dried, the formwork was prepared for casting by installing the bent rebars (Fig. 12), steel mesh and other 3D-printed connections for transportation. Both slabs were cast with Hi-Con UHPC having a characteristic compressive strength (F_{ck}) of 127 MPa. The material was prepared in 300-L batches and mixed for six minutes before being poured into the formwork. The self-compacting properties of the concrete ensured that the intricate details of mould were covered in concrete without vibration. The prototypes were left to air-cure for approximately 24 h before being de-moulded.

Fig. 11 Robotic milling of the styrofoam formworks for RD Slab 01

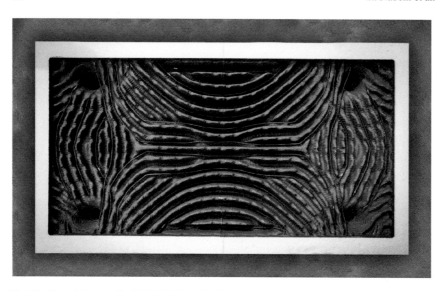

Fig. 12 Coated formwork of RD Slab 01 with 3D bent rebars in MR

11 Results and Discussion

Reducing embodied carbon in reinforced concrete structures is a priority for architectural designers and engineers nowadays. Radical design approaches are required to achieve sustainable targets at a global scale. Construction 4.0 principles and technologies are supporting this paradigm shift by: (i) enabling the automation of complex design operations; (ii) promoting data-driven design with early-stage structural analysis to embed fundamental engineering notions at the concept level; (iii) introducing generative design methods to effectively explore efficient design solutions; (iiii) computational design and fabrication tools becoming more accessible to architects, structural engineers, integrated designers, construction engineers.

The chapter discusses these topics, focusing on experimental activities conducted for designing and engineering ribbed concrete slabs. Several innovative features were presented. Anisotropic RD was employed for the first time to explore structural patterns interactively. Scalar and vector fields in the form of image maps were used to exchange performance data across simulation and modelling software in a synthetic format and drive structural and milling explorations. A software pipeline that connects structural analysis, generative design, robotic fabrication, and MR construction was presented. The pipeline was developed and tested during the SDU CREATE´s summer school in Experimental Architecture (2022) with students from architecture, civil engineering, integrated design, and mechanical engineering who could easily engage with complex design/engineering and fabrication tasks. Two scaled prototypes (Fig. 13) were engineered, manufactured, and successfully tested with non-linear FEA and non-destructive structural tests. The outputs show a previously unseen ribbed slab typology, which combines material and embodied carbon

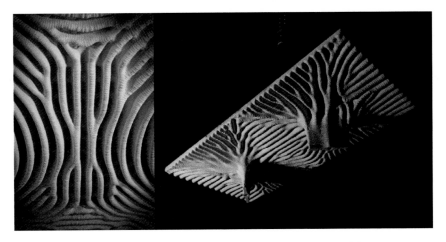

Fig. 13 Close-up view of RD Slab 01 (left) and global view of RD Slab 02 (right)

reduction with distinct aesthetics. Future work will further investigate the opportunities this approach opened, including an evaluation of its acoustics behaviour for future construction applications.

Acknowledgements The work described in this chapter was developed during the Experimental Architecture X Robotic Milling Strategies–SDU CREATE Summer School 2022 (8-21 August).–Organised by CREATE – Led by Assoc. Prof. Dr. Roberto Naboni, University of Southern Denmark (SDU). Teaching team: Roberto Naboni, Alessandro Zomparelli, Luca Breseghello, Anja Kunic, Hamed Hajikarimian. Software Partner: Karamba3D (parametric engineering). Material Partners: Hi-Con (UHPFC), Teknos and Lakkgruppen (coating material). Students: Ahmed Mosallam, Boas Olesen, Cecilia Rask, Christian Jørgensen, Christina Jensen, Elma Ramic, Esben Kay, Eyjolfur Logason, Oscar Oweson, Julia Przado, Kate Heywood, Marie-Louise Merser, Mathias Nielsen, Nanna Larsen, Nicolas Ortiz, Nikolaj Rytter, Omar Rawashdeh, Paula Riehn, Rikke Ejlersen, Sara Sander, Sofie Skov-Carlsen, Steen Hansen, Viktor Hansen.

References

1. Adriaenssens, S., Billington, D.: The ribbed floor systems of Pier Luigi Nervi. Proc. IASS Annu. Symp. **2013**(23), 1–7 (2013)
2. Antony, F., Grießhammer, R., Speck, T., Speck, O.: Sustainability assessment of a lightweight biomimetic ceiling structure. Bioinspir. Biomim. **9**, 016013 (2014). https://doi.org/10.1088/1748-3182/9/1/016013
3. Bischof, P., Mata-Falcón, J., Kaufmann, W.: Fostering innovative and sustainable mass-market construction using digital fabrication with concrete. Cem. Concr. Res., **161**, (2022). https://doi.org/10.1016/j.cemconres.2022.106948
4. Breseghello, L., Naboni, R.: Toolpath-based design for 3D concrete printing of carbon-efficient architectural structures. Addit Manuf. **56**, (2022). Elsevier
5. Cao, Z., Myers, R.J., Lupton, R.C., Duan, H., Sacchi, R., Zhou, N., Reed Miller, T., Cullen, J.M., Ge, Q., Liu, G.: The sponge effect and carbon emission mitigation potentials of the

global cement cycle. Nat. Commun. **11**(1), 3777 (2020). https://doi.org/10.1038/s41467-020-17583-w
6. Flatt, R.J., Wangler, T.: On sustainability and digital fabrication with concrete. Cem. Concr. Res. **158**(May), 106837 (2022). https://doi.org/10.1016/j.cemconres.2022.106837
7. Global Carbon Project (GCP).: Global carbon atlas: CO_2 emissions. (2021). http://www.globalcarbonatlas.org/en/CO2-emissions. Accessed December 2022
8. Gislason, S., Bruhn, S., Breseghello, L., Sen, B., Liu, G., Naboni, R.: Lightweight 3D printed concrete beams show an environmental promise: A cradle-to-grave comparative life cycle assessment. Clean Technol Envir., (2022)
9. Jayasinghe, A., Orr, J., Hawkins, W., Ibell, T.: Boshoff WP (2022) Comparing different strategies of minimising embodied carbon in concrete floors. J. Clean. Prod. **345**, 131177 (2021). https://doi.org/10.1016/j.jclepro.2022.131177
10. Mata-Falcón, J., Bischof, P., Kaufmann, W.: Exploiting the potential of digital fabrication for sustainable and economic concrete structures. RILEM Bookseries **19**(2019), 157–166 (2019). https://doi.org/10.1007/978-3-319-99519-9_14
11. Naboni, R., Kunic, A., Breseghello, L., Paoletti, I.: Load-responsive cellular envelopes with additive manufacturing. J. Facade Des. Eng. **5** (Powerskin), 37–49 (2017)
12. Naboni, R., Breseghello, L.: Kunic A (2019) Multi-scale design and fabrication of the trabeculae pavilion. Addit. Manuf.: Spec. Issue Large Area Addit. Manuf. **27**, 305–317 (2019)
13. Naboni, R.: Cyber-physical construction and computational manufacturing. In: Bolpagni, M., Gavina, R., Ribeiro, D. (eds) Industry 4.0 for the built environment. structural integrity, vol 20. Springer, Cham (2022)
14. Pinsent Masons.: Why the construction industry needs to decarbonise. (2021). https://www.pinsentmasons.com/out-law/analysis/why-the-construction-industry-needs-to-decarbonise (Accessed Dec 2022)
15. Preisinger, C., Heimrath, M.: Karamba—A Toolkit for parametric structural design. Struct. Eng. Int.: J. Int. Assoc. Bridg. Struct. Eng. (IABSE) (2014). https://doi.org/10.2749/101686614X13830790993483
16. Tam, K.-M.M., Coleman, J., Fine, N., Mueller, C.: Stress line additive manufacturing (SLAM) for 2.5-D shells. J. Int. Assoc. Shell Spat. Structures. International Assoc. Shell Spat. Struct. (IASS). **57** (4), (2015)
17. Turing, A.M.: The chemical basis of morphogenesis. philosophical transactions of the royal society of London. Ser. B, Biol. Sci., **237**(641), 37–72 (1952). http://www.jstor.org/stable/92463
18. UN Environment and International Energy Agency.: Towards a zero-emission, efficient, and resilient buildings and construction sector. Global Status Report 2017, (2017)
19. U.S. Geological Survey.: Mineral commodity summaries 2022: U.S. Geological Survey, p 202. (2022). https://doi.org/10.3133/mcs2022

Digitally Designed Stone Sculpting for Robotic Fabrication

Shayani Fernando, Jose Luis García del Castillo y López, Matt Jezyk, and Michael Stradley

Abstract In this chapter, we present case studies in digitally aided marble sculpting for robotic fabrication developed at the Digital Stone Project workshop. The residency brings together artists, architects, designers, researchers and technologists engaging in state-of-the-art digital tools for the realization of innovative works of art in stone. These projects were developed during the Digital Stone Project research residency during 2013 to 2018 and showcase the potential of novel input methodologies to drive creative processes in design, architecture and the arts. The case studies demonstrate both conceptual and technological development in the design process through 3D modelling, scanning and fabrication workflows, developing toolpaths, virtual reality, haptic interaction and reversible construction techniques. The chapter examines the value of robotic technologies in the design and construction process relative to collaborative crafting of the hand and machine. Accommodating for material tolerances and interrogating the implications of computational crafting in relation to Industry 4.0 and exploring the role of the artisan in machine crafted architectural components.

Keywords Digital Stone · Structural Design · Robotic Crafting · Workflows · Sustainability

---

S. Fernando (✉)
DDU-Digital Design Unit, Fachbereich Architektur, Technische Universität Darmstadt, Darmstadt, Germany
e-mail: fernando@dg.tu-darmstadt.de

J. L. G. del Castillo y López
Material Processes and Systems Group, Harvard University Graduate School of Design, Cambridge, MA, USA
e-mail: jgarciadelcasti@gsd.harvard.edu

M. Jezyk
Founder at Foldstar.AI, Hamilton, NJ, USA
e-mail: matt@foldster.com

M. Stradley
Rensselaer School Architecture, Rensselaer Polytechnic Institute, Troy, NY, USA
e-mail: stradm3@rpi.edu

United Nations' Sustainable Development Goals 9. Industry, Innovation, and Infrastructure · 11. Sustainable Cities and Communities · 12. Responsible Consumption and Production

1 Introduction

Exploration in robotic fabrication has seen a remarkable development in the last decade, particularly in the field of architecture. The democratization of Computer-Aided Design (CAD) to Computer-Aided Manufacturing (CAM) tools led to novel inquiries into the affordances of industrial robotic arms in mass-customization workflows, and their role as potential "all-purpose fabrication machines." Such renewed interest led to an explosion of research in the field, enabled by the proliferation of dedicated robotic fabrication shops in architecture schools around the globe as well as specific conferences on the topic [1].

However, it could be argued that, with exceptions, the limitations of conducting robotic fabrication in non-industrial contexts may have biased most of the current research to a particular set of outcomes. On one hand, the additional facility requirements to conduct subtractive manufacturing–ventilation, debris disposal, safety measures–may have favored explorations on additive manufacturing, such as 3D printing or assembly operations. Comparably, the costly tooling necessary to perform intense milling operations on harder materials such as stone or metal could have resulted in work focusing on softer substrates such as wood, foam or fresh clay. Consequently, certain forms of robotic fabrication, such as stone carving, have not had the opportunity to be broadly explored.

Intriguingly, the stone carving industry is itself currently undergoing its own new Renaissance, this time powered by the skill of industrial robots [2]. An increasing number of companies now rely on CAD/CAM workflows and robotic fabrication to accelerate their production and remain competitive in a market where material and labor costs are continuously on the rise [3]. Nevertheless, a large portion of their art-oriented production is geared toward the reproduction of existing physical models, typically relying on 3D scan-to-milling processes. The industrial nature of these fabrication processes, combined with the technical literacy necessary to harness the computational frameworks driving it, constitutes a large entry barrier for designers and artists hoping to explore the creative space afforded by these technologies.

1.1 The Digital Stone Project

The Digital Stone Project [4] and Garfagnana Innovazione [5] are a collaborative team to bring together artists, architects, educators and students from across the globe to the historic Garfagnana region of Italy to work with high technology and ancient

Tuscan stone. Over the course of a month each participant produces a sculpture or prototype that is carved with a 7-axis robot arm and finished by hand.

This collaboration between designers, technologists and the advanced technology facility at Garfagnana Innovazione has exciting implications for the future of architecture, sculpture and digitally aided design and manufacture. The artists and architects collaborated with robotics engineer Gabriel Ferri to create the computational data that drive the robotic cutting arm. This way of working represents the pinnacle of current manufacturing capability joined with computer aided design (CAD).

Originally designed for automotive and aerospace manufacture, these robots have been coopted to re-invigorate the declining stone industry in the Garfagnana region and to reintroduce the skills associated with marble sculpture [6]. The work includes projects made with generative design, 3D scanning, and 3D programs in animation, CAD (computer aided design), and CAM (computer aided manufacture). Participants wrote algorithms to create the designs for carving or developed projects through interactive computer design programs such as Rhinoceros 3D, Grasshopper, Cinema 4D, 3D Studio Max, and ZBrush.

The work produced in the Digital Stone Project residencies reveals how the development of CAD/CAM and robotics digital manufacturing technologies has helped to reduce the disparity between the geometric forms able to be generated using modern design software, compared to the methods of materializing the outcome. This chapter aims to describe the design and manufacturing processes through 6 case studies of prototypes produced during 2013 to 2018. They explore minimal surfaces, non-standard tool paths, interlocking joinery systems and integrated scanning to milling.

During the DSP residencies one of the greatest challenges facing the engineers to fabricate the digital models is translating the artist's idea into stone through the machine. According to Gabriel Ferri, robotics engineer at Garfagnana Innovazione (2022), "since the idea itself is not linked to the material many times it must be adapted to respect the physical limits of the stone. Finding a compromise between the three parts (1) artist's idea, (2) physical limit of the stone and (3) physical limit of the machine, is the main challenge. To address these challenges the digital stone processing facilities has several resources to help enable and develop solutions. They are described in the following.

1.2 Digital Stone Processing

The state-of-the-art facilities at Garfagnana Innovazione have assisted the participants of the residencies to realise their projects efficiently and in a resourceful manner. The main hardware used is described in the following [7]:

Robotized 3D scanner High performance instrument, suitable to scan both outdoors and indoors. Due to its laser technology, it can reconstruct 3D solid models starting from objects with every form, dimension and complexity. The extreme precision in the survey of points and surfaces assure an exact reconstruction of the scanned object. This instrument, born from 100% Italian research (CNR), can be used manually like a common 3D scanner, but also in synergy with their anthropomorphic robots gaining the skill, unique in this sector, of reconstructing clouds of points independently, without further actions needed on the digital realignment of the scanned model.

Anthropomorphic robots Two robotized stations with two high precision anthropomorphic robots (7 interpolated axes) are the state of the art of technology for the stone processing. The vast working area, up to 2.5 m of height, allow the creation of big and full-round works. Besides the large variety of tools for the processing of every type of marble and stone, there is also the possibility to work materials for the rapid prototyping: plastics, polystyrene, resins, etc. These machines, together with their powerful software, can perform complex operations with extreme efficiency: the results of the creation of human or abstract figures are qualitatively excellent. Progressing from the scale model (created on a PC) to the final realization is now more efficient.

CNC work center CNC work center (5 interpolated axes) to work with blocks, solids and marble slabs, granite, natural stone veneer and glass. It makes every kind of drilling, milling, cutting with blades, contouring, shaping, recessing, polishing, carving, engraving, chamfering, 3d writing and processing. With one of the biggest working areas available, this machine can make big lots with precision and high rapidity.

These facilities have remained for the last few years. According to Gabriel Ferri (2022), while no new tools have been developed, new finishing procedures have been created as a result of the workshop residencies. For example, in the work of Jon Isherwood, "He is very attracted to the gesture of manual finishing, so we tried to translate the human gesture into code for the robot in order to replicate the movements that leave the desired mark on the stone [8]." The workflows are directly impacted by the type of project and will be explained in detail in the case studies. As Ferri states, "Perhaps more than the machines in recent years we have witnessed an evolution of the software. And I believe that in the next few years it will evolve even more because as a user I recognize that there are still many gaps and room for improvement." The following will describe case study projects and concepts by participants of the Digital Stone Project workshops. These specific projects were chosen for their innovative design to production methodologies specifically relating to the field of construction and Industry 4.0 in the creation of bespoke architectural components. The projects will be discussed in relation to **project concepts, digital design workflows, tools and techniques**.

2 Minimal Surface Geometries

The first case study project concept uses Minimal Surfaces which are a class of geometric surfaces characterized as having minimal surface area for a given boundary. The classic example is the surface formed by a soap film stretched between a boundary frame [9]. In mathematics, minimal surfaces are defined as 'surfaces with zero mean curvature" [10]. Zero mean curvature can best be understood as any point on a surface where the lines of principle curvature are equal and opposite. It is possible to compute minimal surfaces like the Catenoid and Helicoid based on parametric equations.

Computational optimization and dynamic relaxation techniques can be used to find minimal energy states for these types of surfaces, instead of generating from a mathematical equation. Interestingly there is a strong correlation between the minimal energy states found computationally via optimization and physical, empirically derived forms [11]. It can be observed these minimal surfaces are operating purely in tension. This study is quite similar to catenary curves found in 'hanging chain' models. This makes minimal surfaces also very relevant in the study of tensile fabric structures. It also makes minimal surfaces relevant to the study of forms operation purely in compression, in the form of free-form grids shell structures [12].

The Costa Minimal Surface was discovered relatively recently by Celso José da Costa, a Brazilian mathematician and is used in Case Study 1. The Costa Minimal Surface is rather like a Mobius Strip or a Klein Bottle, where the delineation between the outer and inner surfaces starts to blend. "The Costa surface is a complete minimal embedded surface of finite topology (i.e., it has no boundary and does not intersect itself… Until this surface was discovered by Costa (1984), the only other known complete minimal embeddable surfaces in R^3 with no self-intersections were the plane (genus 0), catenoid (genus 0 with two punctures), and helicoid (genus 0 with two punctures), and it was conjectured that these were the only such surfaces" [13].

2.1 Case Study 1–Costa Minimal Surface (2014)

This project, titled Costa Minimal Surface [14], was designed and fabricated as part of the Digital Stone Project in 2014. Previous research had been conducted in the digital fabrication of tensile structures as well as compression structures. In many of these cases, computational design, simulation and optimization techniques were used to create a digital design, and then computer numerically controlled (CNC) equipment was used to fabricate the final shape. In Case Study 1, a design was created to explore minimal surface geometry using marble instead of in tensile fabric materials. There were three main goals for the design and fabrication of this case study:

Design: From a design perspective, the goal was to explore the interplay between the thinness of these minimal surface 'soap bubble' geometries, contrasted with the heavy materiality of stone. It was also very important to ensure the smooth surface continuity was maintained, to help the observer understand the smooth, sensuous curves of the Costa surface geometry.

Simulation/Computation: Case Study 1 required the development of specialized software to support the simulation, analysis and fabrication of the piece. The goal was to create a parametric design workflow to find the desired geometric form in terms of sculptural qualities in order to achieve the design goals above, then support a seamless process of converting the geometry into CNC code able to be executed on the 7-axis industrial arms.

Fabrication: There was also the technical challenge of milling the stone to follow the smooth, thin surfaces of these geometries. It was unknown exactly how thinly the stone could be milled, and if these doubly curved surface topologies could be expressed in stone at all. It was desired to obtain a slightly translucent material quality in the marble, yet still retain the overall structural integrity of the piece by virtue of the fully in compression structural nature of the minimal surface. How the anisotropic nature of marble would interact with the thin shell geometry was also an unknown.

During the design process for Case Study 1, many mathematical surfaces were explored. The Costa Minimal Surface was found to be most suitable for this type of sculpture, given the constraints on the size of the piece and the available fabrication techniques. The software Mathematica [15] was used to explore the surface domain and generate the geometry (see Fig. 1). Online examples for how to compute the Costa function in Mathematica were referenced [16] (Fig. 2)

A usable geometric mesh needed to be computed from the smooth analytic surface. A grid of points was evaluated in the UV parameter space of the surface and then

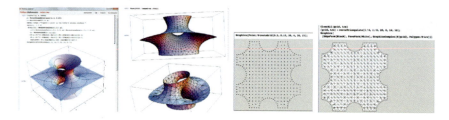

Fig. 1 Left–Mathematica Software was used to plot the surface–Further refinement of the domain of the surface, Middle Right–Computing the UV Grid [16] Right–Triangulation of surface via the UV grid [16]

Fig. 2 Left–Dynamo Computational Design Software was used to parametrically control the design–simulation loop with Mathematica, Right–Plotting the UV points in Dynamo, refining the edge treatments

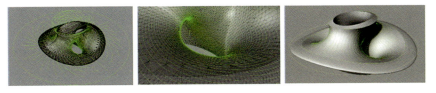

Fig. 3 Left–NURBS curves based on the UV parameterization. Middle–Close up showing relationship between the UV parameterization and the meshed surface geometry Right–The final design geometry

Fig. 4 Top: left–fabrication rough block middle–fabrication detail, right–smooth surface, bottom: left: UV lines on surface, middle- UV lines close up, right: final sculpture

used to construct a mesh. Online examples were referenced and modified extensively [16]. The software Dynamo (an open-source visual programming tool) was used to create the overall parametric design, simulation and fabrication loop. An open-source Dynamo C# software plugin called DynamoMathematica [18] was developed and used to set input parameters and send data to Mathematica. Once the form was simulated, the geometric mesh was sent back to Dynamo. Special functions needed to be developed in Mathematica to compute the normals for this non-trivial surface. [17] The Dynamo script used the normal to allow thickening the meshed surface and creating NURBS curves based on the UV parameterization (green lines below, see Fig. 3).

The milling toolpaths were created and then run on a 7-axis ABB robot in the mountains of Tuscany as part of the Digital Stone Project at Garfagnana Innovazione. Custom CNC G-Code was developed by Gabriel Ferri to allow the UV lines of ruling to be carved into the surface. The piece was finished by hand using various chisels, grinders and sanding equipment. The following images (see Fig. 4) show the fabrication method and final milled sculpture.

3 Interlocking Joinery Systems Based on Catenary Geometry

The following will describe a joinery system exploring Catenary geometry and is inspired and developed by hanging chain models and catenoids, which are explained in the previous case study the Costa Minimal Surface. However, the design intention of this system was to utilize dry joint connections with interlocking wave geometry. While case study 2–Catenary Tales explores the potential tensile forces of the interlocking wave joinery system, Case study 3– Archi-Twist' investigates structural stability through the twisted interlocking wave joint.

The waveform prototype joint design was developed through a process of iterative drawing which was then parametrised in Rhino 3D and GRASSHOPPER. One of the most significant aspects of this geometry is that it is made from 'ruled surfaces' to accommodate fabrication methods with robotic wire cutting. As a comparison study into EPS foam blocks and natural stone blocks [19] two similar geometries were fabricated with Gosford quarries in Sydney. This prototype demonstrates both potential for column and cantilever structural conditions without the use of extra connectors or mortar, demonstrating reversible construction methods.

3.1 Case Study 2–Catenary Tales (2015)

The concept for the development of 'Catenary Tales' (2015) [21] stemmed from an exploration of self-supporting structures in natural stone. The sculptural prototype is made from interlocking wave joinery in varying sizes of modules which engages with the tensile strength of stone. The production process involved the design using 3D modelling software Rhino 3D and manufactured as part of the DSP workshop.

A combination of machined and hand finishes were applied (see Fig. 5) to the sculptural prototype. The overall dimensions are 38 × 36 × 80 cm and weighing at 60 kg in total. The marble type is similar to the colours chosen for Case Study 3– 'Archi-Twist' which include Bianco Acquamarina, Venato Orto di Donna and Bardiglio Imperiale. The darker colours at the base while the lighter Carrara marble at the top which further contributed to highlight the complexity of the geometry and material.

The exhibited prototype 'Catenary Tales' was successful in that through the collaboration of the hand and the machine, the realisation of the initial 3D model was achieved. The specific factors which made it successful include the skill of the craftsman in order to remediate the machine inconsistencies and performance with the material. The skill of the fabrication team and technicians to realise the 3D model into physical reality involved many attempts especially when there were cracks and fractures in the material. The workflow of both the modelling and fabrication process could have further integrated methods of material surveying and scanning. However, the focus was on developing the skill of crafting and carving away material from

Fig. 5 Case study 2 Left–Digital model and assembly sequence, middle–7 axis milling, Right-Final exhibited sculpture

natural stone. Constraints included a limited 25-h machine time for each prototype which affected both the final scale and number of machined parts which were in the end grouped together [20]. The sculptural prototype demonstrates bespoke construction typology with dry joint connections as a contribution towards the innovation of architectural components.

3.2 Case Study 3—Archi-Twist (2017)

The concept of the 'Archi-Twist' prototype designed and manufactured as part of the Digital Stone Project 2017 [22] is a twisted catenary arch comprised of innovative modular interlocking wave joinery based on catenary curvature. The method of geometry generation was developed in an entirely parametric environment. In contrast to using sinusoidal or interpolated curvature (used in the initial wave jointed block geometry), catenary curvature for the interlocking wave amplitude has a higher contact surface area thus facilitating a better interlocking capacity. Furthermore, the 180-degree twist of the bases of the arched structure was modelled for both aesthetic and structural reasons due to the extent of the twisting capacity of ruled surface joints. It not only tested the potential capacity of the wave joint contact surfaces but also provided a stabilising mechanism for the overall macro geometry of the arch [20].

Due to limitations in machine time which were 25 h per project within the workshop setting, the scale and number of blocks to be fabricated had to be reduced from 7 parts to 5. Each part (see Fig. 6) was machined in approximately 5 h including tool changes. However, this was still not enough as a final finished module as there was usually 1–2 cm extra material left to sand by hand using the power tools, grinders and hand sanding paper processes similar to processes used by traditional marble sculptors. The overall dimensions of 'Archi-Twist' were 14 × 66 × 56 cm, weighing at 38 kg in total.

Fig. 6 Case study 3 Left- digital model, middle–fabrication, right–final exhibition sculpture

The above case studies demonstrate the use of bespoke dry joint connections using a specific geometry of the wave jointed blocks which is a developed variation of the osteomorphic block. Modular dry joints in the construction industry have gained acceptance for their versatility and reduced labor costs in comparison to traditional brick and mortar methods. The ease of assembly and disassembly make using connecting blocks for spatial assemblies.

4 Non-standard Tool Paths

A tool path is the direction through space that the tip of a cutting tool follows on its way to producing the desired geometry of the designed work. Nonstandard toolpaths were explored as part of the DSP workshop. The following is an example of a non-standard approach to the development of tool paths and will describe specifically a non-standard method of an engraving toolpath.

4.1 Case Study 4—Orbital Body (2015)

This project, titled Orbital Body, was designed and fabricated as part of the Digital Stone Project in 2015 and exhibited at the Marble Codes exhibition [21]. Orbital Body is a seating element carved from a single block of Bardiglio Imperiale marble and finds novel geometry in the characteristic markings of several different milling passes and tool types. The piece was fabricated through four subsequent machining steps before being hand-finished. As is typical in CNC milling and other fabrication processes, each subsequent machining pass increased in detail and decreased tool size.

This $4 + 1$ step procedure moved through the following machining steps: (1) fabrication of a marble block to match the approximate proportions of the piece's bounding box using a robotically controlled diamond saw, (2) a coarse, flat-end roughing pass which removed the majority of excess material located outside the positive volume of the piece, (3) a ball-end, finishing pass which removed a thin layer of material, closely approximating the curvature and geometry of the digital model, (4) a detailed engraving pass using a ball-end, pencil-neck tool, which inscribed a series of overlapping grooves into the surface of the piece, and (5) the hand-finishing and polishing of the piece to achieve the desired finish and surface texture.

Much of this machining procedure was adapted from standard workflows in CNC-fabrication, however, the detailed engraving toolpath (step 4, above) required a non-standard approach. The engraving toolpaths were produced in Rhino 3D by arraying a series of construction planes along the central mid-line of the piece, intersecting the curved volume of the piece with these planes, and exporting the resulting linework as the engraving toolpath. Originally, the design-intention of the piece was for each engraving toolpath to form a closed loop in space. While this was geometrically possible, the dimensional and rotational limitations of the robot arm made this impossible to fabricate. Working closely with the fabrication engineer, the simulated limit of the robot arm's rotational "reach" was used to adjust the engraving, surgically clipping the engraving toolpaths in order to remove any areas unreachable by the robot arm.

Despite the technical hurdles in realizing the engraving toolpaths, their contribution to the overall piece is substantial aesthetically and functionally. In terms of the work's functionality as a seating element, the overlapping engraving lines of the piece impart tooth and friction to an otherwise smooth, slippery, gloss-finished surface. The depth and spacing of the engraving is deepest/densest across the horizontal surface of the seat and is more shallow/sparse along non-seating areas. As the piece was intended for both indoor and outdoor sites, the engraving toolpaths were designed to function as micro-gutters and help to shed water from the seat surface. These engraved grooves, along with the overall slight convex curvature of the seat, allow the seating surface to shed water quickly and prevent the pooling of water in outdoor environments.

The overall form of the piece seeks to capture qualities of speed and continuity to match the smooth, continuous surface quality of the finely-honed stone. Despite this formal aspiration of continuity and fluidity, the piece had to observe two major functional constraints: balance and weight. Viewed in plan (see Fig. 7), the piece is perfectly rotationally symmetrical. Working in Rhino 3D and Grasshopper, each subtle push and pull of the form is matched symmetrically on the opposite side of the form. This allowed for the sculptural, intuitive manipulation of the overall form during design, while maintaining a consistent center-of-gravity and eventually yielding a seating object which resists overturning. The void at the center of the piece also operates functionally. The removal of material at the center of the piece greatly reduces the weight of the final work, allowing it to be moved or adjusted by two people standing at either end of the piece. In future iterations, the central void of the piece might be extracted as a monolithic block, reducing both machining time and material waste.

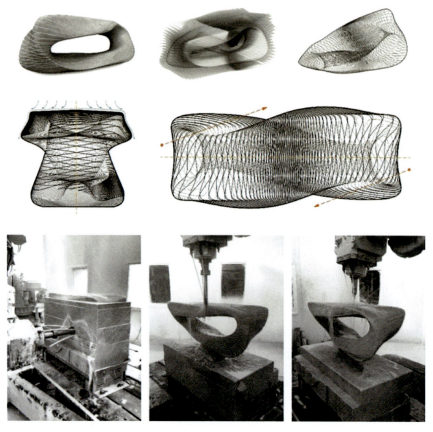

Fig. 7 Case study 4 Top- 3D model showing the development of toolpaths for engraving, Middle–Plan and section views, Bottom–Fabrication using 7 axis milling

5 Interaction Methodologies and Integrated 3D Scanning to Milling

In this research, we borrow methods and logics from traditional sculpting techniques, and explore their power to be computationally augmented, and to include the agency of the designer in the design and fabrication process. The goal was to showcase the potential of novel interaction methodologies to drive creative processes in design, architecture and the arts.

5.1 Case Study 5—Untitled 50,069,744 (2018)

The first sculpture produced for this research is called "Untitled 50,069,744," and it constitutes a case study on the use of digital technologies to computationally augment the 3D sculpting process, inspired by the freedom of physical gesturing. The goal for this piece was to create a system that could capture the fluidity of human motion and freeze it in time into a delicate and graceful manifestation. Additionally, the project sought to push the tectonic capacity of the stone to its limits and investigate how thin the sculpture could be fabricated. It was determined that the form of the object would be generated by digitally tracking a human sculptor's hand, and that these motion steps will be used to generate a self-standing vertical surface to be milled at a specific thickness.

In order to achieve this goal, a bespoke modeling environment was prototyped using Autodesk's Dynamo (see Fig. 8). The system was able to read physical motion from a hand-held motion controller, and use it to generate a three-dimensional thin solid, with its form was inspired by the undulating shape of a piece of fabric draping over itself. The piece was set to be milled out of a 1800 × 800 × 180 mm block of Arabescatto Vagli, a local marble with gray and golden veins, and an elevated degree of translucency through its main white body. A target thickness of 8 mm was estimated by the manufacturer as an ideal compromise between material stability for fabrication and maximizing thinness for translucency.

The milling operations were programmed with commercial CNC software, and the sculpture was milled using an industrial robotic arm and an external rotary 7th axis. As the sculpture was created to be publicly exhibited at the Digital Stone Project exhibition [23] a large industrial robotic fabrication space, only one of the sides was manually polished, whereas the opposite side was left with the raw toolpath markings for the fabrication history of the object to remain more readable (see Fig. 9).

Fig. 8 Left–The interactive, hand-based 3D sculpting environment, Right–initial tests

Fig. 9 Left—Final sculpture model, Right—Milled sculpture in the shop and the exhibition gallery

5.2 Case Study 6—Dereliction (2018)

The second sculpture produced for this research was called "Dereliction," and it is a study on the capacity of computational workflows to be adaptive to the physical world and use preexisting conditions as source input in form generation processes. Additionally, the project showcased the potential of digital technologies to mimic certain forms of craft, and the use of algorithms as creative tools. The main question addressed by this work was how digitally augmented sculpting could be harnessed to reclaim discarded materials from marble extractive processes, giving them a second life by adding value through digital craft, while improving the sustainability of the marble extraction process.

To fulfill this vision, an integrated 3D scan to robotic sculpting workflow was developed. A large-scale laser scanner was attached to the robot flange, and a 360° robot scanning procedure was developed to achieve a full three-dimensional representation of the target stone piece. The scanned point cloud was assembled and rebuilt using commercial 3D-scanning software, resulting in a high-resolution polygon mesh. A custom parametric modeling environment was developed in Dynamo, to accept input polygon meshes, representing the 3D-scanned rock, and output the full robotic procedure to groove a custom pattern on its surface (see Fig. 10). The main contribution of the model is the internal translation of the polygon mesh into a graph data structure, susceptible to be analyzed using propagation algorithms, customizable by the designer. Toolpaths were generated using the Robot Ex Machina framework [24].

For the final piece, a mossy marble boulder abandoned in a nearby creek was reclaimed. The rock was placed on the rotary axis of the robot for scanning, and its

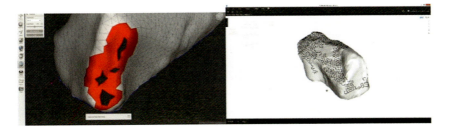

Fig. 10 Left—Mesh reconstruction, Right—toolpath propagation

Fig. 11 Left—Fabrication using 7 axis milling, Right—Milling details

position was fixed to avoid calibration errors. The scanned model was post-processed using the abovementioned workflow, and an error-proof robot procedure was generated. A 5 mm diameter milling bit was chosen, and the procedure was successfully run on the robot, taking approximately 2 h to complete (see Fig. 11). No latter treatment or polishing was applied to the piece.

6 Conclusion and Future Work

The above case studies demonstrate the value of robotic technologies in the design and construction process relative to collaborative crafting of the hand and machine. The collaborative effort accommodates for material tolerances and interrogates the implications of computational crafting in relation to Industry 4.0, while exploring the role of the artisan in machine crafted architectural components. The case studies investigated both conceptual and technological development in the design process through 3D modelling, scanning and fabrication workflows, developing toolpaths, virtual reality, haptic interaction and reversible construction techniques using bespoke geometry and customized toolsets.

However, the trend towards discrete modular construction must be acknowledged as future work specifically for the case studies with dry joint connections. The use of discrete modularity as opposed to bespoke in interlocking joinery systems in this case rationalises the choice of material and geometry. According to Tessmann,

Rossi (2019), "structures are made from discrete and manageable elements that are aggregated in a defined (or rule-based) way. The aggregation becomes a structure if it redirects loads." While both discrete and bespoke modular blocks consume material, discrete geometry prevents excessive waste generation by limiting the number of variations to the block types to be fabricated and assembled. In bespoke systems each module is individually designed and manufactured, often using more material and labor for the design and production process. Discrete systems only use a limited amount of block variations usually limited to two or three hence can use existing waste material and offcuts, depending on the scale of modules. This is significant in today's construction sector in line with the Sustainable Development Goals [26] to build resilient infrastructure, promote inclusive and sustainable industrialization, and foster innovation.

Acknowledgements The authors would like to acknowledge the support of Digital Stone Project president Jon Isherwood, the DSP committee, the staff of Garfagnana Innovazione and Autodesk to support the development of the participants' projects; and the University of Sydney, Australia and Ingenium grant at TU Darmstadt, Germany.

Excerpts and images from Chap. 1 and 5 reproduced from "The Digital Touch. Towards Novel Modeling Frameworks for Robotically-Enhanced Marble Sculpting" by Jose Luis García del Castillo y López, with his permission.

References

1. Brell-Cokcan, S., Braumann, J. eds.: Rob|Arch 2012: Robotic fabrication in architecture, art and design. Springer, (2013)
2. Dorfman, P.: How would michelangelo's sculpture look if he'd had robot apprentices? (2018). [Online]. Available at https://redshift.autodesk.com/robot-sculpture/. Last Accessed 18 April 2022
3. Bubola, E.: We don't need another michelangelo: In Italy, It's Robots' Turn to Sculpt. (2021). [Online]. Available at https://www.nytimes.com/2021/07/11/world/europe/carrara-italy-robot-sculptures.html. Last accessed 18 April 2022
4. Digital Stone Project Homepage https://www.digitalstoneproject.com/. Last Accessed 19 April 2022
5. Garfagnana Innovazione Homepage https://www.garfagnanainnovazione.it/en. Last Accessed 19 April 2022
6. Isherwood, J.: Digital stone project—Exhibition at La Fortezza di Montalfonso, Castelnuovo di Garfagnana. (2013). [Online]. https://static1.squarespace.com/static/5d935416f7d27746a4 fe9afb/t/5db658c1f7a6c635a148f583/1572231364411/2013+DSP+Cat.pdf
7. Garfagnana Innovazione website. [Online]. Available at https://www.garfagnanainnovazione. it/en/polo-tecnologico. Last Accessed 04 April 2022
8. Ferri, G.: Email Interview conducted in 10/02/2022. (2022)
9. Definition of Minimal Surface Weisstein, E. "Minimal Surface." From MathWorld–A Wolfram [Online]. http://mathworld.wolfram.com/MinimalSurface.html. Last Accessed 19 April 2019
10. Hyde, S., Blum, Z., Landh, T., Lidin, S., Ninham, B.W., Andersson, S., Larsson, K.: The language of shape: the role of curvature in condensed matter: physics, chemistry and biology– Chapter 1, p. 19. SBN-13: 978–0444815385 918(1996)

11. Pauletti, R.M., Adriaenssens, S., Niewiarowski, A., Charpentier, V., Coar, M., Huynh, T., Li, X.: A minimal surface membrane sculpture. In: Bögle, A., Grohmann, M. (eds.), Conference: Proceedings of the IASS Annual Symposium 2017 "Interfaces: architecture engineering science. Hamburg, (2017). https://www.researchgate.net/publication/326632807_.
12. Brasz, F.: Soap films: Statics and dynamics. [Online]. (2010). Available at https://www.princeton.edu/~stonelab/Teaching/FredBraszFinalPaper.pdf
13. Kilian, A., Ochsendorf, J.: Particle-spring systems for structural form finding. J Int Assoc Shell and Spatial Struct **46**, 77–84 (2005)
14. The costa minimal surface [Online]. Available at https://mathworld.wolfram.com/CostaMinimalSurface.html. Last accessed 28 April 2022
15. Modern differential geometry of curves and surfaces with mathematica, 2nd ed. ISBN-13: 978–1584884484. CRC, Taylor and Francis, (2006)
16. Costa's minimal surface with minimal fuss—Andrej Bauer [Online]. Available at https://github.com/andrejbauer/costa-surface/blob/master/Costa.pdf. Last accessed 03 Feb 2022
17. Costa minimal surface normal [Online]. Available at https://mathoverflow.net/questions/151733/how-to-compute-the-normals-to-costas-minimal-surface. Last accessed 29 April 2022
18. Dynamo Mathematica C# zero-touch node (Dynamo Plugin) Matt Jezyk [Online]. Available at https://github.com/tatlin/DynamoMathematica. Last accessed 19 April 2022
19. Fernando, S., Reinhardt, D., Weir, S.: Waterjet and wire-cutting workflows in stereotomic practice: material cutting of wave jointed blocks. In: Janssen, M.A.S.P., Loh, P., Raonic, A. (Ed.), Protocols, flows and glitches. 22nd International Conference for Computer-Aided Architectural Design Research in Asia (CAADRIA 2017) (pp. 787–798). Hong-Kong, (2017). http://papers.cumincad.org/data/works/att/caadria2017_018.pdf
20. Fernando, S.: Collaborative crafting of interlocking structures in stereotomic practice. In: Sousa, J.P., Xavier, J.P., Castro Henriques, G. (eds.), Architecture in the age of the 4th industrial revolution—Proceedings of the 37th eCAADe and 23rd SIGraDi Conference—vol 2. University of Porto, Porto, Portugal, pp 183–190. (2019)
21. Isherwood, J.: Marble codes. In Robotic sculpture from Garfagnana, Digital stone project III, Exhibition catalogue (p.7). (2015)
22. Isherwood, J.: Metamorphic resonance. Digital stone project V exhibition catalogue. (2017). https://www.digitalstoneproject.com/previous-residencies
23. Isherwood, J.: Carbo nato di calcio. Digital stone project VI exhibition catalogue. (2018). https://www.digitalstoneproject.com/previous-residencies
24. García del Castillo y López, J.L.: Robot ex machina. In: Proceedings of the ACADIA 2019 Conference, pp. 40–49. (2019)
25. Tessmann, O., Rossi, A.: Geometry as interface: parametric and combinatorial topological interlocking assemblies. ASME. J. Appl. Mech. **86**(11), 111002 (2019). https://doi.org/10.1115/1.4044606(2019)
26. Envision2030 Goal 9: Industry, Innovation and Infrastructure (United Nations) 2021 [Online]. https://www.un.org/development/desa/disabilities/envision2030-goal9.html. Last accessed 29 April 2022

MycoCode: Development of an Extrudable Paste for 3D Printing Mycelium-Bound Composites

Fatima Ibrahim☉, Giorgio Castellano☉, Olga Beatrice Carcassi☉, and Ingrid Maria Paoletti☉

Abstract Additive manufacturing of sustainable and biodegradable materials offers an alternative fabrication paradigm to current composites used in architecture, based on the growth of materials rather than on extraction. This research investigates the extrusion of a mycelial-hemp and clay mix without additional additives and thoroughly evaluates the steps leading to it. The analysis entails four major steps: manual material investigation to figure out the finest ratio between clay and hemp shives for smooth extrudability, understanding hardware and software determinants that impact the resultant printed form, preparation of the material for 3D printing under sterile conditions and printing the Mycelium-clay mix varying properties analyzing the emergent characteristics of the material. Therefore, the research explores the possible combination ratios of a clay substrate with the addition of hemp shives inoculated with *Pleurotus Ostreatus*, without using any additional additives to test the paste's extrudability properties. The research successfully achieved a balanced ratio between mycelium, hemp, clay, and water when the relative percentage of clay and hemp shives were kept at 85–15%, respectively. The investigation also helped deduce that 3D extrusion printing with the Delta WASP 40,100 Clay Printer, with a nozzle diameter of 9 mm, is most optimal when the layer height is one-third of the nozzle diameter and the Extrusion (E) Value and Feed rate (F) is kept constant throughout the printing, in our case at 30 mm and 1500 mm/min, respectively.

Keywords Additive manufacturing · Digital fabrication · Mycelium-clay · Extrudability · 3D Printing · G-codes · Emergent attributes

---

F. Ibrahim (✉) · G. Castellano · O. B. Carcassi · I. M. Paoletti
Dipartimento di Architettura, Ingegneria Delle Costruzioni e Ambiente Costruito, Politecnico di Milano, 20133 Milano, Italy
e-mail: fatima.ibrahim@mail.polimi.it

G. Castellano
e-mail: giorgio.castellano@polimi.it

O. B. Carcassi
e-mail: olgabeatrice.carcassi@polimi.it

I. M. Paoletti
e-mail: ingrid.paoletti@polimi.it

© The Author(s), under exclusive license to Springer Nature Switzerland AG 2024
M. Barberio et al. (eds.), *Architecture and Design for Industry 4.0*, Lecture Notes in Mechanical Engineering, https://doi.org/10.1007/978-3-031-36922-3_29

United Nations' Sustainable Development Goals 9. Build resilient infrastructure, promote inclusive and sustainable industrialization and foster innovation · 11. Make cities and human settlements inclusive, safe, resilient and sustainable · 15. Protect, restore and promote sustainable use of terrestrial ecosystems, sustainably manage forests, combat desertification, and halt and reverse land degradation and halt biodiversity loss

1 Introduction

1.1 Industry 4.0

The current paradigm shift in the Architecture, Engineering, and Construction (AEC) industry, dubbed as the Fourth Industrial Revolution, or Industry 4.0, has created numerous opportunities for innovation and invention. This has also paved the way for the emergence of a more collaborative and cross-disciplinary technological era to emerge. One such innovative research field that has spawned a plethora of sub-research studies is the use of additive manufacturing (AM). This study aims to investigate the marvels of additive manufacturing of bio-based materials that are drawn and built upon the fundamentals of bio-design while also being biodegradable. Although additive manufacturing reduces waste generation, the use of biobased materials ensures that the material is biodegradable at the end of its life cycle, minimizing any significant environmental impacts. Holistically, this research is one of many springing studies that could potentially lead to the replacement of building materials that are detrimental to the environment with more environmentally friendly ones, owing its success to the advancements of technology and digital fabrication.

1.2 State of the Art: Mycelium Fabrication Technologies

In the recent decade, the shift in global perspective due to the excessive contribution to global warming and global carbon emissions encouraged the AEC to explore and use more eco-friendly methods and materials [1]. This has opened a novel avenue: the discourse on bio-based materials and their application in the AEC Industry [2]. Recent research in the burgeoning discipline of biodesign has begun to provide intriguing answers for environmental issues created by the fast population increase and the discard culture that has accompanied it [2]. As Myers [3] suggested, "Biology-inspired methods to design and manufacture" are used in biodesign, and "life creatures [are] essential components in design" [3]. To achieve increased architectural flexibility, construction is being driven towards automation by numerous factors, including a reduction in labor for safety concerns, a reduction in construction time on site, and manufacturing costs [4]. In particular, AM is an interesting process of layer-by-layer material deposition to form 3D model data. It primarily negates the need for excessive formwork while reducing the waste produced [5]. The practice of AM further becomes more environmentally conscious if using sustainable or bio-based materials that are biodegradable or compostable at their end-of-life (EoL) [2]. Engineering natural materials with the themes of sustainability and EoL concerns have become a strategical design approach. Materials made from cellulose, silk protein, eggshell membrane, bamboo, and mycelium composites [6] have recently received considerable attention [7]. In this regard, mycelium-based composites offer renewable and biodegradable options for various design and production processes, including architectural applications [2]. Mycelium is the vegetative portion of fungus, made up of a mass of hyphae [2]. In fungus, hyphae are a long, branching filamentous structure that acts as a growth agent [2]. Each hypha contains one or more cells that divide to accelerate the growth process and has a diameter of 4–6 μm [2]. Mycelium breaks down biopolymers into simpler entities via enzymes released by hyphae, which are subsequently absorbed through active transport. This cellular process allows living creatures to consume carbon-based nutrition [2]. The hyphae emerge from the substrate and into the air, forming a "fluffy or compact coating covering the substrate, known as fungal skin" [8].

Mycelium-based composites have been widely studied and investigated in the previous years [9], which have led to the production of products like Mycelium-based leather by Mycoworks [10], acoustic and flooring panels by MOGU [11], daily use items such as vessels and bowls by Officina Corpuscoli [12], packaging items initially started by Ecovative [13] and furniture pieces by Phil Ross [14] and Eric Klarenbeek [15]. The application of mycelium-based composites is not only limited to products; it has also been explored in the realm of architecture at a pavilion scale by studios and architects that made use of molding techniques such as, the Hy-fi project by The Living Studio [16], Mycotree by ETH Zurich [17], the Growing Pavilion by Company New Heroes E. Klarenbeek [18] and MY-CO space by MY-CO-X Collective [19]. In contrast, Blast Studio's Tree Column [20] and Lund University's Protomycokion [21] are mycelium-based projects that are additively manufactured through a direct ink writing (DIW) 3D printing technique. The application of mycelium-based composites in AEC is a novelty in the last decade, whereas, clay has been used for millennia [22]. This is due to clay's insulating characteristics, heat storage capacity, ability to minimize construction energy use, and high local availability [22]. However, this material is usually associated with traditional and vernacular architecture in undeveloped places [22]. In recent years, the 3D printing of clay and the related parameters have been explored by Benay Gürsoy [23] to analyze the impact of the printer and variables in the G-code on the resultant clay print. Interesting 3D printing projects are present in the literature, such as the Tree Column by the Blast Studio [20] and Protomycokion at the Lund University [21], which uses biopolymers to feed mycelium and create 3D printed objects. Moreover, the conceptual exploration of clay and mycelium with varying substrates such as sawdust and coffee grounds for potential 3D printing in the project Claycelium by IAAC, Barcelona [22] paves the way to combine the clay and mycelium-based potentials in the AM sector. Benay Gürsoy tests the 3D printing of clay and the digital parameters involved in the additive manufacturing process that affects the printed form [23]. However, these projects embed in the extrudable paste the use of additives. Hence, what is missing is a clear guideline on how to develop and prepare an extrudable paste and the 3D printing parameters when a clay, hemp shives, and mycelium composite are combined without additive materials. Therefore, this research offers guidance for embarking upon similar research, by investigating the digital additive fabrication of mycelium clay and by providing a step-by-step explanation of the extrudable mix and printing parameters that influence the printed element.

1.3 Research Gap and Aims of the Research

In this paper, the subject of focus is the additive extrusion of a mixture composed of mycelium, hemp shives, and clay without any extra additives, along with factors that affect the final form during the different steps of 3D printing of the material.

We can discern a symbiotic link between the qualities of clay, a common building material, and mycelium, the living fiber network that connects our plant life across the earth [22]. Mycelium is a living organism that requires nutrients to grow, which are not sufficiently found in clay. However, hemp shives provide ample nutrients to mycelium for its growth, which is why hemp qualifies as an apt substrate for mycelium. A mix of mycelium and hemp cannot be extruded via an extruder or printer due to hemp's water retention characteristic, which results in the poor self-adhesion nature of the mixture and eventually the inability to maintain form [24]. Therefore, owing to clay's viscous, elastic properties, it is incorporated into the mycelium-hemp mix [25]; simultaneously, as mycelium grows through the mix, feeding on hemp shives and piercing through the clay here it acts as a 'natural binder' [7], consequentially holding the entire form together. The fungi-based compound can store biogenic carbon in its biomass and lower the process's environmental impacts [9]. Moreover, the fibers enhance the mechanical properties of the clay [26].

"Tools are never neutral,"–Pérez Gómez, because they "underpin conceptual elaboration" and "the entire process of form production" [27]. In digital manufacturing processes, there exist indeterminacies that may be examined as possible design drivers. Digital fabrication methods demand precise control over the process since specified instructions must be followed before materialization, the shift from a computer model to its physical manifestation is seldom smooth. These technologies control and precision make the transfer from digital to physical easier: intricate and complex digital models may be realized without the need for established hands-on abilities. Despite these benefits, digital fabrication technology does not always encourage innovative design discoveries, or a design approach guided by production [23]. According to Gürsoy [23], beyond the seamless materialization of the digital model, digital manufacturing techniques may continually influence design ideation through emerging tectonic features [23], investigating the ramifications of a digital fabrication method that is not based on imposed and strict formalisms but unique and contextual ones.

Based on the available literature review and room for research, this research explores the possibility of growing mycelium in a clay and hemp mixture, preparing this material for a 3D printer, and exploring its emergent properties as it goes through several facets; such as material preparations manually, digital algorithms, geometries, and generated G-codes, printing limitations such as nozzle sizes, print settings (print speed, extrusion speed, layer height, and nozzle width), printing base (dynamic or static), the material of the print base whether it adheres to the material or not, etc. [2]. Pondering upon the determinants mentioned above, this research investigates the emergent properties of the novel material for 3D printing and the impact it has on the resultant object's form and finishes.

2 Methodology

For a fair investigation, the material preparation step remains of utmost significance as it helps determine the properties of the extrudable paste. *Pleurotus Ostreatus* is a specific mycelium species used in this research [9]. Mycelium requires nutrients to grow, which it usually extracts from lignocellulosic biomass; for this investigation, hemp shives are used as the substrate for the experiments with mycelium. This research method was conducted through material experiments, algorithmic iterations, and a series of 3D printed experiments. A variety of material mixes and samples with different G-codes with varying determinants were extruded with the help of a 3D clay extruder to elaborate various attributes dependent on the material and manufacturing procedure. The study was divided into four stages: (i) manual material testing–determining the optimal ratio for mycelium, hemp shives, and clay for extrudability, (ii) hardware and software configuration- preparation of G-codes, (iii) preparation of material for the 3D printer in sterile conditions and adding mycelium to it, (iv) 3D printing a series of forms with differing G-code attributes, as well as samples with different curve toolpath geometries that have the aim of elaborating printing strategies of the material based on the digital fabrication process.

2.1 Material Investigation

The additive manufacturing of mycelium, hemp shives, and clay allows for more precision of the desired form. The biomass mix is printed, and mycelium is harvested throughout the form to bind the clay and hemp together. As the aim was to have an extrudable mix, an array of physical experiments was carried out experimenting with different ratios of clay and hemp to evaluate the most optimal extrudability with the varying ratios of hemp shives. Clay is especially well suited for 3D printing studies because it allows for extensive modification before, during, and after the printing process. Unlike other materials that can be 3D printed, clay has specific qualities such as viscosity, elasticity, and texture that determine the final printed shape [28]. Apart from those mentioned above, the 3D printer and its extrusion nozzle, the 3D printing parameters, curve toolpaths, and the varying Extrusion value € will also influence the final form as additional parameters. The layer-by-layer deposition of clay is evident in the final form as the layers remain visible despite not being anticipated in the digital 3D model, making it a characteristic of 3D printed clay objects and giving it a tectonic aspect. Moreover, since clay is printed while it is still wet, it can contribute to the sagging of overhangs under gravity as the clay does not harden immediately, making it another emergent property. 3D printing with clay retains the traits of unpredictability in this way [23]. The procedure becomes as crucial as the final product when unanticipated variables must be identified during the process. The accuracy, efficiency, and consistency of 3D printing may no longer be the most enticing features in this new setting. Moreover, the extruded form requires time to dry. As a result, the printed items are exposed to the surrounding conditions and are still susceptible to modification after printing. This brings up new possibilities for a design and manufacturing process incorporating digital and analog techniques [23]. In the current research, no extra additives are used in the mix to singularly explore the properties of viscosity and extrudability of the paste and the growth rate of mycelium with a set ratio of mycelium and hemp. In order to investigate the physical characteristics of the paste to make it suitable for extrusion, the experiments are carried out in two stages. The first one is manual extrusion via a manual syringe to assess whether the mix is extrudable or not. The second stage is the 3D printing of the paste through the Delta WASP 40,100 Clay 3D printer with a 9 mm nozzle diameter [29] with an optimal ratio of clay-hemp mixes and additional computational parameters to evaluate the final form with overhangs. The mixture of mycelium, hemp shives, and water is not extrudable since hemp retains high quantities of water and does not form a cohesive mixture as hemp clumps and does not adhere to each other. Despite hemp's inability to form viscous mixtures, it remains relatively significant to the overall process as it reinforces the clay through the drying process while providing nutrients for the mycelium to grow through the form. For this exercise, hemp shives were ground and sieved by a number 35 laboratory mesh strainer with resulting fibers of length 0.5 mm. The fibers were mixed with varying percentages of clay relative to the weight of hemp to analyze its extrudability. The percentage of clay initially used was 70% clay and 30% of hemp, and the ratios were then altered to achieve the best viscosity for the mixture to make it optimal for printing. The later experiments made use of 75% clay and 25% hemp, followed by 80% clay and 20% hemp, and eventually used 85% of clay and 15% of hemp. Increasing the percentage of clay allowed for more viscosity of the mixture but simultaneously reduced the available substrate (hemp) for the mycelium to feed on.

2.2 Experimental Procedure for Manual Extrusion

Manual extrusions of the clay and hemp mixture were conducted to analyse the viscosity, extrudability, and adherence between layers of the mixture [30]. For this purpose, a standard lab syringe with a 50 ml capacity was used; the tip of the syringe was cut to approximately 9 mm in diameter to simulate the extrusion of a clay printer with a nozzle size of 9 mm. Another significant observation made during the research was the removal of the rubber on the plunger, known as the seal or plunger tip of the syringe, to prevent it from creating an air-tight space within the syringe shaft, as it made it impossible to extrude the paste due to the pressure build-up inside the syringe.

As previously mentioned, a hemp and water mixture are hard to extrude; therefore, it was mixed with clay to achieve the desired consistency. Experiments began by mixing a ratio of 70% clay and 30% of hemp, whereby 10 g of hemp were mixed with 23 g of clay by adding 45 g of water, as shown in Table 1. The percentage of water was proportional to the weight of hemp by keeping it constant at 4.5 times the weight of hemp. It was hard to achieve a printable extrusion due to the clumping of hemp and water as the mixture remained non-adherent, non-viscous, and non-continuous. Moreover, despite applying much manual pressure to extrude the paste, the results were undesirable, and only water came out of the syringe as it was squeezed out of the moist hemp that retained water.

Based on the results from the first round of manual tests, the second round of tests was conducted by increasing the percentage of clay in the mixture from 23 to 30 g (from 70 to 75% of clay), as shown in Table 1. Since there was no more addition of hemp to the mixture, the water content was also kept constant as there was still water retained in the fibers. The ratio between clay and hemp was 75 and 25%, respectively. However, despite increasing the weight of clay, the mixture remained non-extrudable as the water kept separating from the mixture and dripping out through the syringe.

Further experiments were conducted on varying the ratios of clay and hemp as the following set of investigations made use of 80% clay (40 g) and 20% hemp (10 g) with no additional water added (Table 1). This mixture despite being semi-viscous and highly self-adherent was not extrudable as the mixture stuck in the syringe due to the adherence of the mixture within the syringe. Finally, using a mixture of 85% clay (57 g) and 15% hemp (10 g) with water was kept constant at 45 g (Table 1) was optimal as the water was not separating from the mixture, with fine extrudability,

Table 1 Experiments: different Clay-hemp ratio

Material	Clay 70% Hemp 30%	Clay 75% Hemp 25%	Clay 80% Hemp 20%	Clay 85% Hemp 15%
	Weight (g)	Weight (g)	Weight (g)	Weight (g)
Clay	23	30	40	57
Hemp shives	10	10	10	10
Water	45	45	45	45

Fig. 1 Left: first trial—manual extrusion. Experiment: Clay 85% and Hemp 15%. Right: second trial—manual extrusion. Experiment Clay 85% and Hemp 15%

viscosity, adherence, and maintaining its form as shown in Fig. 1. Based on the manual trials conducted, it was deduced that the most optimal ratio for clay and hemp for smooth extrusion and the best layer adhesion was 85% of clay and 15% of hemp with water as 4.5 times the weight of the hemp.

2.3 Material Preparation for 3D Printer

After having evaluated different ratios of clay and hemp shives to achieve the ideal mixture manually, experiments were further conducted with a Clay 3D Printer Extruder- the Delta WASP 40,100 Clay 3D printer, to investigate further the aforementioned hypothesis regarding the attributes of the clay-hemp mixture. The printer had a nozzle of 9 mm used for these experiments and pressure was kept constant at 0.4 MPa, throughout the printing.

In order to conduct the experiments through the 3D printer approximately 3.8 kg of material was prepared, manually to be loaded into the printer. Initially, hemp was ground in a 500 W blending machine to extract 300 g of 0.5 mm hemp shives, the hemp was divided into three portions, 100 g each to make separate batches. To each 100 g of hemp, 567 g (85%) of clay was added, along with periodically adding 450 g of water (Table 2). The mixture was kneaded by hand to achieve a soft and homogeneous consistency.

After creating three separate batches of the clay-hemp mixture, each weighing 1117 g, they were added into three separate autoclavable bags after which the mixtures were sterilized in an autoclave [9] at 120 °C for 1 h [9]. The clay-hemp mixtures after being sterilized were left to cool down for 24 h before adding mycelium to the mix. For each bag, 20% of the weight of the mix without water was taken as the amount of mycelium to be added, therefore, to each bag 133 g of mycelium was added,

Table 2 3D printer material preparation: Clay-Hemp ratio 85%:15%

3D printer, material preparation: Clay 85% Hemp 15% and Water only

Material	Weight (g)
Clay	1700
Hemp shives	300
Water	1350
	Total weight (without mycelium) 3350 g
	Total weight (without mycelium per bag) 1117 g

Table 3 3D printer material preparation: Clay-hemp ratio 85%:15% with Mycelium

3D printer material preparation: Clay 85% Hemp 15% and Water only

Material	Weight (g)
Clay	1700
Hemp shives	300
WaterK	1350
Mycelium (20% of the weight of Clay + Hemp)	400
Total weight (without mycelium) 3350 g	
Total weight (without mycelium per bag) 1117 g	

which was 20% of 667 g of the mix without water (100 g of hemp, 5667 g of clay), as stated in Table 3. Mycelium was mixed well into the clay-hemp mixture under sterile conditions to avoid any contamination. For the printing process the space around the 3D extruder, the filament cartridge of the 3D printer, and the extruder were all disinfected with alcohol both before and after loading the material to ensure maximum sterilization to prevent and reduce the risk of any possible contamination.

3 Geometry Investigation Related to Computational Design

3.1 Computational Design and Its Connection with the Extrudable Paste

To examine the extrudability, layer adhesion, and the deviation from the basic form-different algorithms were generated to examine these properties. The G-codes for the printer were generated via Grasshopper for Rhinoceros 7 [31].

3.2 Computational Design Aims to Exaggerate the Properties of the Paste

All investigations were conducted beginning at a similar point with a cylinder with a radius of 5 cm and a height of 6 cm. Firstly, the layer heights were examined considering the relationship of layer height with nozzle size and the resultant resolution of the form and layer adhesion properties. Three different G-codes with layer heights of 3, 4.5, and 6 mm (Fig. 2) were generated, respectively for the cylinder at a constant Feed rate (F) of 1500 mm/min and an Extrusion value (E) of 30 mm, with the air pressure, was fixed at 0.4 MPa. Researchers suggest, keeping the slice height to be one-third of the nozzle width as a general rule [32]. Jonathan Keep [32] in his publication—'A guide to Clay printing', suggests that the flatter your layer height is in relation to the width of the wall, the more stable your print will be, especially when the wall begins to build out or in [32]. You may prefer a more rounded look to your printed layers, but make sure they are properly packed together, or you will experience delayering during the drying process [32].

For varying other determinants in the experiments, the Feed rate (F) was kept constant, and the Extrusion (E) was altered for the cylinder with the layer height of 3 mm (one-third of the nozzle size-9 mm). Following Benay Gürsoy's investigation [23], for the first G-code, the E value was increased by 10% after thirty percent of the print was completed and in the second G-code the E value was decreased by 20% after thirty percent of the print was completed to understand the impact of the E value with respect to the material's layer adhesion, extrudability and the resultant properties of the form. For the next set of tests, the basic form of the cylinder was sliced at 3 mm, each alternate contour curve was selected, divided into 500 points each, and then each alternate point deviated on the XY-plane outwardly within the parameters of 0 to 15 mm bounds. The resultant points were then interpolated through a third-degree curve and eventually, all the contours were joined together to create a

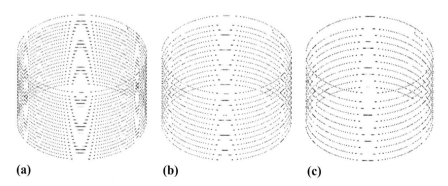

Fig. 2 Cylinder radius 5 cm, height 6 cm. 3(**a**) Contoured Layer Height 3 mm. 3(**b**) Contoured Layer Height 4.5 mm. 3(**c**) Contoured Layer Height 6 mm

curve toolpath (Fig. 3) to be printed at an F value of 1500 mm/min and an E value of 30 mm.

For the final experiment, the Grasshopper script was altered to generate an isocurve toolpath (as shown in Fig. 4) for the cylinder, to deviate from the regular curvilinear toolpath where the Z value gradually changes throughout the print. However, the isocurves were examined keeping the F 1500 mm/min and E value 30 mm.

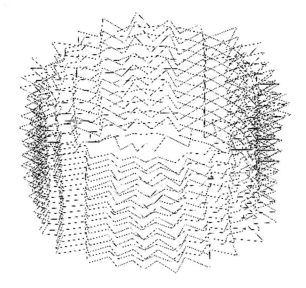

Fig. 3 Curve toolpath generated for uncertain overhangs

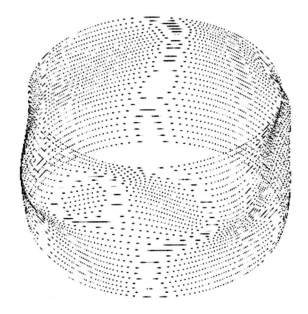

Fig. 4 Isocurve toolpath generated

4 Geometry Investigation Related to Computational Design

The mycelium, hemp shives, and clay paste were successfully printed for all the five G-codes mentioned above, their properties were analyzed, and emergent behaviors of the material were recorded to help further us reach conclusions regarding the properties of the Mycelial-Hemp and Clay mix. A cylinder of radius 5 cm and height of 6 cm was used as a basis to be exploited throughout each experiment. It was also initially recorded that printing on a glass surface resulted in no adhesion between the glass and the first layer of the material, so the material dragged along with the nozzle. However, replacing the glass with a wooden surface allowed the material to be fairly deposited and not drawn along with the nozzle.

4.1 Layer Heights

As displayed in Table 4, different Layer heights for the cylinder were 3D printed; 3, 4.5, and 6 mm, respectively. The toolpath with a layer height of 3 mm remained continuous as opposed to layer heights of 4.5 mm and 6 mm which started showing signs of a non-continuous layer, resulting in loss of form and layer adhesion as the layers kept breaking.

4.2 Varying the Extrusion Value

For the previous set of experiments, the Extrusion value (E) was kept at a constant value of 30 mm; however, for the following two experiments, the E values were increased and decreased to understand the impact of a non-constant E value on the resultant form. As presented in Table 5, Increasing the E value increased the width of the cylinder wall and more amount of material being extruded, which caused the cylinder walls to start caving inwards under self-load. On the contrary, decreasing the E value through the course of the extrusion resulted in non-continuous material deposition and tearing of layers.

4.3 Evaluating Overhangs in the Toolpath

In the last experiments the aim was to understand the properties of the paste by deviating specific points on the XY-plane and interpolated through a third-degree curve to create a curve toolpath. This experiment was carried out to assess the material's behaviour in maintaining the form and layer deposition where the extruder did

Table 4 Extrusion of Mycelial-hemp, Clay mix with variable layer heights

Layer heights	Side view	Top view
3 mm		
4.5 mm		
6 mm		

not deposit layer upon layer on the Z-axis. As shown in Table 6, the material was deposited smoothly, and the lower layers could withhold the shape of the upper layers with overhangs without any significant sagging or breakage of the form.

4.4 3D Printing of Isocurves

Lastly, a curve toolpath was modified from planar curves to periodic planar curves with the Z values changing throughout to understand whether the extrusion paste could withhold the shape or not. This was a step further from the previous experiment

Table 5 Extrusion of mycelial-hemp, Clay mix with increasing and decreasing E values

E value	Side view	Top view
Increasing E value		
Decreasing E value		

Table 6 Extrusion of mycelial-hemp, Clay mix evaluating overhangs

Curve toolpath	Side view	Top view
Overhang bounds (0–15 mm) F = 1500 mm/min E value = 30 mm		

to critically analyze whether the material was suitable for more complex extrusion or not. The material maintained its shape, and the isocurves were visible in the resultant form, with the considerable resolution, as seen in Table 7. The F and E values remained constant at 1500 mm/min and 30 mm, respectively.

Table 7 Extrusion of mycelial-hemp, Clay mix with Isocurves

Curve property	Side view	Top view
Isocurves F = 1500 mm/min E value = 30 mm		

5 Conclusions and Further Studies

The research successfully achieved a balanced ratio between mycelium, hemp shives, clay, and water without adding additives. Where clay and hemp are 85 and 15%, respectively, the amount of water added to the mix is best kept at 4.5 times the weight of hemp, and lastly, the mass of mycelium added should constitute 20% of the weight of the dry mix. After the above experiments were carried out, it was deduced that to keep the highest resolution of the form; it is best to keep the layer height one-third of the nozzle width. In this case, 3 mm layer height remained most suitable.

Altering the Extrusion value (E) during the printing helped us understand that in order to print with a clay 3D extruder like the Delta WASP 40,100 Clay 3D printer, it is best to explore the values of Feed rate (F) and Extrusion Value (E) in proportion to each other and that both values should be kept constant throughout the print to ensure smooth extrusion. From the above experiments, the most suitable Extrusion Value (E) was 30 mm with a constant Feed rate of (F) of 1500 mm/min. Moreover, the material could withhold its form as overhangs were introduced from a range of 0-15 mm without any deformation during and after the 3D printing as the material was able to maintain its shape. The isocurves were visible in the final form, with an acceptable resolution, as seen in Table 10, proving that the material was able to not only maintain the form with deviating points on the curve toolpath in the XY plane but also do the same as points were exaggerated along the Z axis.

The above-discussed ratios of clay, hemp shives, water, and mycelium allow an elastic paste to be extruded without breaking and simultaneously maintaining overhangs without significant changes to the final shape. This research paves the way for further research by questioning which digital parameters (in the G-code) and technical aspects of the 3D printer could be altered to thoroughly study the impacts of 'tools' in the process of 3D printing. Furthermore, exploring and controlling the conditions of the surroundings and the environment of the space the printed form is placed in for the conducive growth of mycelium by establishing measures and conditions that prevent contamination, maintain a sterile environment, and enhance the mycelial growth. Lastly, further studies are needed to understand the deactivation of the digitally fabricated mycelium composite form and investigate the mechanical, acoustic, and thermal properties of the novel extrusion material to reach the most suited architectural scale.

References

1. Lim, S., et al.: Developments in construction-scale additive manufacturing processes. Autom. Constr. **21**(1), 262–268 (2012). https://doi.org/10.1016/j.autcon.2011.06.010
2. Ghazvinian Ali.: A sustainable alternative to architectural materials: Mycelium-based Bio-Composites. (no date). Available at http://thenextgreen.ca
3. Myers, W. (Curator).: Bio design : nature, science, creativity. Museum of Modern Art, (2012)
4. Hebel, D.E., Heisel, F.: Cultivated building materials. De Gruyter, Cultivated building materials (2017). https://doi.org/10.1515/9783035608922
5. Additive manufacturing—General principles—Fundamentals and vocabulary. (2022). Retrieved May 26, 2022, from https://www.astm.org/f3177-21.html
6. van Wylick, A. et al.: Mycelium composites and their biodegradability: An exploration on the disintegration of mycelium-based materials in soil. In: Bio-Based Building Materials, pp. 652–659. Trans Tech Publications Ltd, (2022). https://doi.org/10.4028/www.scientific.net/cta.1.652.
7. Jones, M. et al.: Engineered mycelium composite construction materials from fungal biorefineries: A critical review. Materials and Design. Elsevier Ltd., (2020). https://doi.org/10.1016/j.matdes.2019.108397
8. Appels, F.V.W., et al.: Fabrication factors influencing mechanical, moisture−and water-related properties of mycelium-based composites. Mater. Des. **161**, 64–71 (2019). https://doi.org/10.1016/j.matdes.2018.11.027
9. Carcassi, O.B., Minotti, P., Habert, G., Paoletti, I., Claude, S., Pittau, F.: Carbon footprint assessment of a novel bio-based composite for building insulation. Sustainability. **14**, 1384 (2022). https://doi.org/10.3390/su14031384
10. Mycoworks.: Our Products—MycoWorks. (2022). Available at https://www.mycoworks.com/our-products. Accessed 26 May 2022
11. MOGU.: Home Mogu—mogu. (2022). Available at https://mogu.bio/. Accessed 26 May 2022
12. Corpuscoli, O.: Officina Corpuscoli » The growing lab—Objects. (2022). Available at https://www.corpuscoli.com/projects/the-growing-lab-objects/. Accessed 26 May 2022
13. Ecovative LLC.: Packaging—Ecovative. (2022). Available at https://www.ecovative.com/pages/packaging. Accessed 26 May 2022
14. Mycoworks.: Phil ross grows furniture with mushrooms—Mycoworks. (2022b). Available at https://www.mycoworks.com/phil-ross-grows-furniture-with-mushrooms. Accessed 26 May 2022
15. Fairs, M.: Mycelium Chair by Eric Klarenbeek is 3D-printed with living fungus. (2013). Available at https://www.dezeen.com/2013/10/20/mycelium-chair-by-eric-klarenbeek-is-3d-printed-with-living-fungus/. Accessed 26 May 2022
16. Stott, R.: Hy-Fi, The organic mushroom-brick tower opens at MoMA's PS1 Courtyard|ArchDaily. (2014). Available at https://www.archdaily.com/521266/hy-fi-the-organic-mushroom-brick-tower-opens-at-moma-s-ps1-courtyard. Accessed 26 May 2022
17. Heisel, F. et al.: Design of a load-bearing mycelium structure through informed structural engineering: The MycoTree at the 2017 Seoul Biennale of Architecture and Urbanism. (2017)
18. Klarenbeek & Dros.: Home—The growing pavilion. (2020). Available at https://thegrowingpavilion.com/. Accessed 23 Mar 2022

19. Meyer, V.: My-co space—V. meer. (2022) Available at https://www.v-meer.de/my-co-space. Accessed 26 May 2022
20. Blast Studio.: Tree Column, 3d printed mycelium column from used coffee cups. Available at https://www.blast-studio.com/post/lovely-trash-column. Accessed 23 March 2022
21. Goidea, A., Floudas, D., Andréen, D.: Transcalar design: An approach to biodesign in the built environment. Infrastructures **7**(4), 50 (2022). https://doi.org/10.3390/infrastructures7040050
22. Sheinberg, J.: Claycelium_ Living structures—IAAC blog. (2019). Available at https://www.iaacblog.com/programs/claycelium/. Accessed 26 May 2022
23. Gürsoy, B.: From control to uncertainty in 3D printing with clay. (no date)
24. Amarasinghe, P., Pierre, C., Moussavi, M., et al.: The morphological and anatomical variability of the stems of an industrial hemp collection and the properties of its fibres. Heliyon **8**, e09276 (2022). https://doi.org/10.1016/j.heliyon.2022.e09276
25. Lu, D., Miao, J., Du, X., et al.: A new method of developing elastic-plastic-viscous constitutive model for clays. Sci. China Technol. Sci. **63**, 303–318 (2020). https://doi.org/10.1007/s11431-018-9469-9
26. Rajeshkumar, G., Seshadri, S.A., Ramakrishnan, S., et al.: A comprehensive review on natural fiber/nano-clay reinforced hybrid polymeric composites: Materials and technologies. Polym Compos **42**, 3687–3701 (2021). https://doi.org/10.1002/pc.26110
27. Pérez-Gómez, A.:Persistent modelling. (2012)
28. Chan, S.S., Pennings, R.M., Edwards, L., Franks, G.V.: 3D printing of clay for decorative architectural applications: Effect of solids volume fraction on rheology and printability. Addit.Manuf. **35**, 2020, 101335, ISSN 2214-8604, https://doi.org/10.1016/j.addma.2020.101335
29. WASP Srl.: Clay 3D printer|Delta WASP 40100 Clay—3D printers|WASP. Available at https://www.3dwasp.com/stampante-3d-argilla-delta-wasp-40100-clay/. Accessed 26 Mar 2022
30. Soh, E. et al.: Development of an extrudable paste to build mycelium-bound composites. Mat. Des, **195**, (2020). https://doi.org/10.1016/j.matdes.2020.109058
31. Robert McNeel & Associates.: *Rhino—Rhinoceros 3D*. Available at https://www.rhino3d.com/ Accessed 26 Mar 2022
32. Keep, J.: A guide to clay 3D printing. (2020)

3D-Printing of Viscous Materials in Construction: New Design Paradigm, from Small Components to Entire Structures

Valentino Sangiorgio , Fabio Parisi , Angelo Vito Graziano, Giosmary Tina, and Nicola Parisi

Abstract The advent of industry 4.0 in the construction sector is profoundly changing paradigms that enhance the building construction sector. During the last decade, the experimentation of 3D-printing exploiting *viscous materials* has undergone unprecedented increases by construction companies. Even if reinforced concrete 3D-printing to construct buildings is growing fast, the use of other materials such as *clay* and *raw earth* is not yet affirmed both for the building *components* prefabrication or monolithic constructions. Currently, few 3D-printing applications with clay and raw earth have been experienced by research institutes and companies (e.g. Fablab-Poliba, Italy; Instituto de Arquitectura Avanzada de Cataluña, Spain; WASP company, Italy) for bricks, walls and entire buildings respectively. Beyond practical applications, the academic investigations focused on specific issues only (e.g. structural performances, new design for prefabrication or complex geometry printability). On the other hand, these examples are isolated, and a systematized design paradigm is still missing in the related literature. This chapter aims to define a new design paradigm for 3D-printing with viscous materials. A five-step procedure is proposed to achieve an effective design for both "*small components*" (to be assembled on site) and "*entire structures*" (to be printed in situ). The five steps will guide the reader towards the exploitation of the potential of the technology by experimenting complex shapes and by also respecting the actual limits of the machines. The

---

V. Sangiorgio (✉)
INGEO, University of Chieti-Pescara, 66100 Pescara, Italy
e-mail: valentino.sangiorgio@unich.it

F. Parisi
DEI, Politecnico di Bari, 70125 Bari, Italy

ICITECH, Universitat Politècnica de València, 46022 Valencia, Spain

F. Parisi
e-mail: fabio.parisi@poliba.it

A. V. Graziano · G. Tina · N. Parisi
ArCoD, Politecnico di Bari, 70124 Bari, Italy
e-mail: nicola.parisi@poliba.it

© The Author(s), under exclusive license to Springer Nature Switzerland AG 2024
M. Barberio et al. (eds.), *Architecture and Design for Industry 4.0*, Lecture Notes in Mechanical Engineering, https://doi.org/10.1007/978-3-031-36922-3_30

steps include: (i) Definition of the conceptual design; (ii) Parametric modelling, (iii) Slicing software; (iv) Performance and Printability simulation; and (v) 3D-printing.

Keywords 3D construction printing · Building design · Construction technology · Parametric modelling · Printability simulation · Slicing software

United Nations' Sustainable Development Goals 9. Build resilient infrastructure, promote inclusive and sustainable industrialization and foster innovation · 11. Make cities and human settlements inclusive, safe, resilient and sustainable · 12. Ensure sustainable consumption and production patterns

1 Introduction

Additive Manufacturing (AM) is a rapidly increasing technique that allows the production of three-dimensional objects from a Computer-Aided Design (3D CAD) model through the deposition of layers of material. In contrast with subtractive technologies, the additive process does not involve large quantities of waste materials and gives the opportunity to 3D print complex geometries and shapes [1]. Innovative opportunities in the design and architecture field have been recently created by AM development and the peculiar capabilities of this technology. Plenty of research has been carried out on design for AM and particularly on small-scale applications [2]. In this context, the challenge for designers is to create high-quality products meeting functional design requirements by using AM and considering technological limitations, advantages and capabilities for each process. Indeed, by exploiting such novel technology, designers can generate customized solutions with added value compared to industrial ones [3]. On the other hand, currently, well-defined procedures to exploit the advantages of AM are based on the technician's experience and general design guidelines are missing in the related literature.

This chapter is aimed at defining a new paradigm for the design of 3D-printing *components* (to be assembled on site) and *entire structures* (to be printed in situ). In particular, the chapter proposes five iterative steps to achieve an effective component or building design to be achieved with 3D-printing of viscous materials: (i) Definition of the conceptual design; (ii) Parametric modelling; (iii) Slicing software; (iv) Printability simulation; (v) 3D-printing (Fig. 1). The proposed approach is explained from both a theoretical and practical point of view in three subsections. *Firstly*, a specific paragraph explains in detail the five steps to achieve 3D construction printing from the novel conceptual design to the operational use of machinery. *Secondly*, the five steps are explained from a practical point of view by proposing the design and production of *small building components*. In particular, the generation of 3D-printed clay bricks with complex shapes is presented (Fig. 2, left). *Thirdly*, the proposed method is applied to build a *whole structure* to be printed directly in situ. In this case, the five steps are applied to redefine the conceptual design of the Nubian Vaults to be compatible with a full-scale and support-free 3D-printing production with raw earth (Fig. 2, right).

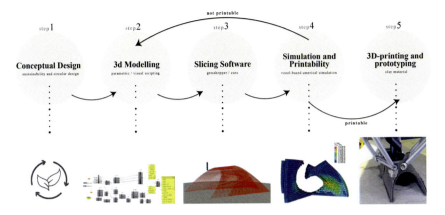

Fig. 1 The five steps to perform an effective design and 3D-printing with viscous materials

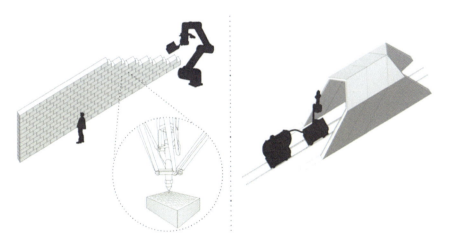

Fig. 2 3D printing of small components (clay bricks, left) and entire structures (Nubian vaults, right)

2 The Five-Step Procedure to Achieve an Effective Design and Prototyping in 3D-Printing

The proposed design paradigm is based on five steps to guide the reader to reach effective design and prototyping in 3D-printing of viscous materials (each step is described in a subsection below). In particular, the *first* step consists of the definition of the conceptual design by identifying the advantages and limitations of a 3D-printing-based production with viscous materials. The *second* step exploits the parametric modelling in order to ensure flexibility of production, customization, parametric analysis and optimization. The *third* step regards the slicing of the 3D model by setting all the 3D printing parameters (including both technology and materials). The *fourth* step is performed in order to analyze the performance of the component and simulate the 3D printing process to verify the effective printability of the designed model. If the printability simulation gives negative results, it is necessary to repeat steps 2, 3 and 4 until the combination of the design, modelling, technology and materials can produce a perfectly printable prototype. The *fifth* and last step consists of the effective 3D printing of the designed object.

2.1 Definition of the Conceptual Design

The definition of conceptual design for additive manufacturing with viscous materials is based on 3 fundamental points: (a) Review of available AM technologies with a focus on capabilities and limitations; (b) Definition of the aim of the project and selection of the most suitable technology for viscous materials for entire structures or components to be assembled; (c) awareness of general limitations of 3D-printable geometries.

(a) In the first stage of conceptual design, an accurate analysis of the current AM processes is required. Seven AM processes are identified in the framework of additive manufacturing standards development by the international standard ISO/ASTM [3]: Directed Energy Deposition, Powder Bed Fusion, Sheet Lamination, Binder Jetting, Material Extrusion, Vat. Photopolymerization and Material Jetting. In the field of architecture and construction, the most relevant and widespread technique is the material jetting processes with viscous materials (e.g. Contour Crafting [4], Xtree [5] and Cybe [6]). Cementitious materials have been the most explored in AM for being cheaper than other materials and characterized by suitable mechanical properties. Nevertheless, there are AM technologies aimed at raw earth and clay deposition [7]. Consequently, the guideline for designers is to select a set of available technologies by considering the following advantages of applying an AM in construction: (i) the possibility to build more complex, functional and customized geometries, composed by parts well integrated and more easily assembled; (ii) use of innovative mixtures as a combination of materials with different properties; (iii) automation of production with reduction of manual work (specifically for entire structures), (iv) possibility to optimize of the used material by enhancing the life cycle sustainability and reducing resource consumption, waste production and pollution [8].

(b) In the second stage, a specific AM system can be selected and applied, according to design objectives and required performances. Anyhow, the first selection of the specific AM system depends on the size of the element to be produced. Currently, two different functional approaches can be identified: (i) application of large-scale systems for the fabrication of entire structures on-site; (ii) application of small/medium-scale systems to produce small building components to be assembled. Secondly, the best AM technology can be selected on the basis of the availability, costs and geometric limitations of 3D printing.

(c) In the third stage of the conceptual design process, all the geometric limitations of 3D printing must be known and taken into consideration. The construction of angled or standard overhangs, bridges, bores and channels, excessively thin wall thickness and too small elements might not be correctly printable, strictly depending on the specific AM system and its characteristics. The thickness of the nozzle, the extrusion method and the material employed highly affect the geometry of overlapping layers. Once all the geometrical limitations are known, the designer can effectively realise the concept of the component or construction. In the end of the conceptual design, the draft of the component is defined together with the target performance to be reached (e.g. functional, structural, and thermal characteristics in order to satisfy minimum regulatory standards and user comfort).

> Note that the designer generally has fewer geometric limitations in conceiving various small components to assemble at forming large-scale structures.

2.2 Parametric Modelling

Parametric modelling belongs to the family of computer-aided design (CAD) concerning the use of computer-based software to aid in design processes. On the other hand, if compared with other tools, parametric modelling is able to build geometry by exploiting mathematical equations operated with visual scripting. This process provides the ability to change the shape of the model's geometry immediately when specific dimension values are modified. Consequently, once the conceptual design of the model is ready, it is necessary to identify all the dimension values that need to be parameterized (e.g. thicknesses, lengths, heights, curvatures, and parameters that define any internal fillings). The aim of parametric modelling is of basic importance since it establishes the flexibility and adaptability of the ideated component and allows fast customization, analysis and optimization of the final product. In addition, the achieved parametric 3D model can operate in synergy with specific performance analysis software. For this purpose, an iterative process can be set up to adjust the dimension values and improve the performance until the printability is reached and the target performances are satisfied (Sect. 2.4). A correct execution

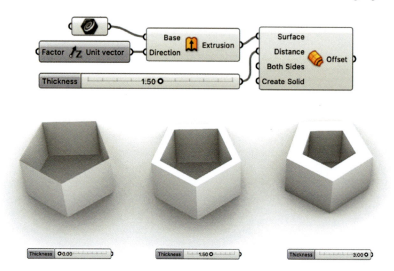

Fig. 3 Example of a parametric model with a parameterized thickness (Grasshopper software)

of the parametric model makes the subsequent phase of the iterative adjustment of dimension values simpler and more effective.

Operatively, the parametric modelling is carried out by using specific software (e.g. Rhinoceros and Grasshopper). In these tools, 3D modelling is achieved by using specific *components* (hereafter written in *italics*) to be dragged onto a canvas and adequately connected between them. Figure 3 shows how an example of a parametric model is generated starting from a pentagonal shape. The pentagon is firstly extruded (with *Extrude* component) in the Z cartesian direction and secondly, an *offset surface* is applied by parameterizing the thickness (one-dimension values only can be varied in this example). In the end of the process, the parametric model allows to easily change the thickness dimension values. In particular, the figure shows three configurations obtained by setting different thicknesses of 0, 1.5 and 3 respectively.

2.3 Slicing

For the 3D printing process, analogously for other digital fabrication technologies, a connection between the digital environment and the real physical one is necessary. This connection can be achieved through computer-aided manufacturing software (CAM) that generates a toolpath (in an alphanumeric language) to provide the machine with the processing instruction. For additive technologies, the CAM procedure to convert a 3D model into specific instructions for the printer is called "slicing". Indeed, the process operates a "slicing" of the three-dimensional model to obtain a set of layers re-creating the geometry. The output of the *slicing* is called G-Code and

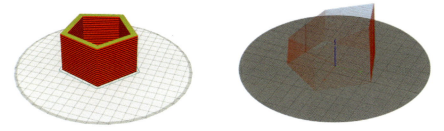

Fig. 4 Automatic (left) and parametric (right) slicing of a simple 3D model

consists of instructions for the 3D printer that can create the object in the physical environment by producing successive layers of material. During the slicing process, numerous printing parameters can be defined such as the layer height (related to final component resolution), the pattern of the internal fill or the parameters related to the printing supports.

In this section, two possible approaches are specified to set the slicing: (i) the automatic slicing by using specific software (e.g. Cura, Simplify3D or Slicer3D) and (ii) the parametric slicing to directly draw the printer toolpath.

The software to achieve automatic slicing are very fast and user-friendly but are also limited to specific algorithms. In fact, even if this software allows to change of numerous printing settings, the tool path is automatically generated by the program and cannot be entirely decided by the user. On the other hand, "parametric" slicing overcomes the drawback of automatic slicing and gives the user full control of the procedure. In this way, the creation of the G-Code is entirely controlled by the user without randomly generating infill or tool path as in the case of the more traditional slicing software. To provide an example, by using this tool it is possible to create toolpaths that are not planar to the printing plane (an option not yet available in slicing software). Figure 4 shows the slicing process obtained for an example geometry (the same of Fig. 3) obtained both with an "automatic" (left) or "parametric" (right) slicing that allows not planar layers.

2.4 Performance and Printability Simulation

The performance analysis is fundamental to achieve functional, structural, and thermal characteristics in order to satisfy minimum regulatory standards and user comfort. For this purpose, another benefit of parametric modelling is the ability to create ready-to-use input files for powerful performance simulation software (e.g. Abaqus). Typically, such software are based on finite element method (FEM) analysis that can be used to obtain solutions on a broad spectrum of engineering problems including thermal, structural and acoustic investigations. In the proposed five-step approach, the FEM is used in synergy with parametric modelling by setting an iterative procedure. The objective is to adjust the parameter of the model until the

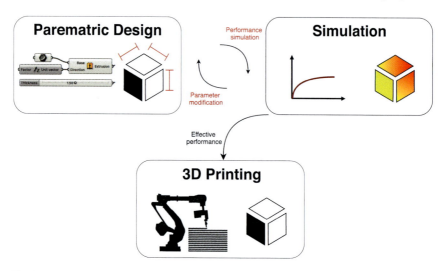

Fig. 5 Iterative process for performance simulation and parameter adjustment

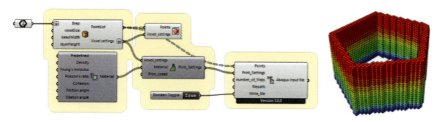

Fig. 6 The connection between Grasshopper and Abaqus software for FEM analysis

minimum performances defined in the conceptual design are satisfied. Such operation is simpler and faster if the parametric model allows changing all the key parameters influencing the performance of the component. Figure 5 schematizes the iterative process for performance simulation and parameter adjustment.

In addition, Fig. 6 shows the components used in Grasshopper to connect the model with Abaqus software and run a printability simulation (a ready-to-use file for Abaqus is generated in Grasshopper). The resulting FEM analysis concern a printability simulation of two different configurations of the thickness (same example of a pentagon extruded shown in the previous sections).

2.5 3D-Printing and Prototyping

The last step of the proposed procedure concerns the effective 3D-printing of the designed object. Even if the previous phases are carried out in a workmanlike manner, negligence in the final phase can compromise the quality of the result. Indeed, the products derived from 3D printing of viscous materials can be classified in three categories: (i) High-quality products meeting aesthetic, functional and structural requirements; (ii) Medium quality products not meeting aesthetic requirements with details not appreciable; (iii) Insufficient quality products not meeting aesthetic, functional and structural requirements, totally or partially collapsed.

To this aim, the current subsection provides a useful overview of the critical issues that can arise throughout the modelling and printing process. In particular, Fig. 7 shows 4 main fields that can generate criticalities: (I) Printing material, (II) Printer settings, (III) Design or (IV) Slicing. Every field is divided into more specific features affecting the results of a 3D printing process. In turn, each feature is associated with a specific effect derived. In the last part of the figure (on the far-right side), possible solutions to correct the wrong features are proposed. In the following discussion, a letter (**bold in brackets**) associates every specific feature (in **bold**) of the figure with the connected description.

Material features: The most important feature of the material employed is its viscosity (**a**). This feature leads to formal three-dimensional variations of the product and to its natural sagging as soon as the layers are deposited on top of each other. In this case, to solve a problem of a "too fluid" material it is possible to induce heat during the printing process or including additives in the mixture can be necessary to accelerate

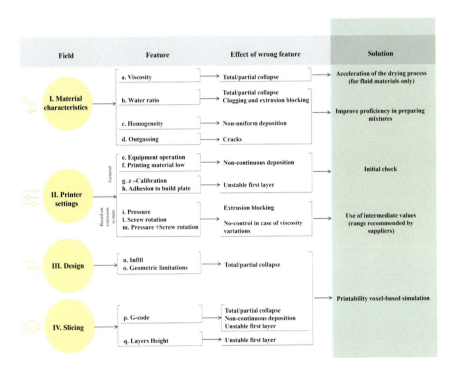

Fig. 7 Overview of possible criticalities during 3D printing

the hardening process. In preparing mixtures, the water ratio is also crucial (**b**). Indeed, the material can result too solid or liquid according to the quantity of water introduced and respectively being less extrudable or tending to collapse immediately after extrusion. The mixture needs to be homogeneous and outgassed as well (**c, d**). Different consistencies and the presence of trapped air in the mixture may result in the printing of a product characterized by superficial defects such as cracks or that tends to collapse. Wrong features, in this case, can be avoided by preparing a mixture in a workmanlike manner with a suitable quantity of water, well-pressed and mixed.

3D printer settings: during the printing process, equipment failures (**e**) or material depletion (**f**) can occur. If the operations stop for some reason, the parts already printed get wasted or recycled and the 3D-printing process has to restart from the beginning in order to be continuous. Moreover, the origin on the z-axis needs to be well calibrated (**g**) and the extruded material must adhere to the build plate (**h**). Difficult adhesion is due to the different nature of viscous materials compared with the materials of which the plate is made. Wrong settings cause unstable deposition of the first layer with the consequent collapse of the entire extruded structure. Equipment full checks before launching a 3D printing process are considered to be indispensable. For what concerns the specific extrusion system employed, it is necessary to set proper values for piston pressure (**i**), screw rotations number (**l**) or both (**m**) in hybrid systems technologies. In case of viscosity variations and to prevent extrusion blocking, the use of intermediate values among ranges typically recommended by suppliers is suggested to be able to intervene on these settings during printing. To give an example, if the maximum pressure allowed by the machinery is being used and it is necessary to have a greater quantity of extruded material (e.g. the material is too dense) it is not possible to make this correction during the printing process by further increasing the pressure.

Design: the project of the infill structure determines the quality of the product. Inadequate infill regions (**n**) as well as the extrusion of "not printable" geometries (**o**) can lead to the total or partial collapse of the structure.

Slicing: errors in G-code (**p**) can produce an unstable deposition of the first layer and non-continuous deposition or total/partial collapse of the structure. Furthermore, layer height (**q**) needs to be well configured in order to avoid the fabrication of low-quality products in which the details are not appreciable.

In general, an effective FEM printability simulation (proposed in Sect. 2.3) should detect errors and prevent total or partial collapse during the 3D-printing processes.

3 The Five Steps Applied to Clay Bricks with Complex Infill

In the building sector, the possibility of creating **small components** to be assembled on-site can be very effective both for achieving customised design and high performances. The flexibility of the 3D printing can be useful to get complex geometries as in the case of the hi-performance clay bricks of Sangiorgio et al. [9]. In particular, this section shows how the proposed five iterative steps have been applied to obtain clay bricks with complex internal filling geometries [9].

3.1 The Conceptual Design of Complex Bricks

The first stage of the conceptual design concerns the review of available AM technologies. Among the available technologies, in the case of 3D printed brick, the preliminary review focuses on Material Jetting exploiting clay material. During the first screening, the range of 3D printers of the Wasp company (Italy) is selected for the closeness of the supplier to the used laboratory (FabLab Poliba, Bari, Italy) and for economic convenience. In the second stage, since the proposed application concern the production of small/medium-scale building components, the machine Delta Wasp 40,100 for clay is selected. Indeed, such a machine represents a good compromise among availability, costs and geometric limitations of 3D printing (typical of a delta printer) [10]. In the third stage, the conception of the new printable bricks is defined by considering the limits of the selected machine. Beyond the limits of the selected 3D printer, two ideas guided the design: (i) The observation of the traditional and widespread external shapes and internal wall thickness of the brick; (ii) The use of periodic minimal surfaces to generate the complex internal configuration of the bricks.

Note that the periodic minimal surfaces are geometries well known for the high mechanical performances also used in ceramic 3D printing. To this aim, such geometries are selected to be integrated into clay bricks [11].

3.2 Bricks Parametric Modelling

The purpose of the parametric modelling in the case of 3D printed bricks is twofold: (i) reach flexibility and adaptability of the elements allowing for a quick change of different periodic minimal surfaces as filling geometries; (ii) connect the parametric modelling with an accurate FEM analysis software in order to simulate the brick printability.

For the first purpose, the obtained visual script allows for quick modify the dimension of the brick (length, width and height), the number of minimum surfaces that can be inserted in the brick, and the thicknesses of the infill and of the external walls. In particular, the script is divided into four clusters of components in order to achieve *external shell generation* and thickness, *infill generation* exploiting the minimal surfaces, *infill thickness* and *finalization* of the brick respectively (Fig. 8). For the second purpose, the connection of the parametric modelling to the simulation software is achieved with Abaqus (FEM analysis software) through a plug-in named *VoxelPrint* for Grasshopper [12]. The components of *VoxelPrint* allow to create a voxelization of the designed geometries (convert the geometry to a set of identical

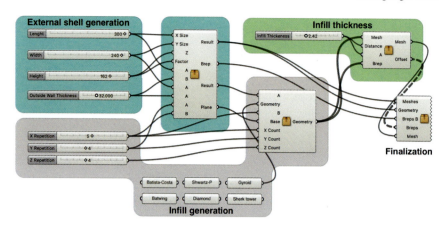

Fig. 8 Visual scripting (with clustered components) to generate complex bricks

finite elements) and consequently achieve a ready-to-use input files for simulation in Abaqus [9].

3.3 Slicing Software

In the 3D printing of the proposed clay bricks, a specific path of the extruder is not necessary. Consequently, among the two slicing processes described in Sect. 2.3 the simplest approach, based on common slicing software such as Cura, is chosen.

The key parameters selected for the slicing include a layer thickness of 1 mm printed by a nozzle of 2 mm diameter and a printing speed of 30 mm/s.

3.4 Simulation to Verify Infill Printability

With the aim of achieving a printable brick with a complex infill exploiting minimal surfaces, an iterative process is set to adjust the dimension of the parametric model until the cells of the brick result printable from simulation. The parameters are varied by respecting the limitation of the Italian regulatory [9]: the minimum thickness of the external walls (external shell) is 1 cm while the thickness of the internal walls is considered to be at a minimum 0.8 cm. In addition, the external shell of this first prototyping is 15 × 12.5 × 9 cm.

Starting from the lowest values of internal fill thickness (0.8 cm) and the number of cells (which cause the total collapse of the bricks) the iterative process led to an increase in the number of cells up to 4 × 5 × 5 and an internal thickness of 3 mm. Indeed, to provide an example, Fig. 9 shows the analysis of the internal geometry

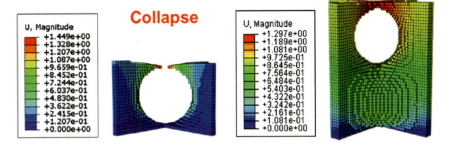

Fig. 9 Brick cell collapse or cell perfectly printable

of "Sherk Tower" that collapses with a thickness of 1.5 mm and became perfectly printable with a thickness of 3 mm.

3.5 3D-Printing and Prototyping

In conclusion, the prototypes of the bricks are printed by using a Delta Wasp 40,100 for clay by respecting all the suggestions proposed in subsection 2.5, specifically for obtaining a suitable viscosity of the printing material. To sum up, Fig. 10 shows the Slicing in the software Cura, the 3D printing and the final brick.

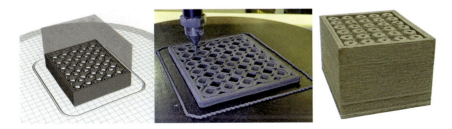

Fig. 10 Slicing in the software Cura, 3D printing and final brick

4 Design of a Support-Free 3D Printed Nubian Vaults

Large-scale 3D-printing of viscous materials is becoming relevant in the research sectors of space and post-disaster architecture for giving the opportunity to extrude local earthen resources. The aim of this section is to propose a preliminary approach for 3D printing structures entirely made of raw earth. Due to technological limits in 3D-printing overhangs [13], structures are typically combined with traditional flat roofs in which just vertical walls or columns are additive manufactured. 3D-printing of roofs is a challenging process, and temporary or permanent supports may be necessary. However, the use of supports made of materials different from the locals should be avoided in extreme environments. In fact, the design and transport of formworks could be expensive and the realization hard or dangerous for workers. In the course of history, several techniques of masonry have been developed for building massive structures without the use of supports, such as Nubian vaults, Catalan vaults, and Persian vaults [14]. Considering the similarities between additive manufacturing and masonry, especially in compositional and structural fields [13], a further goal of this section is to determine a suitable strategy for closing 3D-printed structures by applying masonry constructive principles with a focus on the Nubian vault.

4.1 The Conceptual Design of Nubian Vaults

The Nubian vault is a characteristic structure of the ancient region of Nubia composed by raw-earthen bricks held together by mortar. The bricks were assembled to form arches inclined at an angle of 45 degrees in relation to the level of ground. In this structure, formworks are not needed since each added layer is supported by that previously printed and already stable. Such construction principle has been reworked to be used in 3D printing with raw earth. In particular, the prototype has been conceived for being realized in the rural area of "Malandra Vecchia" (Abruzzo, Italy) and is inspired by the traditional earthen architecture present in its surroundings. In the forms, the project reminds of the "hut archetype" (Fig. 11, left). The innovative technological "hut" is thought to be made of inclined layers. Two components can be observed: (1) the part 1 is characterized by hut-shaped layers tilted at 45°, (2) the part 2 works as support for the tilted part and is composed by layers extruded horizontally.

4.2 Vault Parametric Modelling

Also in this case, the aim of the parametric modelling is allowing quick changes in the model in order to optimize it and conduct a FEM analysis for the printability simulation.

Fig. 11 Components of the nubian vault prototype and FEM printability simulation

Parameters consistently affecting the performances of the prototype are: (i) width of the span, (ii) height of the ridge line, (iii) height of the eaves line, (iv) angle of inclination of the hut-shaped layers, (v) angle of inclination of the sidewalls, (vi) thickness of infill, external walls and roof.

4.3 Parametric Slicing Process

The printing process is of basic importance in this case, consequently, a parametric slicing process is set in order to draw an effective toolpath. Indeed, this method allows the creation of 45° inclined toolpath and enables different printing settings such as layers height and infill regions for each part of the prototype.

4.4 Performance Simulation

In order to verify the performance of the inclined toolpath, a suitable FEM simulation (Abaqus software) is applied by investigating the deformation trends of the layers (according to the executed slicing). The right part of Fig. 11 shows the deformation of the upper layer (heavier in the central part of the arch) that remains acceptable to ensure printability.

4.5 3D-Printing and Prototyping of the Vault

In the last phase, the effects of the correct slicing method can be noticed by printing a prototype and verifying the application of Nubian vault principles during production. Figure 12 shows the small-scale tests to simulate both the 3D printing with horizontal and inclined layers performed at "Fablab Poliba" Digital Fabrication Laboratory and by using a WASP Delta 40,100. The prototype realized with horizontal layers shows several printing criticalities (it collapses on the top and does not allow for

Fig. 12 The prototypes with different slicing methods: horizontal and 45° layering

Fig. 13 Render of the application of the 3D printed Nubian Vault in Malandra Vecchia (Italy)

the full closure of the structure) while the inclined layers are perfectly printable. In conclusion, Fig. 13 shows a rendering of the application of the Nubian vault in Malandra Vecchia, a fraction of the municipality of Casalincontrada, (Abruzzo, Italy).

Acknowledgements This research was funded by the European Union—European Social Fund—PON Research and Innovation 2014-2020, FSE REACT-EU.

References

1. Pacillo, G.A., Ranocchiai, G., Loccarini, F., Fagone, M.: Additive manufacturing in construction: A review on technologies, processes, materials, and their applications of 3D and 4D printing. Material Design & Processing Communications **3**(5), 253–256 (2021)

2. Vaneker, T., Bernard, A., Moroni, G., Gibsona, I., Zhangd, Y.: Design for additive manufacturing: Framework and methodology. CIRP Ann. Manuf. Technol. **69**(2), 578–599 (2020)
3. ISO/ASTM 52900: Standard terminology for additive manufacturing—General principles — Terminology. In ASTM International, West Conshohocken (2015)
4. Khoshnevis, B., Brown, M. E.: Techniques for sensing material flow rate in automated extrusion. United States Patent WO/2009/070580 (2008).
5. Jipa, A., Dillenburger, B.: 3D Printed Formwork for Concrete: State-of-the-Art, Opportunities, Challenges and Applications. 3D Printing and Additive Manufacturing 00, 1–22 (2021).
6. Camacho, D.D., Clayton, P., O'Brien, W.J., Seepersad, C., Juenger, M., Ferron, R., Salamone, S.: Applications of additive manufacturing in the construction industry—A forward-looking review. Autom. Constr. **89**, 110–119 (2018)
7. Stampante 3d per case Crane WASP. https://www.3dwasp.com/stampante-3d-per-case-crane-wasp/. Last accessed 19 Jan 2022
8. Peng, T., Kellens, K., Tang, R., Chen, C., Chen, G.: Sustainability of additive manufacturing: An overview on its energy demand. Addit. Manuf. **21**, 694–704 (2018)
9. Sangiorgio, V., Parisi, F., Fieni, F., Parisi, N.: The new boundaries of 3d-printed clay bricks design: printability of complex internal geometries. Sustainability **14**(2), 598 (2022)
10. Bell, C.: 3D printing with delta printers. Apress, Berkeley, California (2015)
11. Restrepo, S., Ocampo, S., Ramírez, J. A., Paucar, C., García, C.: Mechanical properties of ceramic structures based on Triply Periodic Minimal Surface (TPMS) processed by 3D printing. J Physics: Conference Ser. **935**(1), 012036. IOP Publishing, Bristol (2017)
12. Vantyghem, G., Ooms, T., De Corte, W.: VoxelPrint: A Grasshopper plug-in for voxel-based numerical simulation of concrete printing. Autom. Constr. **122**, 103469 (2021)
13. Carneau, P., Mesnil, R., Roussel, N., Baverel, O.: Additive manufacturing of cantilever—From masonry to concrete 3D printing. Autom. Constr. **116**, 103184 (2020). https://doi.org/10.1016/j.autcon.2020.103184
14. Cowan, H.J.: A history of masonry and concrete domes in building construction. Build. Environ. **12**, 1–24 (1977). https://doi.org/10.1016/0360-1323(77)90002-6

A Study on Biochar-Cementitious Composites Toward Carbon–Neutral Architecture

Nikol Kirova and Areti Markopoulou

Abstract Concrete is currently the second most consumed material in the world, with it's core ingredient–cement, emitting 900 kg of CO_2 into the atmosphere with each ton produced. Carbon sequestering amendments can help tackle the negative impacts of the concrete construction sector. As the concrete construction industry along with the computational design tools and manufacturers evolve, they move towards new materially and structurally informed construction methods aiming at carbon neutral solutions. The research investigates the use of biochar (carbonised bio-waste) as an aggregate for sustainable cementitious composites. The literature on the topic suggests that the main limitation of biochar as a concrete amendment is the reduction in mechanical properties associated with an increase in biochar content. This chapter, however, approaches this as a design challenge for maximising carbon sequestration while reaching optimal structural performance. Based on the novel biochar-cementitious composites developed in IAAC (the Institute for Advanced Architecture of Catalonia), the work further investigates and defines new design principles for traditional building elements so that they can obtain a carbon–neutral or negative footprint. The programmability of the material mix, for instance, combined with additive manufacturing and computational design tools makes it possible to design and manufacture functionally graded architectural elements whose properties vary based on the mechanical or qualitative performance, among others. The research chapter uses three case studies and examines fabrication strategies as well as new techniques for material allocation and performance-driven design toward carbon-negative materially informed building elements. Following the properties and origins of biochar, a novel approach for building structures acting as "carbon sinks" is proposed.

Keywords Biochar · Cementitious composites · Carbon–neutral · Carbon sinks · Material-driven design

---

N. Kirova (✉) · A. Markopoulou
Institute for Advanced Architecture of Catalonia, 08005 Barcelona, Spain
e-mail: nikol.kirova@iaac.net

A. Markopoulou
e-mail: areti@iaac.net

United Nations' Sustainable Development Goals 9. Industry, innovation and infrastructure · 11. Sustainable cities and communities · 12. Responsible consumption and production

1 Introduction

The increase in carbon dioxide (CO_2) emissions has been linked to rising global temperatures for nearly a century [52]. In 1938, Guy Callendar connected carbon dioxide increases in Earth's atmosphere to global warming which led to the formulation of the "Carbon Dioxide Theory of Climate Change" by Gilbert Plass in 1956 [34]. In the 1950s, a dramatic increase in the burning of fossil fuels to make electricity, oil for vehicles, produce steel and manufacturing overall vastly accelerated the rate of CO_2 being pumped into the atmosphere. Since then, atmospheric CO_2 has been increasing at an exponential rate [41]. The construction sector may not be the only contributor to this phenomenon, however, it is among the biggest.

The buildings and construction sector accounted for 36% of final energy use and 39% of energy and process-related CO_2 emissions in 2018, 11% of which resulted from manufacturing building materials and products such as steel, cement and glass [52]. Among those construction materials concrete stands out with its high production rate, being the second most used material in the world, which greatly increases its associated environmental costs. Compared to other conventionally used construction materials, concrete has a rather low embodied energy and embodied carbon but due to its excessive use, it is now considered to have a detrimental impact on the environment [14].

According to sustainability research, the world has exceeded four out of seven planetary boundaries. The areas of climate change, biodiversity loss, nitrogen cycle and land use have left the so-called safe operating space for humanity [47]. The built environment is a major contributor not only to CO_2 emissions but also to resource depletion. Regardless of the urgency to act and reduce the carbon footprint of the construction sector, concrete is still the second most consumed material, with three tonnes per year used by every person in the world [51].

The main ingredient of concrete is cement. Cement requires limestone and gypsum to be heated to about 900 degrees Celsius so that water molecules can be removed, making a powdery material that when needed can be chemically activated with the addition of water to form a durable solid, binding aggregated to produce what we now call concrete. For each ton of cement produced, 900 kg of CO_2 is emitted into the atmosphere making the cement industry alone responsible for approximately 5% of global anthropogenic CO_2 emissions [12].

Due to the global urbanisation trend, the housing demand in cities is rising. It is estimated that 60% of the new urban areas needed in 2030 have not yet been built [22]. It is also suggested that the built environment will nearly double by 2050. Aside from the fact that the concrete industry suffered a decline in 2020, the demand for concrete is projected to continue rising in the near future.

As previously discussed, the increase in CO_2 emissions has been directly linked to climate change and temperature rise. A point is reached where climate change cannot be mitigated solely by the reduction of the production and use of materials such as cement. There is an urgency to develop novel methods and approaches in the field of architecture and design.

Various frameworks, policies and international agreements have been put in place as acting measures to prevent further temperature rise and resource depletion as well as provide a better and healthier living environment. In 2015, the Paris Agreement, a legally binding international treaty on climate change, was adopted by 196 parties. Its goal is to limit global warming to well below 2, preferably to 1.5 degrees Celsius, compared to pre-industrial levels. To achieve this long-term temperature goal, countries aim to reduce the emissions of greenhouse gases such as CO_2 [10]. In the same year, the United Nations adopted the Sustainable Development Goals (SDGs), a universal development agenda, which goals need to be fulfilled by the year 2030 and by all UN countries worldwide [22]. Among the 17 goals depicted in the SDGs agenda, this chapter is going to address the following: Goal 9: Industry, innovation and infrastructure; Goal 11: Sustainable cities and communities; Goal 12: Responsible consumption and production.

Furthermore, in December 2019, the European Commission presented for the first time the Green Deal, which sets out how to make Europe the first climate-neutral continent by 2050 [16]. On the 14th of July 2021, the European Commission adopted a set of proposals to make the EU's climate, energy, transport and taxation policies fit for reducing net greenhouse gas emissions by at least 55% by 2030, compared to 1990 levels. Achieving these emission reductions in the next decade is crucial to Europe becoming the world's first climate-neutral continent by 2050 and making the European Green Deal a reality [16].

Various other reports by the European Commission suggest how to move towards a more sustainable future. For instance, adopting Advanced Manufacturing Technologies was proposed as a particularly beneficial research area, especially with regard to the construction sector where additive manufacturing or 3D printing has been suggested as an interesting field of research [15]. Another potential indicated innovation area is Advanced Materials. It has been indicated that lightweight materials can play a crucial role in Europe as new materials replace the old ones with more optimised and environmentally friendly alternatives [53]. Furthermore, with regard to the construction industry, it has been suggested that agricultural waste can be used to mitigate the large environmental impact of the industry and potentially link the industries. It is proposed that waste material from the end of the life cycle from the agricultural field can become useful material at the beginning of the life-cycle construction materials.

The "Advancing Net Zero" (ANZ) campaign conducted by the World Green Building Councils is a global programme working towards total sector decarbonisation by 2050. There are various ideas pushed forward within the programme such as energy efficiency and carbon offsetting. A net-zero carbon vision acknowledges the time value of carbon emissions from materials and construction. One of the proposed ways to achieve the Net-Zero vision is through radical cross-sector collaboration similar to the one proposed by the European Commission [56]. Collaboration between industries can aid circularity in design and economy.

Building on the idea of cross-industry collaboration, the "From Farm to Facade" EU initiative proposes exactly that by combining end-of-life products with new materials, the agricultural and construction sectors can mutually benefit [37]. For example, if waste produced by the agricultural sector is used as a product in the construction sector not only does waste obtain a newfound value and is therefore upcycled but raw materials are not extracted and therefore depleted in the same manner. This scenario aids both sectors to become more sustainable, producing less waste and decreasing their energy consumption and loss.

In 2016 a report by the European Environmental Citizen's Organisation for Standardisation (ECOS), included biochar among other waste materials that can be used as a fertiliser in the agricultural industry. Up until recently, there were no links between biochar and the construction sector, but this has changed. Research publications on the non-agricultural uses of biochar have increased exponentially since 2017.

2 Background

2.1 Cementitious Materials with Reduced Environmental Impact

As previously stated, there is evidence that the concrete construction sector has a large negative impact on the environment. To mitigate and reduce this impact, various solutions are investigated and developed both from the material engineering side and from the construction technology side. Research is being carried out to develop mitigation strategies to control CO_2 emissions from the cement manufacturing sectors while retaining concrete's performance. These strategies include changes in the raw materials, the introduction of carbon-negative amendments, novel fabrication techniques, and material reduction, among others.

In this context, life cycle assessment and the effectiveness of alternative bio-based ingredients in cement manufacturing are also investigated. Additionally, notable studies on the efficacy of waste and ash-based amendments in cement have been reported. Although the addition of biobased amendments such as fly ash or biochar are cost-effective, they may reduce concrete's performance. Studies have found that the addition of biomass to cement causes its degradation due to the alkaline nature of the cementitious material which results in a decrease in durability [55]. In ash-based amendments, the presence of organic and inorganic impurities can be linked to the further depletion of cement's properties [28, 40].

There are several types of industrial waste or by-products that have pozzolanic properties and therefore can substitute cement or be added to concrete to reduce the amount of cement needed. Some of the most used by-products are: blast furnace slag which derives from the production of iron and steel (in blast furnaces); silica fume—a by-product of the manufacturing of silicone; and fly ash—a by-product of the coal industry.

Regarding blast furnace slag, research shows that partial substitution of cement with slag can improve durability and reduce the carbon footprint creating a more sustainable alternative to traditional concrete. However, both blast furnace slag and silica fume are not commonly used in concrete buildings but rather in infrastructure in harsh environmental conditions. Furthermore, comparing the costs associated with the use of slag and silica film to the costs of Portland cement it is still more economically favourable to use Portland cement.

Fly ash is one of the most adopted industry amendments for concrete. It has significant environmental benefits. With the right quantity of fly ash, the structural performance of concrete is improved but with the wrong balance, the decrease in strength is significant. Fly ash is mostly used in infrastructural elements rather than buildings mostly because it is better suited for precast rather than on-site construction but also because it has a degree of unpredictability. The main limitation of fly ash is that it does not allow air to penetrate through the material which makes it not suitable for cold climates and winter construction. It is also linked to the industry that produces it which is the coal industry. The coal industry is very harmful in terms of air pollution. Fly ash is abundant in developing countries that are still relying on coal for heating and energy production but in places such as Europe sufficient amounts of fly ash are not available. Making it harder to integrate into the construction sector.

Cement replacement and sustainable concrete amendments are slowly penetrating the concrete market. One of the main limitations of the slow adoption of these material systems is the associated economical costs and dependability on external, often highly polluting, industries. The research proposes an alternative solution for a sustainable concrete amendment that derives from the agricultural industry. It is proposed that biochar can be used within cementitious materials in order to decrease the carbon footprint of concrete and cementitious mortars. Biochars are produced throughout the globe by the local agricultural industries making them an abundant product in comparison to some of the previously discussed industrial by-products.

2.2 Biochar in Architecture

Organic waste or by-products from plants and/or animals, in particular waste from the agricultural industry or forestry management, have been underutilised for decades. Their primary use is for fuel production by direct combustion [3, 25]. Biochar is the solid residue obtained from the controlled thermal decomposition (pyrolysis) and gasification of biomass under limited oxygen [5, 54]. Notable control parameters are heating rate, temperature, and feedstock type (Tomczyk et al. 2020). The quality and type of feedstock affect the structure, chemical composition, and yield of the produced biochar. Biochar is black in colour, rich in carbon, has a large specific surface area, and has a porous structure. Aside from carbon, compounds such as ash, nitrogen, oxygen, and sulphur have been identified in biochar [27].

Biochar production has increased significantly in the last decade. Companies such as Carbofex, a Finish biochar producer, are developing new technologies for biochar production with minimal environmental impact. Biochar has been produced mainly as a by-product from agricultural waste and feedstock management, as well as forestry management. Biochar is commonly used for soil remediation [29, 46], carbon sequestration [49], energy storage and conversion purposes [26], etc. The most common use of biochar, currently, is as a fertiliser. The architectural and construction applications of biochar are underdeveloped and have been explored more rigorously only in the past several years.

The properties of biochar have been explored for the first time in an architectural context by the Ithaka Institute. As demonstrated by previous research the addition of biochar can lower the thermal transfer and enhance the water absorption of cementitious mortars and other materials. These qualities create an ideal condition for isolating buildings and regulating excess humidity. A study from the Ithaka institute has demonstrated, on a building scale, how the use of biochar plaster can regulate humidity, provide carbon negative finishing, and enhance thermal insulation while maintaining breathability [44]. It is also suggested that when biochar is applied in brick form with a thickness of up to 20 cm it can be used as a substitute for Styrofoam as an insulative material. Furthermore, if biochar is to be mixed with clay and lime instead of cement at the end of life, the material can be directly used as a compost closing the cycle of the construction material.

"Made of Air" was founded in 2016 by architects Allison Dring and Daniel Schwaag, who have developed a type of thermoplastic that can be easily moulded and holds large quantities of biochar-derived organic waste mixed with sugar cane. In April 2021, thermoplastic was installed on a building for the first time. An Audi dealership in Munich was clad in seven tonnes of hexagonal panels. It was assessed that the entire facade cladding accounts for 14 tonnes of carbon that would be sequestered until the recycling of the cladding tiles [23].

Another example of a biochar-derived product that has application in architecture comes from a U.S.-based company called "Interface". They produced their first carbon-negative carpet tile in 2021. The company specialises in carpet backings made of biochar and other composites. Because the company's product is on the market there was little information found on the material composition and carbon footprint.

The state-of-the-art analysis shows that currently there are no biochar-derived building elements or structural elements for the architectural sector. All the products mentioned above are related to either interior use or exterior cladding systems. However, it is interesting to note that all of the products reached the market after 2020. The interest in biochar and its ability to sequester carbon is rising and more and more companies, industries and researchers are looking into the properties of the material.

The research proposes that there are further applications of biochar cementitious composites that can be used within the architectural and construction sectors to reduce the carbon footprint of structural building elements.

2.3 Biochar Cementitious Composites: Performance

The following section discusses several categories of biochar research. An extensive literature review on biochar in architecture from a material science and engineering perspective is used to set the ground for the research. The main categorisation is done in three categories based on material performance: Environmental Performance; Mechanical Performance; Qualitative Performance. Following, several case studies of existing architectural applications of biochar are presented and analysed. Furthermore, three applied research projects where the author has contributed to the state of biochar-cementitious materials research are described and analysed in detail.

2.3.1 Environmental Performances

Carbon Sequestration

Carbon sequestration, including geological, oceanic, and terrestrial ecosystem sequestrations, plays an important role in mitigating global warming. Embedding carbon sequestering techniques and materials in industries such as the construction industry can aid in meeting environmental goals with minimal cost. Among many construction materials that are associated with carbon sequestration, indubitably timber stands out. Wood is one of the most environmentally friendly construction materials. Recently, it has gained a lot of research interest in reaching higher design freedom with the aid of digital fabrication methods. Even so, concrete is still the number one globally used construction material due to the associated design freedom, cost efficiency and ubiquity. Making it important to investigate carbon sequestration possibilities of concrete and cementitious mortars. Biochar can help the concrete construction industry meet the current environmental demands due to its long-term carbon sequestration potential. Furthermore, as cement or a fine aggregate replacement in concrete or mortars it does not only sequester carbon but additionally decreases the consumption of cement; hence, saving energy and reducing the discharge of CO_2 in cement production (Li et al. n.d). Several studies on cement mortars demonstrate the carbon sequestering potential of biochar when investigated as an amendment to the composite [16, 17, 20, 21, 33, 42, 48].

Most studies focused on carbon sequestration propose treatments to improve the performance of the material. One of the discussed options is pre-treating biochar with a 0.1N HCl solution. It is suggested that this technique can improve the carbon sequestration ability of biochar-cement mortars [56]. The research shows that this treatment provides additional benefits to the compressive strength of the mortars, with up to 15% of biochar the compressive strength at 28 days of curing decreases by only 4Mpa (from 44.7 to 40.6). Furthermore, at only 5% biochar concentration the data show that the mortars have a 75% CO_2 sorption, giving evidence of the carbon sequestration potential. Another study looks in detail at the difference between saturated and unsaturated with CO_2 biochar in mortar mixture [56]. It concludes that the use of saturated biochar triggers off carbonation, which affects the strength development and porosity of mortar. Biochar saturated with CO_2 can control the ingress of water into mortar primarily due to fine particle size, but the water absorption is drastically increased. Therefore, it is recommended to use unsaturated biochar as a means of sequestration of stable carbon in cement composites. It was interesting to find out that in a previous publication by the same research team it was discussed that saturated $CO2$ biochar can have an enhanced carbon sequestration potential [56]. Such evolution of the research perspective is encouraging to researchers in the field of sustainable cementitious materials through the perspective of biochar. This shows that there is a need for further and more in-depth research before biochar-cementitious composites become marketable to the industry in a convincing manner.

Furthermore, it is important to point out that the type of organic material used to produce biochar would produce biochar with different carbon content. There are two main factors to be taken in account: the amount of carbon inherent in different types of feedstocks and the production process. The International Biochar Initiative classifies the organic carbon content into the following three categories [9]:

- Class 1 = >60% (e.g., from plant and tree waste)
- Class 2 > 30% < 60% (e.g., from poultry manure mixed with organic bedding material, paper sludge)
- Class 3 > 10% < 30% (e.g., from cow manure, sewage sludge).

This classification proposes that it is more desirable to use plant or tree waste feedstock to produce biochar. Another aspect to be taken into account is the transportation of the material. The closer the source is to the manufacturing facility and/or site the lower embodied carbon and therefore the higher carbon offset is generated.

The above-discussed studies on biochar as a carbon sequestering material in cementitious mortars demonstrate not only the growing research interest in the field but also that it is backed up by data showing the potential that the integration of biochar in those material systems has. These studies are part of a larger body of research that aims to combat the carbon emission of concrete construction while making as little change to the construction method as possible. I believe that to make a real and disruptive impact on the industry we have to consider not only how to improve the material but also how to design better with the improved material.

Vegetation Compatibility

As previously stated, soil fertiliser products are one of the main widely recognised applications of biochar. There is persistent evidence that charcoal made of biowaste can improve the fertility of soils. The first notions of the positive impact that biochar has on vegetation have been associated with the "Terra preta" soil type. *"Terra preta"* has a characteristic black colour due to its high charcoal content. It was created by farming communities between 450 BCE and 950 CE in the Amazon Basin to combat the low fertility of the local soil. The biophilic properties of biochar have been recorded and studied ever since.

In the field of architecture, the biophilic properties of biochar have not been widely explored; however, there are some studies that demonstrate that biochar improves the vegetation compatibility of vegetation concrete. It was demonstrated that in pervious or vegetation concrete, biochar can accelerate plant growth and provide a healthy microbial environment at 5% content by weight [58]. Additionally, biochar pervious concrete samples show both greater compressive strength and splitting tensile strength when the biochar content is below 6.5%, above that concentration these strengths are compromised [39]. These studies outline that there is still room for future research in order to better understand how biochar can be used to promote growth and biodiversity in the built environment. Furthermore, it is suggested that the vegetation compatibility and even possible growth-enhancing properties can be beneficial to the end-of-life of biochar-cementitious composites.

Mechanical Performances

Compressive Strength

The compressive strength of biochar cementitious materials has been rigorously investigated within the past decade. It has been demonstrated that the mechanical properties differ based on the biochar type, content, and pyrolyzing temperature. The various studies show different trends when it comes to the effects biochar addition has on the compressive strength of concrete and mortars. Overall, most studies agree that a high concentration of biochar compromises the strength of the composite while low concentrations (below 2%) can improve the strength due to better hydration. It is also agreed that the effect of the type and origin of the biochar has a great effect on the mechanical properties of the cementitious composite and calls for further investigation and mapping out of these differences. For example [2], investigated the compressive strength of biochar produced from rice husk, poultry litter, and paper mill sludge. Concrete blocks with 0.1%, 0.25%, 0.5%, 0.75%, and 1% biochar in relation to the total volume of the composite were produced. Concrete with biochar obtained lower compressive strength than conventional concrete. After 28 days of curing the study found that the 1% addition of biochar improved the strength of the concrete. The researchers encouraged the use of a higher percentage of biochar in concrete production without surpassing 1%. By further investigation, these results were explained by the improved hydration during the curing process.

The study of [8] supported the observations of [2]. The compressive strength of concrete with the addition of biochar derived from dry distiller grains was investigated. In particular, the study looked at the effects of replacing sand and aggregate within the composite with 1.2% and 3% of biochar. The findings showed a minor increase in the mechanical properties after the addition of biochar. The addition of 3% biochar yielded maximum strengths of 21 MPa (when replacing sand) and 22 MPa (when replacing aggregates).

In a study by [19], the optimum concentrations of biochar for increasing the compressive strength of cement mortar were found to be 1% and 2%. When compared to standard mortar, a significant increase of 22% and 27% was observed. Following that, biochar addition beyond 2% resulted in a reduction in strength.

Furthermore, [4] reported maximum strength on the concrete samples with added 5% bagasse biochar. In contrast, [32] observed a decrease in compressive strength when as high percentages of biochar as 5% were added. It was suggested that this decrease was caused by the high-water retention capacity of the biochar. Based on these findings, we can conclude that the compressive strength of the concrete is highly influenced by the feedstock used in the production of biochar [45], also agreed that the properties of the feedstock material have a great impact on the compressive strength of the composite. By comparing the outcomes of the above-mentioned studies, we can conclude that the feedstock, pyrolysis temperature, and particle size can considerably affect the mechanical properties of biochar/cementitious composites, regardless of the biochar content.

Some alternative research focusing on high-performance concrete shows that biochar may not be suitable for this application. [11] investigated the effects of replacing quartz powder with saw dust-based biochar at 2% and 5% in ultra-high-performance concrete. The results after 28 days of curing showed a reduction in the compressive strength by 13% and 14% respectively. The authors suggested that this is due to the biochar's poor properties in comparison to the other constituents (silica fume, silica sand, and quartz powder) in the ultra-high-performance concrete.

Furthermore, [31] investigated how biochar and MgO will affect the compressive strength of concrete when used in a synergetic matrix. A weed tree-based biochar was used and added at 2%, whereas the MgO was added at 4 and 8%. The study showed that MgO in higher quantities reduced the strength of the composite but the addition of biochar to the MgO counteracted that reduction in the long run and therefore had an overall positive impact. Adding 2% of biochar to the 8% MgO resulted in an increase of 6% in comparison to the control sample after over 3 months of curing.

One of the few studies where higher concentrations of biochar have been examined is [35]. The study investigated the effects of biochar addition to cementitious mortars in three different proportions (5%, 10%, and 20% of cement weight). The compressive strength of the biochar mortars at room temperature was recorded at 35, 39, 28, and 16 MPa for 0%, 5%, 10%, and 20% biochar addition, respectively. Similarly, to other studies, the conclusion is that the addition of biochar in higher quantities than 5% leads to a decrease in compressive strength. Furthermore, it can be concluded that even though the effects of feedstock type, pyrolysis temperature as well as particle size have an undeniable effect on the mechanical properties of biochar-cementitious composites, high concentrations of any form of biochar will always lead to a decrease in the compressive strength. The compressive strength of concrete is directly linked to its composition. It is demonstrated that by introducing biochar into the mixture the compressive strength decreases. In this research, the above state correlation between compressive strength and biochar concentration will be used as a design driver for functionally grading the amount of biochar in cementitious composites.

Tensile Strength

Concrete is known for its weak tensile strength which is improved by the addition of reinforcement. Flexural strength is a metric that represents the tensile strength of homogeneous materials such as concrete and mortar. The rule of thumb is that the flexural strength of concrete is about 10–20% of the compressive strength. There are studies examining the tensile strength of biochar-cementitious composites. Cracks in the cement structure may form because of factors such as stress concentration, drying, and shrinkage, which may increase under loading [1]. As a result, microcracks may develop into macro cracks or become connected to the adjacent microcracks, forming branches, and resulting in failure. Another important consideration is the brittleness of the concrete materials, which has a direct impact on the tensile strength of the cement. Biochar could be used as an efficient particle reinforcement to increase the tensile strength of cement.

Most of the studies that investigate the effect of biochar addition on the tensile strength of the biochar-cementitious composite agree that with about 1% addition of biochar the tensile properties are improved [15, 35]. It is suggested that the higher specific surface area of the biochar is the main reason for the increase in flexural strength, which contributed to the enhanced interaction with the matrix. It is also evident that the type of feedstock used had a great impact on the tensile enhancement. For instance, studies with hazelnut shells, cherry pits and peanut shells proved more beneficial in comparison to coffee powder and sawdust [28, 30]. It is worth noting that, based on the particle size, type, and preparation of biochar, different concentrations are required to optimise different mechanical properties. The concentration also affects the amount of water needed to improve the performance of concrete.

Fire Resistance

Studies on the fire performance of concrete have shown that, compared to building materials such as steel and wood, concrete demonstrates the best fire resistance [13]. The addition of reinforcements further increases the thermal stability of the concrete. However, according to previous research, the size and form of the reinforcements may not have a significant influence on thermal stability [43]. In addition, reinforced concrete exhibits an increase in strength up to 450C and decreases as the temperature is elevated beyond 600C (Jackiewicz-Rek et al. n.d). When produced at high temperatures of pyrolysis, biochar creates strong C–C covalent bonds that make it very stable and insusceptible to temperature elevation. Even though fire resistance is not the focus of the research, it is important to attain a deep understanding of the impact biochar has on various properties of cementitious composites. Based on the literature review it is suggested that this will improve the fire resistance of cementitious composites such as concrete.

2.3.2 Qualitative Performances

Thermal Conductivity

The thermal conductivity of concrete is varying from 0.62 to 3.3 W/(mK) for standard concrete, or 0.4–1.89 W/(mK) for concrete with lightweight materials [6]. Based on a study of nine concrete samples containing varying quantities of biochar it has been established that biochar has a positive effect on the insulating properties of the concrete. It is suggested that this effect is due to the high porosity and the low thermal conductivity of biochar (Berardi and Naldi, 2017). Even though there the research showed some inconsistency with thermal conductivity of samples with biochar concentration below 5% the effects of poor distribution seem to be negated at higher biochar loading levels, demonstrated in the samples containing 10% and 12% by weight.

Other research also backs up the findings and shows that increasing the aggregate volume fraction increases the thermal conductivity of the concrete sample [36]. However, these studies are typically performed at much lower water to cement ratios, so the excess water content that the biochar demands may have an unexpected impact on the insulating properties. Generally, the thermal conductivity of a material decreases as the material density increases. However, it is possible that this could also come down to the build-up of the biochar.

Moreover, the thermal conductivity of the concrete samples is consistent across the temperature range. This shows that the biochar would be effective in insulating buildings in different weather conditions. Biochar addition has a positive effect on improving the thermal insulating properties of the concrete [8]. The results of this study show that biochar addition improves the thermal resistance of the concrete, even at lower concentrations. However, the insulation would still need to be used in building applications with the concrete incorporating biochar, but the decreased thermal conductivity would undoubtedly contribute to improved thermal efficiency and limit thermal bridging effects.

An intriguing study published in the "Science of the Total Environment" journal suggests a new perspective of biochar as a building material with improved hygrothermal properties [36]. It was demonstrated that biochar-mortar can be used as a functional building material to improve the hygrothermal performance of the building envelope. Two types of biochar were compared within a biochar-mortar composite with a mixing ratio of the biochar from 2 to 8%. The thermal conductivity of biochar-mortar composites was decreased as the biochar addition increased. Furthermore, the water vapour resistance factor of biochar-mortar composites increased by 50.9% compared to the reference specimens. The study proposed that biochar-mortar composites can contribute to humidity control without facilitating mould growth and additionally improve the thermal insulation of cementitious mortars.

Humidity Regulation

As biochar is created through the process of pyrolysis it commonly has high porosity and surface area on a microscopic level. This can vary based on the conditions of pyrolysis and type of feedstock but often the produced biochar has high water retention capacity due to its porosity. As previously stated, the most common use of biochar is in the agriculture industry, where it serves as a fertiliser that retains moisture while preventing mould growth. However, the inherent relation between biochar and water opens the possibility for other applications in the fields of water purification and construction.

Looking into biochar applied in previous concrete, studies show that by increasing biochar content, the water adsorption increases, and the porosity of the specimen decreases which helps in regulating the growth of microorganisms [57]. Furthermore, when used in pervious concrete pavement blocks, biochar can help reduce surface runoff, purify water pollution, and mitigate the urban heat island. An investigation on how to prolong the evaporative cooling effect of paving blocks by incorporating biochar particles, as hygroscopic filler, show that replacing a small quantity of cement by biochar could effectively improve the evaporative cooling performance [50].

Acoustic Dampening

Sound absorption is the loss of sound energy when sound waves come into contact with an absorbent material such as ceilings, walls, floors and other objects, as a result of which, the sound is not reflected back into the space. Acoustics comfort is an increasingly important topic between architects and designers due to its role in boosting productivity and reducing anxiety among other benefits. One of the metrics used to measure sound absorption is the sound absorption coefficient (SAC). The SAC for plain cast concrete is about 0.02, indicating that about 98% of the sound energy is reflected by the surface. Even though acoustics is a complex field of study, where many parameters have to be accounted for, designing for specific acoustic conditions can be tricky. The three main variables to be taken into account when designing spaces for acoustic comfort: are surface geometry, materiality and depth of spatial separator.

Biochar as an amendment for cementitious composites is shown to improve the SAC over a range of frequencies [8]. The study compares biochar to activated carbon, and even though both the addition of biochar and activated carbon had a noticeable impact on the sound absorption across the entire frequency range. It is suggested that the sound energy dissipation within the interconnected pore networks in the concrete created by the biochar addition, was responsible for the high sound absorption coefficients. Even more encouraging is that biochar seemed to have the same effect as the activated carbon in this regard, with both the 10% and 15% samples showing near identical curves as that of the concrete with the 7.3% activated carbon. While this could be due to the higher concentrations of biochar, it would be expected that the considerably higher surface area and associated porosity of the activated carbon would result in higher sound absorption properties.

3 Digital Matter and Intelligent Constructions

The architectural applications of biochar have been studied in several projects developed in the Digital Matter and Intelligent Construction (DMIC) research studio taking place in the first year of the Master of Advanced Architecture in the Institute for Advanced Architecture (IAAC). The DMIC studio tackles the environmental challenges of the building sector with research on the implementation of advanced digital technologies of computational design, material computing, human–computer interaction, and artificial intelligence coupled with the latest tools in digital and robotic fabrication. The studio introduces a model of materially responsive and circular architecture that presents possibilities for designing novel performances and dynamic metabolisms in the building industry.

Starting from allocating problems and opportunities in the local context, each year students aim to identify potential resources that can be upcycled into performative and adaptive building components for different architectural features. Students dive deep into analysing material opportunities that are embedded into the fabrication process, creating a coherent and informed narrative of the material life-span while using computation techniques and digital fabrication in order to establish workflows to design more sustainable and contextualised buildings. The design studio researches the implementation of computed, active or zero emissions material systems.

Biochar is studied as a carbon-negative matter that can augment and engage certain material performances. During the development of the below-discussed projects, the focus was on investigating the different applications of biochar in terms of a material matrix, fabrication techniques and architectural design. Various mixtures were developed and tested for mechanical properties, thermal insulation and humidity regulation. Fabrication techniques from casting to robotic material allocation were explored in order to allow design freedom and multi-material optimisation through a functionally graded design process. Additionally, the design that manifests from the material properties was explored and proposed within the conclusions of each research project.

3.1 Research Introduction and Significance

As discussed in the previous chapter, biochar as an amendment for cementitious composites has been thoroughly explored from a material engineering perspective. The addition of small quantities of biochar within concrete has many beneficial properties such as carbon sequestration, improved vegetation compatibility, fire resistance, thermal insulation, humidity regulation and acoustic dampening. However, by introducing biochar to the composite matrix the mechanical properties decline. In particular, it is demonstrated that the quantity of biochar is proportional to the decrease in both compressive and tensile strength. To deal with this, the research looks into the adoption of functionally graded material and design approaches that would allow for the optimal allocation of lower or higher quantities of biochar in correspondence to the structural needs. The design possibilities of the material system are mostly unexplored, especially in the context of maximising the carbon sequestration potential aiming to mitigate the environmental impact of concrete construction.

The research looks into three case studies where the material system has been applied through a performance-driven design framework in order to demonstrate the potential and implications of the use of biochar in concrete building construction. The research hypothesis is that through the adoption of biochar-cementitious composites, within the construction industry, buildings can be transformed from carbon sources to carbon sinks. The following research projects investigate the material developments, fabrication possibilities and architectural applications of biochar-cementitious composites through an empirical study based on physical and digital experimentation.

3.2 Research Projects

3.2.1 Cast in Carbon (2019)

"Cast In Carbon" is a research project developed over the period of six months in 2019. This investigation started the research on biochar in the DMIC in IAAC. The project studies the impact of large quantities of biochar within material systems that can be used in the construction sector, initially in clay and cement composites but eventually narrowed down to cementitious composites. The project was inspired by the idea that the carbon cycle can be altered as excess CO_2 is removed from the atmosphere and sequestered within building structures and envelopes. The main premise is that as carbon is stored in buildings it converts them from "carbon sources" into "carbon sinks" holding carbon for up to decades. The research focuses on the physical properties of biochar as a standalone material and as a part of the material system, replacing environmentally harmful construction materials in existing scenarios.

Biochar is the material obtained when organic matter undergoes thermal decomposition under the limited supply of oxygen at relatively low temperatures (<700). It is a very stable carbon-rich material, which can sequester carbon for thousands of years. On average 1 kg of finished biochar can sequester up to 1.8 kg of carbon dioxide. A life cycle assessment is conducted to further understand the net embodied carbon of biochar (Fig. 1). The feedstock from which biochar is produced can affect the properties and carbon sequestration potential of the final material system. For instance, hardwood feedstock has a higher value of negative embodied carbon in comparison to softwood feedstock. This is due to the amount of carbon that the tree photosynthesises during the natural growth process. As biochar is commonly produced from forest management feedstock, the embodied carbon of the feedstock together with the processing, transport and pyrolysis conditions have to be taken in consideration. It is estimated that softwood biochar will have a carbon neutral embodied carbon value whereas hardwood biochar will have a negative embodied carbon value.

Biochar can be obtained from pyrolysis of biomass in the temperature ranges of 200-900 °C. The biomass is heated inside a pyrolysis kiln up to the required temperature with a limited supply of oxygen. The by-products obtained from this process are biochar, bio-oil and syngas. The temperature at which the pyrolysis occurs plays an important role in the physical and chemical properties of the obtained biochar (Fig. 2) [35–41]. Higher temperature (900 °C) results in lower density and microporous structure whereas lower temperature (200 °C) results in higher density and macroporous structure. The density and microporous structure of the biochar has a direct impact

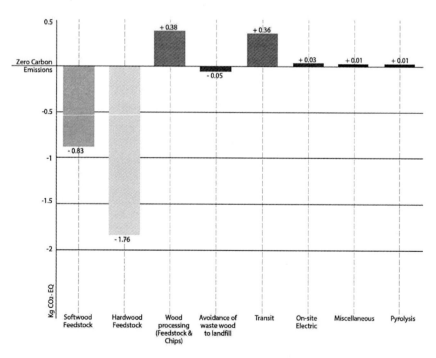

Fig. 1 Embodied carbon in production processes to obtain biochar alongside the embodied carbon of softwood and hardwood-based biochar

on its performance including the thermal conductivity and absorption of both gases and water. The programmability of the properties of biochar with the conditions of pyrolysis opens the possibility to modify and create specific performances based on the context or minimise the embodied energy of the material.

In the first phase of the material, research biochar was explored within two composite bases: cement and clay. A range of specimens was produced to be tested for compressive strength with a hydraulic press. It was observed that the clay composites were less stable and exhibited large shrinkage. At the time, the testing equipment in the lab was limited to 100 PSI pressure. All cement-based mixtures with less than 60% of biochar have strength above the threshold in comparison to the clay-based mixtures where the samples with less than 50% of biochar were above the threshold (Fig. 3).

Furthermore, it was demonstrated that the addition of biochar decreases both the weight and compressive strength of the material. In comparison to the studies in the subchapter "Mechanical Performances" where the higher concentration of biochar within the composites was 20%, this study explored large quantities of biochar reaching 90% [35]. However, the trend is consistent as the addition of biochar above 10% has proven to be detrimental to the mechanical properties of the composites. Based on this first experiment the research continues to investigate cement-based composites as they appear more suitable for the construction sector.

Further tests on the developed biochar-cementitious mortars (BCMs) with varying proportions of biochar are conducted looking into the following performances of the material system: compressive strength, humidity retention and thermal conductivity.

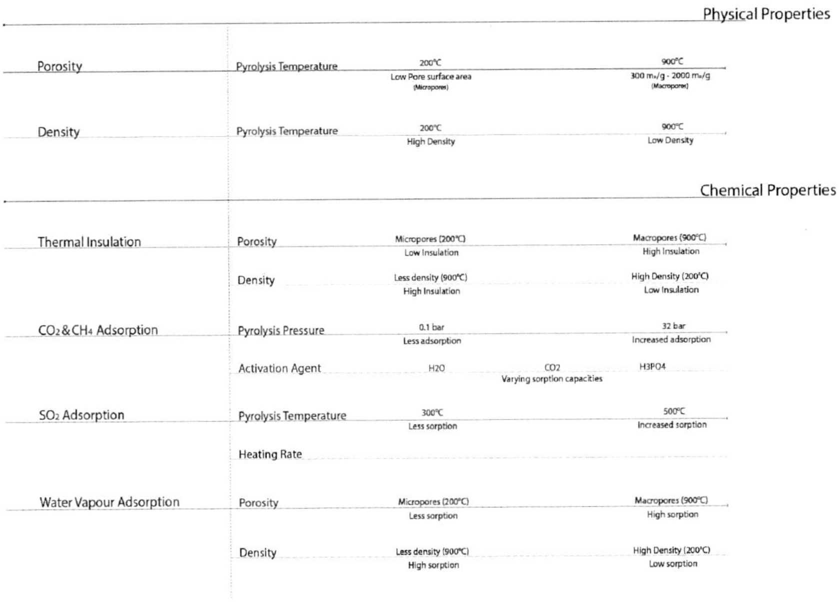

Fig. 2 Programming biochar properties based on pyrolysis conditions

Fig. 3 Cement composites

Multiple specimens of 50/50; 40/60; 30/70 proportions of cement and biochar as well as 100% cement control specimens are produced and left to cure for 28 days.

The following experiment is conducted and compressive strength measurements are collected using a hydraulic press at the Universitat Politècnica de Catalunya material testing laboratory. Due to imperfections on the edges of the specimens the test is not conclusive. However, it is determined that the compressive strength decreased exponentially between the pure cement specimen and the 50% biochar specimen and it continued to decrease as the biochar quantity reached 70% (Fig. 4). Further tests have to be carried out and more samples have to be tested to get a better estimate on the actual compressive strength of the composites.

Specimens from the same series are used for a water absorption test. The test is conducted within a control chamber with a humidifier over 2 days. The weight of the samples is measured thirteen times over the testing period and the relative humidity within the chamber is recorded (Fig. 5). It is evident that the amount of biochar is proportional to the decrease in weight; however, no distinct effect of the increase in relative humidity on the weight of the specimens with biochar is observed. The

A Study on Biochar-Cementitious Composites Toward Carbon–Neutral ...

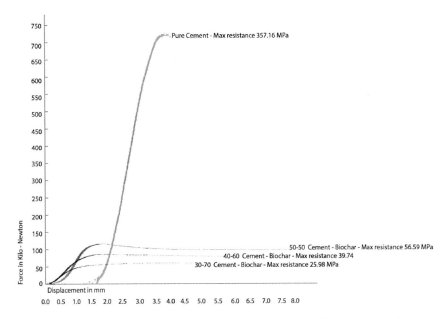

Fig. 4 Compressive strength of the developed samples as recorded during student investigation

specimen of 70% biochar exhibited slightly higher water absorption in comparison to the rest.

Furthermore, a second test on the water absorption is conducted following a different method. The speciements are left in water for 2 h and their change in weight is measured again for 2 day. It is observed that it takes 1 h for the samples with biochar to fully hydrate whereas the pure cement samples continue hydration for the full 2 h. The 50/50 BCM specimen shows the largest water absorption and retention in comparison to all other material mixtures (Fig. 6).

The last test from the series is on thermal conductivity. Rectangular specimens with 200 × 200x20mm dimensions are developed with the same material composition. The specimens are tested using a custom thermally insulated apparatus where one side of the sample is exposed to heat and measurements are taken from both sides of the specimen (Fig. 7). The biggest temperature difference between the front and back faces of the samples is measured in the 40/60 BCM. The test is inconclusive as only one set of samples is tested however, based on the results obtained it is concluded that biochar within the material mixture can improve thermal insulation.

In conclusion, the set of experiments demonstrated that the higher percentage of biochar within the cementitious composites resulted in lowered mechanical properties, improved water absorption and thermal insulation. Further research is conducted to reflect the limitation of the material system, specifically the lack of sand, fibres and aggregates within the composition.

The "Cast in Carbon" research project proposes a modular, functionally graded design approach taking advantage of the material properties without compromising

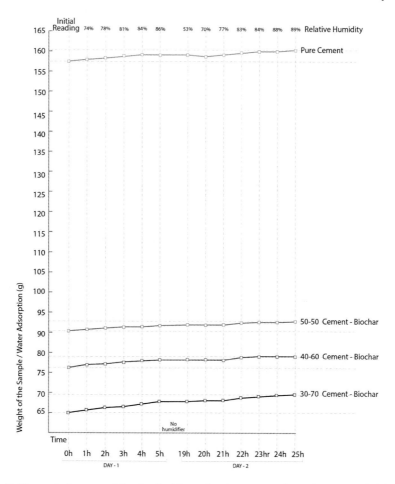

Fig. 5 Change in the mass of the developed samples due to water absorption as recorded during student investigation

the structural integrity of the structure or building element. Two main architectural applications are studied: a vault pavilion and a curtain brick facade.

Comparing the embodied carbon, weight, strength and cost of standard clay brick with BCM bricks of 50/50; 40/60 and 30/70 proportions, shows that the BCM bricks score better on all parameters making them both more structurally performative and carbon sequestering (Fig. 8). The brick system is further developed with a study on interlocking possibilities (Fig. 9). The interlocking system is explored as an option that would minimise the need of extra mortar.

Looking at the architectural design proposals, the vault pavilion is suggested as a compression only structure taking into account that as concrete without reinforcing the developed material systems performs only in compression (Fig. 10). The vault is designed using parametric design tools, in particular Rhino's Grasshopper and the

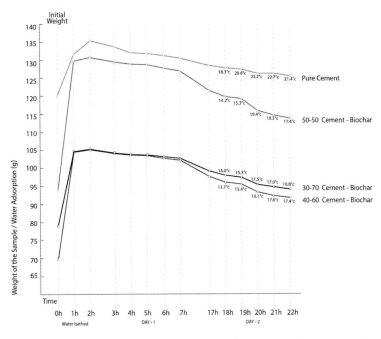

Fig. 6 Comparison of water absorption measurements of four material compositions based on water submersion test

Karamba plug-in. The vault is functionally graded based on structural analysis. The second design application is a curtain brick facade following a similar principle but utilising other material performances such as thermal insulation and humidity regulation (Fig. 11). A final architectural vision of the first iteration of biochar architectural language with both the vault structure and the curtain brick facades is presented as a conclusion of the investigation (Fig. 12).

3.2.2 Terra Preta (2020)

The second research project, "Terra Preta", was developed in 2020. The research builds upon the "Cast in Carbon" investigation and explores how biochar can be used within a geopolymer rather than an Ordinary Portland cement (OPC) material system as well as the possibility of natural fibre integration within the material matrix as tensile reinforcement. Unlike Ordinary Portland (pozzolanic) cement, geopolymers do not form calcium-silicate-hydrates (CSHs) for matrix formation and strength but utilise the polycondensation of silica and alumina precursors to attain structural strength. The main constituent of geopolymers source of silicon and aluminium which are provided by thermally activated natural materials (e.g. kaolinite) or industrial byproducts (e.g. fly ash or slag) and an alkaline activating solution which polymerizes these materials into molecular chains and networks to create a hardened binder. It is

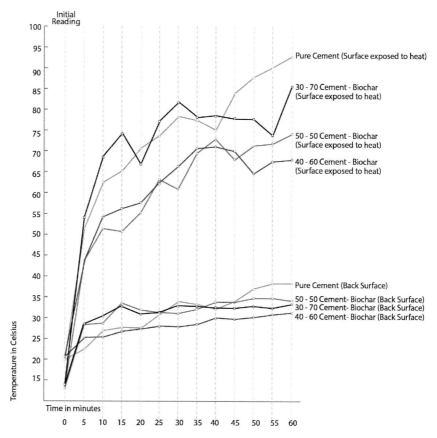

Fig. 7 Comparison of thermal conductivity measurements of four material compositions. Data was collected with a custom apparatus

also called alkali-activated cement or inorganic polymer cement. It is considered a less environmentally harmful alternative to OPC as it has a fraction of the embodied carbon of ordinary cement.

Embedding fibres within the Biochar-Geopolymer composites (BGCs) was the focus of the "Terra Preta" investigation. Natural fibres were selected as a cost-efficient, sustainable alternative to carbon fibre or polypropylene fibres. Various natural fibres, such as hemp, jute and flax, were studied and hemp was selected for further investigation due to its superior tensile strength.

Specimens (200 mm × 50 mm × 30 mm) for flexural strength test with a three-point bending testing apparatus were fabricated with raging biochar quantity from 0 to 80% (Figs. 13 and 15). Both geopolymer and ordinary cement material mixtures were fabricated for comparison. Two strategies for integrating the hemp fibres were studied: layered and multi-directional. In the layered approach, two layers of continuous fibre in the longitudinal direction located 5 mm above the button and 5 mm below

A Study on Biochar-Cementitious Composites Toward Carbon–Neutral …

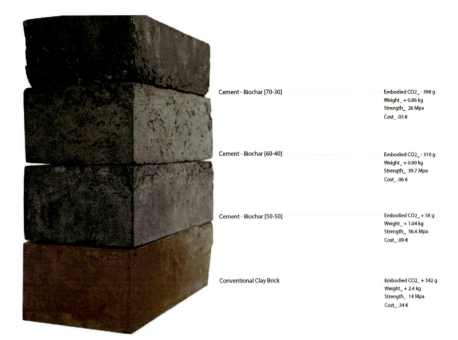

Cement - Biochar [70-30]
Embodied CO2_ - 398 g
Weight_ + 0.86 kg
Strength_ 26 Mpa
Cost_ .03 €

Cement - Biochar [60-40]
Embodied CO2_ - 310 g
Weight_ + 0.90 kg
Strength_ 39.7 Mpa
Cost_ .06 €

Cement - Biochar [50-50]
Embodied CO2_ + 58 g
Weight_ + 1.04 kg
Strength_ 56.6 Mpa
Cost_ .09 €

Conventional Clay Brick
Embodied CO2_ + 342 g
Weight_ + 2.4 kg
Strength_ 14 Mpa
Cost_ .34 €

Fig. 8 Comparison of ordinary clay brick and BCM bricks

Fig. 9 Interlocking brick study

Fig. 10 Vault pavilion architectural proposal

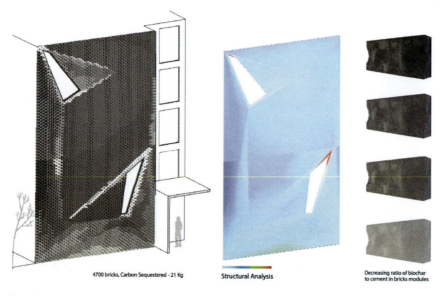

Fig. 11 Curtain brick facade architectural proposal

the top of the specimens are pre-tensioned on the mould before casting. Extra sets of fibres in the lateral direction were positioned through the middle of the specimens using the same strategy (Fig. 14).

The project was developed during the first wave of COVID-19 and the fabrication laboratory was closed, leaving the physical testing to the findings disclosed in Fig. 14. Based on the experiments it is suggested that the flexural strength of specimens with layered hemp fibres is higher than the multi-directional fibres. The BGCs exhibited

Fig. 12 "Cast in Carbon" architectural vision for carbon sequestering architecture

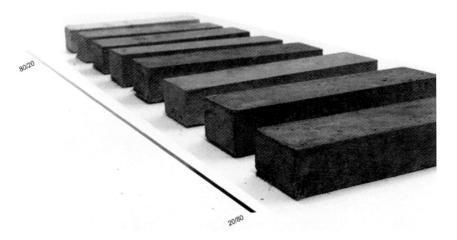

Fig. 13 Specimens with a range of biochar proportions within geopolymer-based composites

higher strength than the BCMs but more tests have to be conducted to verify the findings.

The architectural vision, proposed by the groups of students, is of a monolithic structure with a large mass, maximising the carbon sequestering capacity of the BGCs. Rhino's Grasshopper was used to generate the bulky geometries using the structural analysis and optimization plug-in Millipede. This computational design process was coupled with the Monolith plug-in which allows for a volumetric material distribution and was used to allocate a functionally graded BGC material on

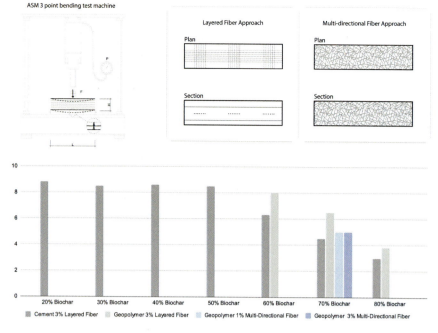

Fig. 14 Comparison of the flexural strength of BCMs and BGCs of specimens with and without hemp fibre

Fig. 15 Specimens after flexural strength testing

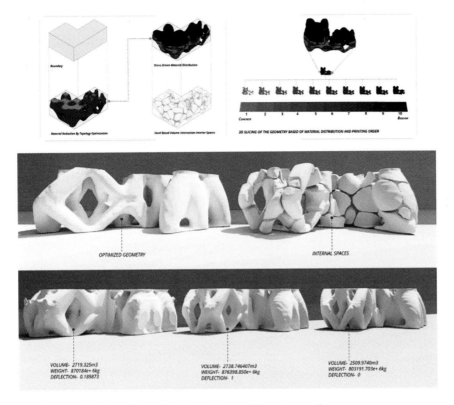

Fig. 16 "Terra Preta" architectural vision and computational approach

the optimised geometry (Fig. 16). The process is to be developed further to obtain more informed design solutions. After six months of investigation the research was closed. The architectural vision served as a demonstrator of the radically different architecture that can be generated with a materially-driven approach towards carbon sequestering buildings.

3.2.3 Carbon Copies (2021)

The last case study on the topic is titled "Carbon copies". The research project is focusing on the development of an automated bi-material concrete casting technique within a functionally graded design that provides a sustainable and material-efficient construction method compatible with the industry utilising a carbon negative BCM material system. The research builds up on the previously discussed research within the "Cast in Carbon" and "Terra Preta" case studies.

The research proposes that biochar is used as a partial substitute for cement, thus facilitating material savings and substantially lowering the amount of embedded carbon in concrete structures. A system for robotic bi-material allocation within a standardised formwork is developed based on a study of structural and material optimization of standardised building elements. The aim is to minimise the use of cement by allocating it exclusively where it is structurally required and incorporating a biochar-based mix to fill the rest of the formwork, thus lowering or neutralising the embodied carbon of structural elements.

The project's methodology revolves around 3 three essential fields of exploration: materiality, fabrication and design. The material aspect informs fabrication and design elements by providing necessary information about their properties, such as consistency, viscosity, flow rate and curing time. This knowledge subsequently influenced the decisions made during fabrication and design phases. The investigation went through the stages of material exploration: manufacturing and studying the properties of BCMs; manual casting: fabrication and evaluation of multi-material optimised functionally graded structural elements; automated casting: integration of robotics into the fabrication process to achieve better control over multi-layered material deposition. The combined research on all topics showed a potential to adopt the developed fabrication workflow for the full-scale manufacturing process of prefabricated concrete structural elements.

In order to understand the principal material behaviour and properties of the composite a setup of seven manually casted prototypes are created. In the first material experiment, biochar and OPC are the materials used to create BCMs similar to the material system developed in "Cast in Carbon". The material production workflow is based on grinding and finely screening the biochar. The biochar powder is mixed with cement and liquefied with water to be able to cast the mass in moulds for curing. The base sample consists of 100% cement. In steps of 10%, the amount of cement is reduced and the proportion of biochar is increased. As biochar is absorbing water it is necessary to increase the amount of water to be able to achieve the same viscosity of cement (Fig. 17). The needed amount of water to hydrate the composite extends proportional to the amount of biochar. The water absorption leads to a strong binding behaviour of the biochar composite. After the curing time, the models are compared in terms of weight. The increased amount of biochar results in a lighter composite. Due to the higher water ratio, the biochar composite has a longer curing time.

The setup for the second material exploration experiment consists of adding sand to the biochar cement mixture. The content of sand is kept constant at a ratio of 10%. The amount of cement was reduced in steps of 10% (Fig. 17). Based on that exploration a graded layered sample is created to investigate the binding behaviour of the different mixtures. Compared to the cement and biochar composites adding sand to the mixture results in a lighter material.

The third experiment lies in investigating the composites behaviour with adding fibres to the mixture, as a continuation to the "Terra Preta" research project. The used fibres are hemp and sisal in different lengths of 5, 10 and 20 mm. As concrete has good capabilities of working in compression the fibres are added to the mixture to be able to react to tension forces. A horizontal prototype in a working principle of a beam is casted. This beam consists of three zones. The upper layer is the compression zone where the most amount of cement is used. The neutral middle layer consists of

A Study on Biochar-Cementitious Composites Toward Carbon–Neutral ... 567

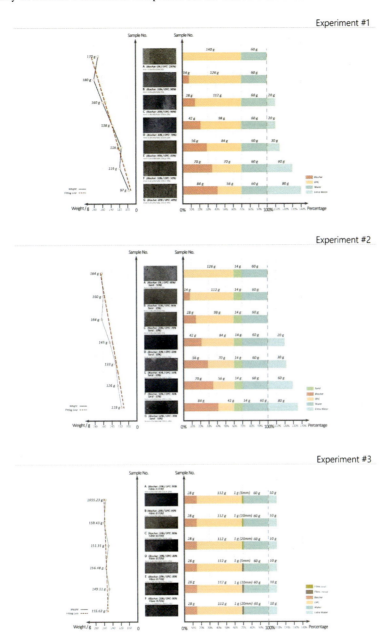

Fig. 17 Material system development based on three key material matrix experiments

50% biochar. The third layer is the tension zone where the fibres are added to the mixture.

The samples were not tested for compressive or flexural strength. The main evaluation criteria was the weight of the samples. It was concluded that the sample with the highest ratio of biochar resulted in higher demand in water content but lowest mass.

The developed computational workflow introduces a complete building element manufacturing process for structural elements that is based on data acquired from conducted material experiments and the precedent research projects: "Cast in Carbon" and "Terra Preta". Based on the optimization data acquired from such plugins as Karamba and Millipede, a customised G-code for robotic arm ABB140 was programmed to prefabricate these building elements for subsequent assembly on-site of construction (Fig. 18).

Throughout the investigation the goal was to produce a physical prototype based on the structural and material optimization of a scaled down beam sample. Based on the input parameters for computational protocol, such as target density, cell size, load and support cases, a target optimization result was acquired, clearly demonstrating the distribution of two materials for casting process that still uses conventional formwork, but allocates both cement and biochar-based mixes exclusively where it is structurally required (Fig. 19).

Initially, manual casting was used to test bi-material allocation with a BCM and cement-based mortar mixture. It is essential that the interface between the two mixtures is studied in detail and the fabrication process allows for precision and control of the material allocation. For this purpose, a set of fabrication methods were

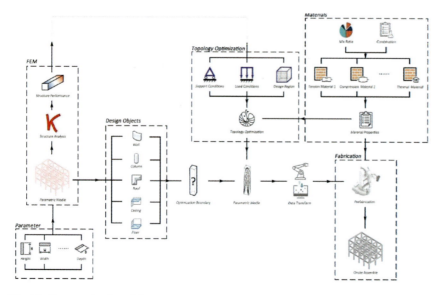

Fig. 18 Computational workflow

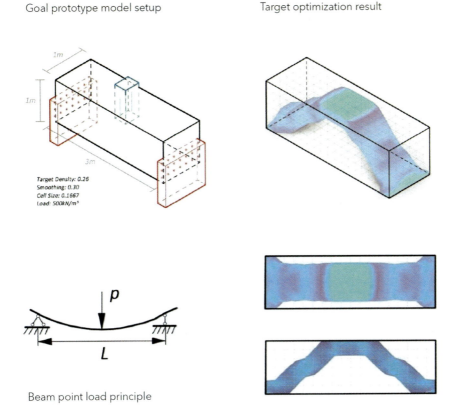

Fig. 19 Topological optimisation used to inform the material distribution of a scaled down beam to be used in the fabrication process

developed, executed and evaluated within a holistic approach. The proposed fabrication strategies were distributed in two categories: physical boundary (techniques that require the use of guiding separators for clear material deposition) and seamless casting that investigated binding behaviour between two mixes (Fig. 20).

Through evaluation of the physical prototypes of all of the explored fabrication methods, it was concluded that the best results were achieved using the pouring techniques (Fig. 21).

Based on the results of manual casting, the stage voxel grid technique helped to achieve the most accuracy of material pouring. To overcome the technique's limits in the project required the integration of robotics into the fabrication process that would provide better control over multi-layered bi-material deposition. The project revisited the topic of material exploration in order to create two suitable mixture compositions for the robot and in addition investigated the aspects of material deposition timing, robot tool path, end effector configuration and prototype reinforcement options (Fig. 22).

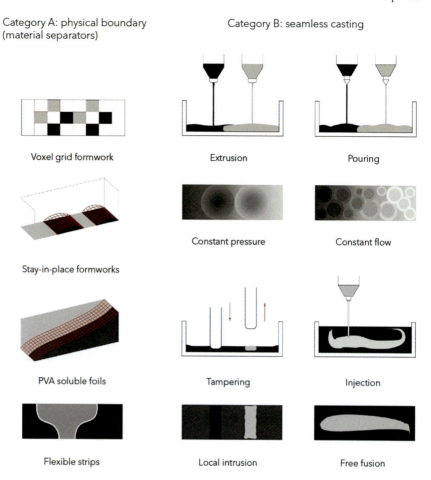

Fig. 20 Diagrammatic representation of manual casting fabrication techniques

The first iteration end effector was implemented with a Y-connector that would allow interchangeable bi-material outputs through a single nozzle. After several experiments, it was discovered that due to the fact that mixtures have different densities they tend to disrupt the overall flow in the Y-connector loop, which proves them to be sensitive to pressure. Furthermore, segregation in the biochar-cement mixture was happening due to large biochar particles that absorb water and accumulate at the bottom of the cartridge thus blocking the nozzle. Lastly, a biochar-cement mixture containing 40% of biochar was not suitable for robotic fabrication due to its high viscosity and quick water absorption. As a consequence, two new mixtures containing 20 and 30% of biochar were put to trial to test their performance. The mixture containing 20% biochar was defined as a new working mixture.

Based on the performance of the first end-effector iteration the second robotic fabrication setup went through significant improvements. More accurate pressure

A Study on Biochar-Cementitious Composites Toward Carbon–Neutral …

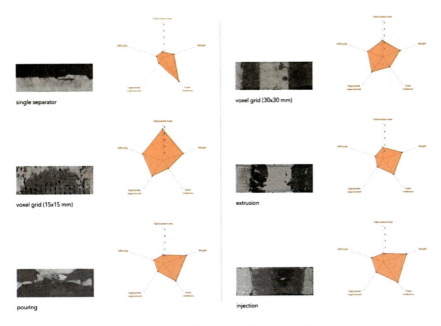

Fig. 21 Prototypes of the tested manual fabrication techniques and their evaluation

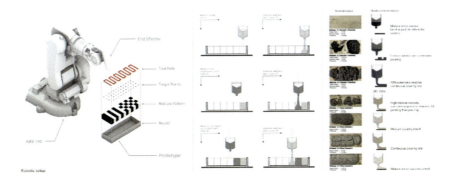

Fig. 22 Robotic fabrication experimental set-up and initial tests

control was achieved by providing the setup with two independent pressure control terminals for each working mixture. The initial end-effector was redesigned for faster assembly and disassembly to make sure that each pouring cycle operates within the same time frame between the end of the mixing stage and the start of each layer of material deposition. Most importantly, it was decided to use two separate nozzles for each working mixture, so the pressure inside the cement cartridge would not affect the material output of the biochar-cement cartridge. Finally, a mediator device (pinch valve) was implemented into the setup to refine material deposition timing by blocking the flow at a specific point, thus distributing equal amounts of material deposition for each voxel (Fig. 23).

Reinforced concrete frame building structures are composed of a network of columns and connecting beams that form the structural 'skeleton' of a building. Initially, the optimisation process was applied to the entire building but this method was not proven ineffective and a second approach was taken where the building

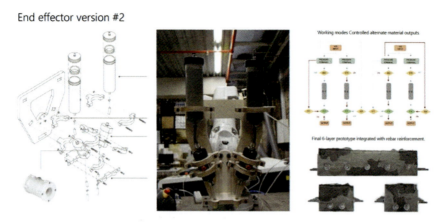

Fig. 23 Final robotic fabrication set-up

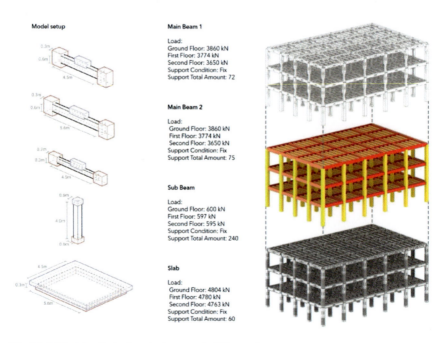

Fig. 24 Material optimisation strategy on a building scale

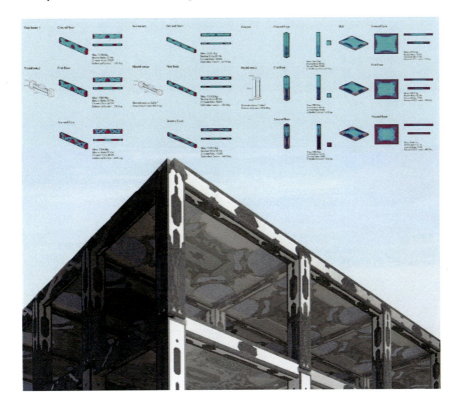

Fig. 25 *"Carbon Copies"* architectural vision of bi-material optimised structural elements

elements are extracted and optimised individually. Material optimization was applied to four structural elements: main beam, sub-beam, column and slab (Fig. 24). The elevation of the elements was also taken into account as the higher floors have to take fewer loads and therefore can have a predominantly biochar mixture.

3.3 Conclusions

The global environmental challenges are calling for novel solutions and sustainable practices in the fields of architecture and construction. In order to respond to the SDGs agenda, the research looks into strategies for integrating biochar as an aggregate for carbon–neutral cementitious composites within a materially driven design process. The aim is to propose innovative solutions for construction materials and processes towards sustainable architecture.

The research chapter uses three case studies and examines fabrication strategies as well as new techniques for material allocation and performance-driven design toward carbon-negative materially informed building elements. The material systems are developed over three distinct research projects under the same agenda, starting with investigating how biochar can be integrated within clay and cement mixtures. The second interaction looks into the possibility of moving away from OPC to geopolymer-based composites and how natural fibres can be incorporated into the mixture. The final project creates a mixture for robotic fabrication for bi-material allocation of BCMs. Due to the identified impact of the addition of biochar on the structural performance of the mixture as well as the carbon sequestration potential, it is proposed that a multi-material allocation would allow for an increase in the carbon sequestration without compromising the structural properties. This aspect is crucial for the development of both the material system and the fabrication process.

Fabrication techniques from manual to digital and eventually to robotic fabrication were studied and tested. The first project explores the modular and discretised approach to functionally graded structures and building elements. This allows for mass fabrication of modules with different properties that can be arranged based on computational optimisation of performances such as thermal insulation, humidity regulation and structural integrity. The "Cast in Carbon" case study demonstrates the possibility of obtaining architectural solutions that are tailored to multi-performance architectural elements and structures within a functionally graded material system.

The second architectural application is on monolithic structures. This proposal focuses on design without considering the fabrication process which limits it to a speculative outcome. The final case study "Carbon Copies", explored the intersection of material, design, and fabrication, focusing on structural performance. The strength of the final proposal is in the pragmatic approach that investigated how robotic fabrication can be used to obtain bi-material deposition. The three case studies show the evolution of the research from a material and fabrication perspective, concluding with a BCM material system that can be applied to a functionally graded design informed by computational simulation and optimisation processes. A range of architectural applications are proposed and even though the research is still in development it demonstrates a possible change in the design practice that is driven by sustainable material systems.

References

1. Ahmad, S., Tulliani, J.M., Ferro, G.A., Khushnood, R.A. Restuccia, L., Jagdale, P.: Crack path and fracture surface modifications in cement composites. Frattura ed IntegritaStrutturale 9, 524–533 (2015). https://doi.org/10.3221/IGF-ESIS.34.58
2. Akhtar, A., Sarmah, A.K.: Novel biochar-concrete composites: Manufacturing, characterization and evaluation of the mechanical properties. Science of the Total Environment 616–617, 408–416 (2018). https://doi.org/10.1016/j.scitotenv.2017.10.319
3. Antar, M., Lyu, D., Nazari, M., Shah, A., Zhou, X., Smith, D.L.: Biomass for a sustainable bioeconomy: an overview of world biomass production and utilization. Renew. Sustain. Energy Rev. **139**, 110691 (2021)
4. Asadi Zeidabadi, Z., Bakhtiari, S., Abbaslou, H., Ghanizadeh, A.R.: Synthesis, characterization and evaluation of biochar from agricultural waste biomass for use in building materials.

Construction and Building Materials **181**, 301–308 (2018). https://doi.org/10.1016/j.conbui ldmat.2018.05.271
5. Cha, J.S., Park, S.H., Jung, S.-C., Ryu, C., Jeon, J.-K., Shin, M.-C., Park, Y.-K.: Production and utilization of biochar: a review. J. Ind. Eng. Chem. **40**, 1–15 (2016)
6. Choi, W.C., Yun, H. do, Lee, J.Y.: Mechanical Properties of Mortar Containing Bio-Char From Pyrolysis. J. Korea Institute Struct. Maintenance Inspection **16**, 67–74 (2012). https://doi.org/10.11112/jksmi.2012.16.3.067
7. Cosentino, I., Restuccia, L., Ferro, G.A., Tulliani, J.M.: Type of materials, pyrolysis conditions, carbon content and size dimensions: The parameters that influence the mechanical properties of biochar cement-based composites. Theoret. Appl. Fract. Mech. **103**, 102261 (2019). https://doi.org/10.1016/j.tafmec.2019.102261
8. Cuthbertson, D., Berardi, U., Briens, C., Berruti, F.: Biochar from residual biomass as a concrete filler for improved thermal and acoustic properties. Biomass Bioenerg. **120**, 77–83 (2019). https://doi.org/10.1016/j.biombioe.2018.11.007
9. Das, S.K., Ghosh, G.K., Avasthe, R.: Valorizing biomass to engineered biochar and its impact on soil, plant, water, and microbial dynamics: a review. Biomass Conversion and Biorefinery (2020). https://doi.org/10.1007/s13399-020-00836-5
10. Delbeke, J., Runge-Metzger, A., Slingenberg, Y., Werksman, J., 2019. The Paris agreement, Towards a Climate-Neutral Europe: Curbing the Trend. https://doi.org/10.4324/9789276082569-2
11. Dixit, A., Gupta, S., Pang, S.D., Kua, H.W.: Waste Valorisation using biochar for cement replacement and internal curing in ultra-high performance concrete. J. Clean. Prod. **238**, 117876 (2019). https://doi.org/10.1016/j.jclepro.2019.117876
12. Dr. Fatih Birol, 2019. World Energy Outlook 2019 エ. World Energy Outlook Series.
13. Drzymała., T, Jackiewicz-Rek, W., Gałaj, J., Šukys, R.: Assessment of mechanical properties of high strength concrete (HSC) after exposure to high temperature. J. Civ. Eng. Manag. **24**, 138–144 (2018). https://doi.org/10.3846/jcem.2018.457
14. Estokova, A., Vilcekova, S., Porhincak, M.: Analysing Embodied Energy, Global Warming and Acidification Potentials of Materials in Residential Buildings. Procedia Engineering **180**, 1675–1683 (2017). https://doi.org/10.1016/j.proeng.2017.04.330
15. European Commission, 2022. Advanced Technologies for Industry, Advanced Manufacturing Technology [WWW Document]. URL https://ati.ec.europa.eu/technologies/advanced-manufa cturing-technology
16. European Commission, 2019. European Commission - The European Green Deal, European Commission - Press re.
17. Gupta, S., Kua, H.W.: Factors determining the potential of biochar as a carbon capturing and sequestering construction material: critical review. J. Mater. Civ. Eng. **29**, 04017086 (2017). https://doi.org/10.1061/(asce)mt.1943-5533.0001924
18. Gupta, S., Kua, H.W.: Application of rice husk biochar and thermally treated low silica rice husk ash to improve physical properties of cement mortar. Theoret. Appl. Fract. Mech. **104**, 102376 (2019). https://doi.org/10.1016/j.tafmec.2019.102376
19. Gupta, S., Kua, H.W., Pang, S.D.: Biochar-mortar composite: Manufacturing, evaluation of physical properties and economic viability. Constr. Build. Mater. (2018). https://doi.org/10.1016/j.conbuildmat.2018.02.104
20. Gupta, S., Kua, H.W., Tan Cynthia, S.Y.: Use of biochar-coated polypropylene fibers for carbon sequestration and physical improvement of mortar. Cem. Concr. Compos. **83**, 171–187 (2017). https://doi.org/10.1016/j.cemconcomp.2017.07.012
21. Gupta, S., Muthukrishnan, S., Kua, H.W.: Comparing influence of inert biochar and silica rich biochar on cement mortar – Hydration kinetics and durability under chloride and sulfate environment. Constr. Build. Mater. **268**, 121142 (2021). https://doi.org/10.1016/j.conbuildmat.2020.121142
22. Guterres, A., 2020. The Sustainable Development Goals Report 2020, United Nations publication issued by the Department of Economic and Social Affairs.

23. Hahn, J., 2021. Atmospheric CO2 is "our biggest resource", says carbon-negative plastic brand Made of Air [WWW Document]. URL https://www.dezeen.com/2021/06/24/carbon-negative-plastic-biochar-made-of-air-interview/
24. Jackiewicz-Rek, W., Drzymała, T., Kuś, A., Tomaszewski, M., n.d. DURABILITY OF HIGH PERFORMANCE CONCRETE (HPC) SUBJECT TO FIRE TEMPERATURE IMPACT.
25. Lee, S.Y., Sankaran, R., Chew, K.W., Tan, C.H., Krishnamoorthy, R., Chu, D.-T., Show, P.-L.: Waste to bioenergy: a review on the recent conversion technologies. BMC Energy (2019)
26. Li, S., Bi, X., Tao, R., Wang, Q., Yao, Y., Wu, F., Zhang, C.: Ultralong cycle life achieved by a natural plant: miscanthus × giganteus for lithium oxygen batteries. ACS Appl. Mater. Interfaces **9**, 4382–4390 (2017)
27. Liu, R., Xiao, H., Guan, S., Zhang, J., Yao, D.: Technology and method for applying biochar in building materials to evidently improve the carbon capture ability. J. Clean. Prod. **273**, 123154 (2020). https://doi.org/10.1016/j.jclepro.2020.123154
28. Liuzzi, S., Sanarica, S., Stefanizzi, P.: Use of agro-wastes in building materials in the Mediterranean area: A review. Energy Procedia **126**, 242–249 (2017). https://doi.org/10.1016/j.egypro.2017.08.147
29. Matovic, D.: Biochar as a viable carbon sequestration option: global and Canadian perspective. Energy **36**, 2011–2016 (2011). https://doi.org/10.1016/j.energy.2010.09.031
30. Mensah, R.A., Shanmugam, V., Narayanan, S., Razavi, S.M.J., Ulfberg, A., Blanksvärd, T., Sayahi, F., Simonsson, P., Reinke, B., Försth, M., Sas, G., Sas, D., Das, O.: Biochar-added cementitious materials—a review on mechanical, thermal, and environmental properties. Sustainability (Switzerland) **13**, 9336 (2021). https://doi.org/10.3390/su13169336
31. Mo, L., Fang, J., Huang, B., Wang, A., Deng, M.: Combined effects of biochar and MgO expansive additive on the autogenous shrinkage, internal relative humidity and compressive strength of cement pastes. Constr. Build. Mater. **229**, 116877 (2019). https://doi.org/10.1016/j.conbuildmat.2019.116877
32. Mrad, R., Chehab, G., 2019. Mechanical and microstructure properties of biochar-based mortar: An internal curing agent for PCC. Sustainability (Switzerland) 11. https://doi.org/10.3390/su11092491
33. Nair, J.J., Shika, S., Sreedharan, V.: Biochar amended concrete for carbon sequestration. IOP Conf. Ser. Mater. Sci. Eng. **936** (2020). https://doi.org/10.1088/1757-899X/936/1/012007
34. NASA, 2022. Climate Change Evidence: How Do We Know?
35. Navaratnam, S., Wijaya, H., Rajeev, P., Mendis, P., Nguyen, K., 2021. Residual stress-strain relationship for the biochar-based mortar after exposure to elevated temperature. Case Studies in Construction Materials 14. https://doi.org/10.1016/j.cscm.2021.e00540
36. Park, J.H., Kim, Y.U., Jeon, J., Yun, B.Y., Kang, Y., Kim, S.: Analysis of biochar-mortar composite as a humidity control material to improve the building energy and hygrothermal performance. Sci. Total Environ. **775**, 145552 (2021). https://doi.org/10.1016/j.scitotenv.2021.145552
37. Polytechnic University of Bari, 2021. From farm to façade: finding new uses for agricultural waste [WWW Document]. URL https://ec.europa.eu/regional_policy/en/newsroom/news/2021/04/04-12-2021-from-farm-to-facade-finding-new-uses-for-agricultural-waste
38. Precast Concrete Market Size, I.R., 2021. Precast Concrete Market Size, Share & Trends Analysis Report By Product (Structural Building Components, Transportation Products), By End-use (Residential, Infrastructure), By Region, And Segment Forecasts, 2021 - 2028.
39. Qin, Y., Pang, X., Tan, K., Bao, T.: Evaluation of previous concrete performance with pulverized biochar as cement replacement. Cement Concr. Compos. **119**, 104022 (2021). https://doi.org/10.1016/j.cemconcomp.2021.104022
40. Ramamurthy, K., Harikrishnan, K.I.: Influence of binders on properties of sintered fly ash aggregate. Cement Concr. Compos. **28**, 33–38 (2006). https://doi.org/10.1016/j.cemconcomp.2005.06.005
41. Rebecca Lindsey, 2021. Climate Change: Atmospheric Carbon Dioxide, Climate.gov.
42. Rockwood, D.L., Ellis, M.F., Liu, R., Zhao, F., Fabbro, K.W., He, Z., Derbowka, D.R.: Forest trees for biochar and carbon sequestration: production and benefits. In: Applications of Biochar for Environmental Safety, pp. i, 13 (2020). https://doi.org/10.5772/intechopen.92377

43. Rudnik, E., Drzymała, T.: Thermal behavior of polypropylene fiber-reinforced concrete at elevated temperatures. J. Therm. Anal. Calorim. **131**, 1005–1015 (2018). https://doi.org/10.1007/s10973-017-6600-1
44. Schmidt, H.-P., 2014. The use of biochar as building material [WWW Document]. URL www.biochar-journal.org/en/ct/3
45. Sirico, A., Bernardi, P., Belletti, B., Malcevschi, A., Dalcanale, E., Domenichelli, I., Fornoni, P., Moretti, E.: Mechanical characterization of cement-based materials containing biochar from gasification. Constr. Build. Mater. **246**, 118490 (2020). https://doi.org/10.1016/j.conbuildmat.2020.118490
46. Sizmur, T., Quilliam, R., Puga, A.P., Moreno-Jiménez, E., Beesley, L., Gomez-Eyles, J.L.: Application of biochar for soil remediation. In: Agricultural and Environmental Applications of Biochar: Advances and Barriers, pp. 295–324. Soil Science Society of America, Inc., Madison, WI, USA (2015)
47. Steffen, W., Richardson, K., Rockström, J., Cornell, S.E., Fetzer, I., Bennett, E.M., Biggs, R., Carpenter, S.R., de Vries, W., de Wit, C.A., Folke, C., Gerten, D., Heinke, J., Mace, G.M., Persson, L.M., Ramanathan, V., Reyers, B., Sörlin, S.: Planetary boundaries: Guiding human development on a changing planet. Science (2015). https://doi.org/10.1126/science.1259855
48. Suarez-Riera, D., Restuccia, L., Ferro, G.A.: The use of Biochar to reduce the carbon footprint of cement-based. Procedia Struct. Integrity **26**, 199–210 (2020). https://doi.org/10.1016/j.prostr.2020.06.023
49. Tan, R.R.: Data challenges in optimizing biochar-based carbon sequestration. Renew. Sustain. Energy Rev. **104**, 174–177 (2019).
50. Tan, K., Qin, Y., Du, T., Li, L., Zhang, L., Wang, J.: Biochar from waste biomass as hygroscopic filler for pervious concrete to improve evaporative cooling performance. Constr. Build. Mater. 287, 123078 (2021). https://doi.org/10.1016/j.conbuildmat.2021.123078
51. Thibaut Abergel, Brian Dean, John Dulac, I.H., 2018. Global Alliance for Buildings and Construction, 2018 Global Status Report, United Nations Environment and International Energy Agency.
52. United Nations Environment Programme, G.A. for B. and C., 2020. 2020 Global Status Report for Buildings and Construction: Towards a Zero-emissions, Efficient and Resilient Buildings and Construction Sector - Executive Summary.
53. van de Velde, E., Kretz, D.: Advanced technologies for industry. Product watch : flexible and printed electronics (2021). https://doi.org/10.2826/242713
54. Vigneshwaran, S., Sundarakannan, R., John, K.M., Joel Johnson, R.D., Prasath, K.A., Ajith, S., Arumugaprabu, V., Uthayakumar, M.: Recent advancement in the natural fibre polymer composites: a comprehensive review. J. Clean. Prod. **277**, 124109 (2020)
55. Wei, J., Meyer, C.: Degradation mechanisms of natural fiber in the matrix of cement composites. Cem. Concr. Res. **73**, 1–16 (2015). https://doi.org/10.1016/j.cemconres.2015.02.019
56. WorldGBC, 2019. Bringing embodied carbon upfront: Coordinated action for the building and construction sector to tackle embodied carbon. World Green Building Council 35.
57. Xie, C., Yuan, L., Tan, H., Zhang, Y., Zhao, M., Jia, Y.: Experimental study on the water purification performance of biochar-modified pervious concrete. Constr. Build. Mater. **285**, 122767 (2021). https://doi.org/10.1016/j.conbuildmat.2021.122767
58. Zhao, M., Jia, Y., Yuan, L., Qiu, J., Xie, C.: Experimental study on the vegetation characteristics of biochar-modified vegetation concrete. Constr. Build. Mater. **206**, 321–328 (2019). https://doi.org/10.1016/j.conbuildmat.2019.01.238

DigitalBamboo_Algorithmic Design with Bamboo and Other Vegetable Rods

Stefan Pollak and Rossella Siani

Abstract Algorithmic design software is widely acknowledged as a tool to manage complex design tasks and to enhance material optimization, structural performance, ergonomic needs or similar aspects. The present paper investigates how these tools can be applied to projects that use an important amount of non-standardised, natural materials. The use of renewable and locally sourced materials is becoming mandatory if we accept the challenge of providing an appropriate built environment for a growing world population. A special focus is given to vegetable rods such as giant reed and bamboo. Building tradition provides uncounted examples of how humankind employs natural fibres to erect or ornate its shelters. Some of them can inspire new uses to be applied in contemporary architecture. The aforementioned digitally controlled design processes are normally meant to feed so-called computer aided manufacture processes. Such methods generally need highly standardised materials. The use of renewable materials in such a framework is often impossible due to intrinsic irregularities of natural resources. Can this gap be bridged? The present paper illustrates the design-and-build technology *DigitalBamboo* thought to conciliate the two realms of natural building materials and algorithmic design control. The method has been conceived for experimental projects made of Italian bamboo in the form of strips but can be applied to other vegetable fibres or rods and to other geographical contexts. The investigated technology includes appropriate communication tools to bridge the divide between designer and builder. The illustrated technology is based on manual assembly of digital data and includes ways of transposing geometric entities into topological textures, physical nodes and structures.

Keywords Natural materials · Algorithmic design · Appropriate fabrication · Gridshell · Assembly maps · Nodes

---

S. Pollak (✉)
AK0 – Architettura a kilometro zero ETS, Rome, Italy
e-mail: stefan.pollak@akzero.org

R. Siani
University of Parma, Parma, Italy
e-mail: rossella.siani@unipr.it

© The Author(s), under exclusive license to Springer Nature Switzerland AG 2024
M. Barberio et al. (eds.), *Architecture and Design for Industry 4.0*, Lecture Notes in Mechanical Engineering, https://doi.org/10.1007/978-3-031-36922-3_32

United Nations' Sustainable Development Goals 9. Build resilient infrastructure, promote inclusive and sustainable industrialization and foster innovation · 11. Make cities and human settlements inclusive, safe, resilient and sustainable · 12. Ensure sustainable consumption and production patterns

1 Introduction

The fourth industrial revolution allows architectural design along with its workflow control, the construction itself, and the performance monitoring to be more and more implemented by computation and automatised processes [1].

DigitalBamboo is an experimental research that applies algorithmic design approaches to bamboo, a natural material with interesting structural and expressive features.

The inquiry is part of the broader research line *DigitalNature* [2], which investigates the meeting points between innovation and tradition, between advanced design and locally sourced natural materials.

The main challenge of this research is to adapt Industry 4.0 processes, which generally imply numerically controlled machines or robots [3] and highly standardised products, in order to make them cope with natural materials in their raw form. An original method is proposed to deal with this delicate step.

Using a material in its natural form, with a few processing steps to reach the assembly phase, reduces the impact of the product life cycle assessment [4, 5], both in the initial pre-production phase and in the disposal phase.

Traditional building models with regrowing, natural materials are implicitly circular and provide benefits both for the environment and the local economy. Such good practices can increase their effectiveness and multiply their positive effects on the ecosystem if associated with more complex control tools.

The increase in process-complexity that comes along with the combination of digital tools with physical production means can allow for an improvement in performance and hence for a better management of resources. This opens new perspectives with respect to sustainability, a concept that is transforming itself from a merely conservative approach as defined in the Brundtland Reports [6] or in concepts like *Décroissance Sereine* [7–9] towards a more active attitude as witnessed in the 17 UN Sustainable Development Goals. Similarly to what happens in mature living organisms, where a growth in complexity takes the place of a physical growth, this can be seen as a passage from a quantitative towards a qualitative growth [10].

The use of algorithmic generative design combined with construction processes that make use of natural materials endeavours a specific niche of this perspective.

2 Physical and Digital

2.1 Fast Growing Plants as Building Material

In a circular perspective, the use of vegetable material is crucial. Canes like giant reed or bamboo are largely available in many regions of the world and have a rapid growing cycle, i.e. if compared to timber. Reed canes like *Arundo donax* can be found all around the Mediterranean Sea and in the Middle East. Similar species grow in other continents. The areas where bamboo is native include almost all tropical regions of Asia, Africa and Latin America, but most of the 1.600 known species can grow up to a latitude of 41° North and South [11]. Vegetable rods are commonly used as building materials. [12–18]

The present research focuses on bamboo grown in Italy, mainly *Phyllostachis viridiglaucescens*. This species can provide culms of up to 11 m of length. Its diameters appear in a range between 40 and 80 mm with a wall thickness of mature material that reaches 4 to 5 mm.

The diameters vary along the culm with a maximum at approximately 1 m from the ground and then a constant decrease towards the top. This constrains the cane's bending behaviour, which typically has larger radii in its lower part and the possibility of tighter bends in the upper part. The result is a characteristic asymmetric arch.

Bamboo strips are an alternative to round culms. To source them whole canes can be divided with a splitter, a cutting tool with 3, 4, 5 or more blades, according to the diameter and to how broad the strips shall be. The bending behaviour of strips has less constraints compared to entire canes which allows for freely designed shapes. Freshly cut, 3 cm broad, green strips of *Phyllostachis viridiglaucescens* can bend with a radius as small as 35 cm.

2.2 Digital Design and Production Tools

Algorithmic design [19–22] (also defined parametric) makes use of specific, so-called generative software, whose use is growing in industry 4.0. The final design conformations emerge from a set of pre-established relations and the results vary as the parameters vary in a form finding process [23–28]

Generally, the algorithmic design work is associated with CAD/CAM processes, and thus allows to carry out mechanised production processes with a numeric control [29]. Such computer numeric control machines (CNC) can process various materials, even of natural origin, but with one common feature: the format of the materials is standard.

In the case studies presented here the bamboo material is used in its natural form, in whole rods or strips, and as such the format varies in thickness, length, weight, as well as for the numerous irregularities it presents on the external surface.

The need to use bamboo in its natural form has triggered the creation of a specific construction process, capable of translating digital data into very precise manual operations, so as to combine the advantages of parametric performance control with a constructive model which is close to tradition. This allows us to incorporate consolidated solutions or to involve specific locally available skills.

2.3 The Case of Triaxial Bamboo Strip Gridshells

Gridshells are lightweight constructions made of linear elements and nodes that collaborate in order to reach an efficient structural performance [30]. Strips of split bamboo can be used as building material for such works.

This study examines methods of digital design to control shapes and performances of such gridshells (Fig. 1).

The bamboo strips are organised in three layers according to their orientation (horizontal, left and right). The three families of axes cross in a mesh of nodes, each node joins three strips, one from each layer.

The digital model contains all the needed information in terms of proportion, size and scanning of the parts. The tolerance between the digital and the physical model is in the range of a few millimetres; a tolerance that does not affect the building's performance but, on the contrary, allows for more leeway while dealing with the natural material's peculiarities.

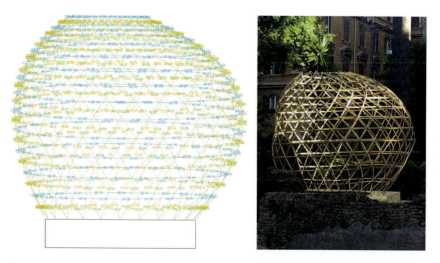

Fig. 1 *Pagurus urbanus pacificus*, a leisure pavilion in a public garden in Rome as a virtual model and in built reality

3 Digital and Empirical Information Management

3.1 Morphogenetic Design

The digital design is developed with *Rhinoceros* [31] combined with the plugin *Grasshopper* [31], a generative software for parametric design. The design phase, or in other words, the development of the virtual project [3] is divided into two steps: the definition of an algorithm and the application of differentiated parameters for the single case [23, 24].

The virtual project represents, like the genotype of an organism, a range of formal possibilities that determine the specific case, the phenotype (Fig. 2).

The morphogenesis of a formal composition is in relation with the most suitable parameters for the project's logic and function. Along with the freely chosen parameters that shape the design, the algorithm can be fed with information on the material's physical constraints such as bending radii, torsion data as well as structural parameters, environmental factors, functional or even cultural data.

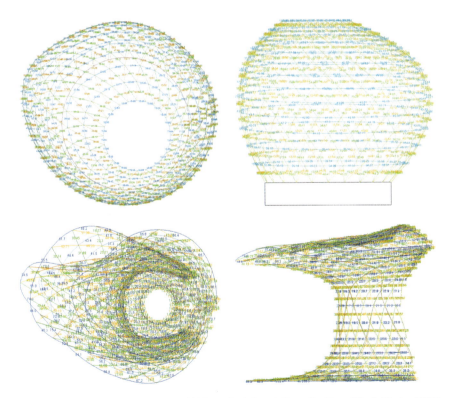

Fig. 2 Two projects (*Pagurus*, Rome 2019 on the left and *Jinen* for Tono Mirai, Venice 2021) compared. The top views highlight the origin of the two designs from a common genotype while the side views clearly show the phenotypic peculiarities

3.2 Curvature and Torsion as Parameters

The data on minimum curvature constrain the composition's formal possibilities. The local datum affects the global behaviour, determining variations on the composition.

The curvature analysis (Fig. 3) refers to the curvature of individual strips, organised by warping or to the surface that determines the overall shape of the structure. The analysis of single strips allows us to investigate possible criticalities of different variants. Once bending radii that are inappropriate for the available material are detected, the design can be modified with a reiterated feedback process until a suitable result is reached.

The analysis of the surface curvature is represented by a graph with a colour gradient which, in our example, goes from red (maximum) to yellow (minimum) and defines the degree of curvature of the surface in relation to the minimum and maximum of the specific template (Fig. 4). It is a less precise tool compared to the previous one, because it does not provide absolute but only relative data. It is not

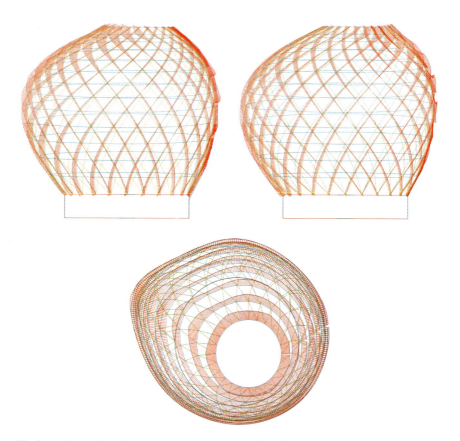

Fig. 3 Analysis of bending curvatures within a gridshell configuration

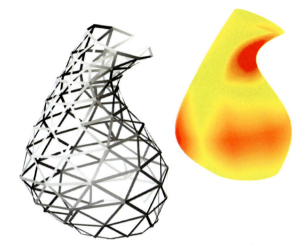

Fig. 4 Surface curvature analysis. Red highlights areas with excessive bending stresses

fruitful to use it in the definition phase of the project, because it does not determine the points that exceed a certain degree of curvature, but it is very effective in the phase of detail design.

The singular (non-standard) nature of bamboo makes it possible to distinguish the strips in relation to the variations in performance that come along with their dimensional features. Having a simplified scheme that identifies the points that require a greater bending effort, allows to better distribute the material with respect to flexibility and extend the optimization process throughout the whole construction phase.

The bamboo strips, in addition to curvature, are subjected to twisting. As for the bending behaviour, the study phase can highlight critical points of torsion and thus warn the designer. In some cases the torsional effect can enhance the material's natural resistance within the gridshell. Most criticalities can be faced with empirical adjustments. As an example, longer screws can be used for those nodes where the torsion is expected to be higher. The loser spacing between the three layers of strips allows for a reduction in torsion. Manual dexterity and some experience in handling the material can empirically solve such situations.

3.3 Shape Optimization

Generative software allows for shape optimization in relation to endogenous and exogenous parameters.

Endogenous parameters include the limit figures for curvature and torsion. While in the process of analysing such data, as described above, is a matter of verification, the same information packages can drive the optimization approach in a process of morphogenesis. The implied numerical values can make a solution emerge that can be considered as optimal with respect to the examined data set.

When gravitational forces and exogenous weight forces are combined with the material's features, structural morphogenesis is possible.

Other plug-ins, *Kangaroo* [32] and *Karamba* [33], are added to the Rhinoceros + Grasshopper [31] software system, which simulate the physical behaviour of structures. The structure's final shape emerges from the forces it is subjected to. In bamboo gridshells, this structural optimization work allows for experimentation with complex yet stable compositions.

Climatic factors such as sun or wind patterns can determine a morphogenesis for the optimization of environmental comfort. In this case the bamboo gridshell becomes the supporting structure of shading systems or corridors for cross ventilation.

3.4 Quantity Design

Working with parametric software allows us to keep a large amount of numerical data related to the project under control in real time. This can include material quantities.

For example the software can help in defining the total length of the bamboo strips, from which the number of whole rods is easy to obtain. Further data include the actual number of strips with its specific length. These data are also organised in groups according to the shell's wrap. Knowing the length of each strip allows for a resource optimisation while choosing the material to use. It also helps in defining the additional amount of overlap material to consider for those strips that can't be built from one single piece. The number of nodes, another information that can be assessed in real time, corresponds to the number of connectors needed.

Keeping the quantities under control while designing also allows to adjust the outcome according to changes in material availability or budget.

Designing with real time quantity control speeds up the construction process, ensures greater precision and helps to reduce waste.

3.5 Digital Models for Triaxial Bamboo Strips Gridshells

The digital model of the bamboo gridshell is generated by a surface, which follows the logic of morphogenesis and formal optimization and is then divided into a triaxial pattern that discretizes the surface in triangles and vertices. In the physical translation, the pattern's edges correspond to bamboo strips with a width between 25 and 30 mm and a thickness of approximately 5 mm, while each vertex represents a node or connection.

As the morphogenetic process includes specific parametric data on the material's behaviour, specific constraints come as a result. With respect to the described bamboo strips, the distance between two distance points varies in a range between 10 and 50 cm. A smaller distance would make the assembling process difficult while meshes with more than the said maximum distance could lead to buckling effects and affect the object's overall stability.

Even in the various compositions, the three axes have a similar organisation: the first axis is composed of arches or closed circles, the other two cross along the first with an opposite inclination.

The model is represented in a simplified way through the segments between the nodes in which each strip is discretized. The overall form is already readable, however in order to have a graphic representation which is more consistent with the final result, strip thickness or colour information can be added.

3.6 The Translation from Digital to Manual

In order to translate the virtual model into built reality some additional passages are needed, especially with respect to how the flux of information from the software to the building site is managed. The steps of this process are: definition of the virtual model, prefabrication, assembly. The prefabrication process consists in pre-perforating each strip with its connection holes in the exact position.

The virtual model already contains the indication on where to drill, a geometric entity that corresponds to the distance between two specific nodes. It is actually possible to visualise this information as a cloud of figures; a suggestive but not easy to read representation. The need for readability comes along with the fact that the natural material requires to be processed by humans. An assembly map with a specific level of abstraction solves this gap (Fig. 5).

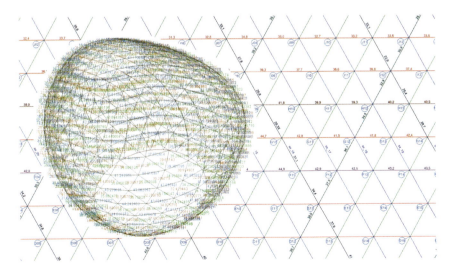

Fig. 5 Numeric assembly map. The number cloud (left) represents the actual shape while the abstract plane representation makes the manufacturing possible

A special script allows to project each distance-figure onto a grid of equilateral triangles just for the sake of graphic order. For every strip to be used in the final work, this simplified representation makes it easy to deduce the following information:

- _total length,
- _alphanumeric code attributed to the strip,
- _distance between each node,
- _alphanumeric code attributed to every node.

This last information is crucial for the passage from prefabrication to final assembly where nodes with the same name have to be joined with a connector element as shown in Fig. 7.

3.7 Nodes and Joints as Control Entity

Bamboo strips can mainly be connected in two different ways. If interwoven in patterns that are tight enough, it is possible to generate stiff curved shapes that can stand only through friction. Such weaving patterns can be bi-axial (warp and woof) or tri-axial with a pattern of hexagons and triangles (Fig. 6).

As an alternative, bolts or other punctual connections can be used to join different layers of strips. According to the design's overall character, iron bolts or timber pins can serve as connectors to hold a grid pattern in place (Fig. 7).

In both cases it is important to consider the right bolt or pin diameter which has to cope with the strip's width in order to balance connection strength with the fact that the strip shall not be injured too much. The connector's length varies according to the degree of curvature and torsion in the various positions of the shell surface.

The described nodes are reversible which allows disassembling and reassembling the structure several times, an additional benefit in terms of circularity.

Fig. 6 Woven shells with bi-axial (left) and tri-axial (right) patterns

Fig. 7 Iron bolts (left) or timber pins with hemp rope (right) as node-connectors

Triangular patterns are the most used due to their intrinsic stability. At the same time triangular meshes are also easy to control with algorithms. In the virtual model the physical node is nothing else than the intersection of three axes.

In the built reality, the three orientation axis lay on three overlapping layers which requires an offset adjustment of the virtual model that takes the strip thickness into account.

3.8 Construction

The gridshells are assembled manually with simple tools: saws, drills, screwdrivers and similar carpentry tools. The precise execution of the indications contained in the assembly map shown in Fig. 5 makes the designed shape appear progressively.

At the end of a correct execution the final design exactly corresponds to the morphogenetically designed virtual model (Fig. 8).

4 Conclusions

The use of locally sourced natural material in architecture is getting of crucial importance if we want to shape appropriate spaces for a growing world population. Facing the increasing complexity that comes with global interconnection and always more rapid new scientific notions, calls for control tools in all design stages.

Methods that can make the non-standard, natural peculiarities cope with algorithmic (or parametric) design tools have to be found. *Digitalbamboo* is a first attempt to make these two realms come closer to one another in a perspective of defining a new architecture that is natural and digital at the same time.

Fig. 8 *Jinen* for Tono Mirai (design: Tono Mirai; technical implementation: Salvatore, A., Siani, R., Pollak, S. - Venice 2021) and *Pagurus urbanus pacificus* (Rome 2019) built

References

1. Barberio, M., Colella, M.: Architecture 4.0 Fondamenti ed esperienze di ricerca, Maggioli Editore (2020). ISBN: 8891639004
2. Siani, R.: Materiali Naturali – Progettazione Generativa. Dall'antitesi alla sintesi. In: Perriccioli, M., Rigillo, M., Russo Ermolli, S. Tucci, F. (editors) "Design in the Digital Age. Technology Nature Culture | Il Progetto nell'Era Digitale. Tecnologia NaturaCultura", Maggioli Editore (2020). ISBN 978–88–916–4327–8
3. Figliola, A., Battisti, A.: Post-Industrial Robotics. Exploring Informed Architecture. Springer Singapore (2021). ISBN: 978–981–15–5277–9
4. McDonough, W., Braungart, M.: Dalla culla alla culla. Come conciliare tutela dell'ambiente, equità sociale e sviluppo, Blu Edizioni, Torino (2003)
5. Baldo, G.L., Marino, M., Rossi, S.: Analisi del ciclo di vita LCA: materiali, prodotti, processi. Ed. Ambiente, Milano (2005). ISBN 9788889014295
6. World Commission on Environment and Development: Our Common Future. Oxford: Oxford University Press. p. 27 (1987)
7. Latouche, S.: Decolonizzare l'immaginario. Il pensiero creativo contro l'economia dell'assurdo, ed. EMI (2004)
8. Latouche, S: Petit traité de la décroissance sereine. Mille et Une Nuits (2007)
9. Latouche, S.: Mondializzazione e decrescita. L'alternativa africana, edizioni Dedalo (2009)
10. Capra, F, Henderson, H: Qualitative growth. In: Outside Insights. London, Institute of Chartered Accountants in England and Wales, October (2009)
11. Dunkelberg, K. et. alt.: Bambus/ Bamboo. n° 31 of IL (Mitteilungsreihe des Instituts für leichte Flächentragwerke), Stuttgart (1988)
12. Liese, W.: Bamboo preservation and soft rot - Report to the Government of India. FAO-EPTA
13. Janssen, J.J.A.: Building with bamboo. Intermediate Technology Publications, London (1987)
14. Ghavami, K.: Application of bamboo as a low cost energy material in civil engineering. In: Proceedings of the Third CIB-RILEM Symposium, materials for low cost housing. Funavit, Mexico city, Mexico (1989).
15. Gauzin-Müller, D.: Architecture en fibres végétales d'auhourd'hui. Grenoble, Museo / CRAterre in partnership with amàco. (2021)
16. Minke, G.: Building with Bamboo - Design and Technology of a Sustainable Architecture. Basel, Birkhäuser. (2012/2022)
17. Krausse, J.: Formen nach dem Vorbild der Natur. Interview by Sabine Kraft & Schirin Taraz-Breinholt. Arch+, n° 159/160. Aachen, Arch+ Verlag. (2002).
18. Velez, S., von Vegesack, A., Kries, M.: Grow Your Own House. Weil am Rhein, Vitra Design Museum (2013)
19. Lolli, G.: Definizioni di algoritmo. In: Matematica e Calcolatori, Le Scienze, quaderni n.14 (1984)
20. Berlinski, D.: The adventure of the algorithm: the idea that rules the world. Harcourt (1999)
21. Oxman, R., Oxman R.: Theoris of the Digital in Architecture. ed. Routledge New York (2014)
22. Tedeschi, A.: AAD Algorithms-Aided Design: Parametric Strategies using Grasshopper. Paperback (2014)
23. Deleuze, G.: Difference and Repetition. Columbia University Press, New York (1968)
24. De Landa, M.: Deleuze and the genesis of form. Universitätsverlag Winter Gmbh (2000)
25. Otto, F., Rasch, B.: Finding Forms – towards an architecture of the minimal. Axel Menges, Stuttgart (1995)
26. Otto, F., Schaur E. et. al.: Natürliche Konstruktionen - Formen und Konstruktionen in Natur und Technik und Prozesse ihrer Entstehung. DVA, Stuttgart (1982)
27. Otto, F.: Netze in Natur und Technik - Nets in Nature and Technique. IL 8. Stuttgart, Institut für Leichte Flächentragwerke (1976)
28. Nerdinger, W.: Frei Otto, das Gesamtwerk – Leicht bauen, natürlich gestalten. Basel, Birkhäuser (2005)

29. Kolarevic, B.: Architecture in the Digital Age: Design and Manufacturing. Spon Press, London (2003)
30. Pugnale, A., Sassone, M.: Morphogenesis and structural optimization of shell structures with the aid of a genetic algorithm. Journal-International Association For Shell And Spatial Structures **155**, 161 (2007)
31. LNCS Homepage, htttps://www.rhino3d.com/ last accessed 2021/09/21
32. LNCS Homepage, http://kangaroo3d.com
33. LNCS Homepage, https://www.karamba3d.com

Virtual Reality Application for the 17th International Architecture Exhibition Organized by La Biennale di Venezia

Giuseppe Fallacara, Ilaria Cavaliere, and Dario Costantino

Abstract This paper aims to investigate the use of Virtual Reality (VR) as a support for expositions and cultural events through the presentation of a case of study related to the 17th International Architecture Exhibition organized by La Biennale di Venezia. The idea for this experimentation was born during the period of Covid-19 pandemic, in which it was impossible to travel freely. The goal was to make part of the exposition available to be visited virtually all over the world, in hubs equipped with VR headsets. Thanks to a collaboration with the organizers of the exposition, a VR app has been developed in order to allow people to visit the Giardino delle Vergini, which for several years has hosted the Italian Pavilion and where this year were placed the installations of prof. Giuseppe Fallacara and his research team together with the works of other international firms. Ethic matters have been taken into consideration during the app development. The VR app has been developed non to be a mere reproduction of the original site, but to be an alternative experience of visit. This work can bring two apparently contradictory advantages: on one hand the differences between virtuality and reality can encourage people to travel and visit the exposition in Venice; on the other hand, barriers of place and time are overcome. Therefore, everyone can visit the Giardino delle Vergini, even people who can't move.

Keywords Virtual reality · Extended reality · Virtual architecture · Biennale di Venezia · Virtual tour

United Nations' Sustainable Development Goals 8. Promote sustained, inclusive and sustainable economic growth, full and productive employment and decent work for all

---

G. Fallacara (✉) · I. Cavaliere · D. Costantino
Dipartimento di Architettura, Costruzione e Design, Politecnico di Bari, 70124 Bari, Italy
e-mail: giuseppe.fallacara@poliba.it

I. Cavaliere
e-mail: ilaria.cavaliere@poliba.it

D. Costantino
e-mail: dario.costantino@poliba.it

1 Introduction

This article aims to describe a work made for the 17th International Architecture Exhibition organized by La Biennale di Venezia, titled *How will we live together*?

The exhibition had to be held in 2020 but it had to be postponed due to the pandemic, therefore it was inaugurated in May 2021. The pavilion called *Porzione d'infinito*, designed by prof. Giuseppe Fallacara and by the architect Maurizio Barberio, was selected for the Italian Pavilion and was placed in the Giardino delle Vergini, at the Arsenale of Venice.

All the problems and the restrictions connected with Covid-19 brought the authors to reflect upon the theme of accessibility of museums, expositions and cultural events and encouraged to think about potentialities of VR and the benefits that this technology could bring to tourism and culture disclosure. It has been demonstrated that Virtual Reality applications help stimulate potential tourists to actually visit physically the places virtually explored and enhance the appeal of less known sites [1].

This topic is linked to the eighth goal of the United Nations' Sustainable Development Goals and deals with decent work and economic growth, since tourism is an important part of one state's economy.

1.1 What is Virtual Reality?

As stated in [2], Most popular definitions of virtual reality make reference to a particular technological system. This system usually includes a computer capable of real-time animation, controlled by a set of wired gloves and a position tracker, and using a head-mounted stereoscopic display for visual output.

For Greenbaum Virtual Reality is an alternate world filled with computer-generated images that respond to human movements. These simulated environments are usually visited with the aid of an expensive data suit which features stereophonic video goggles and fiber-optic data gloves [3].

Even though Virtual Reality seems to be considered as a brand new technology, it is not. As a matter of fact, the history of Virtual Reality, or better said Extended Reality (XR)—which is the group of technologies that includes Virtual Reality, Augmented Reality (AR) and Mixed Reality (MR)–, starts in 1832, when the first attempt of enhancing the concrete reality was made thanks to the invention of the stereoscopy by Sir Charles Wheatstone. That said, the evolution of the virtual technology as we know it today starts from the '60s, when Morton Heilig invented the first example of immersive cinema called Sensorama. Of course, Heilig made just a first attempt that was too heavy, complex and expensive to be spread, but at the same time it was extremely inspiring and became the starting point of an almost sixty years process that has not ended yet.

After Sensorama there were many other experimental tools indeed, such as the Sword of Damocles in 1965 by the Turing Award winner Ivan Sutherland, who was also the inventor of Sketchpad. In this case, too, the technology was raw and uncomfortable and the headset was heavy enough to require a particular coupling system hooked to the ceiling (which is the reason why it was called as it was).

Over time these flaws were corrected more and more, and researchers started to focus also on new ways of showing digital worlds, since Virtual Reality environments consisted of wireframe rooms only. Therefore, the first example of Google Street View was made with the name of Aspen Movie Map in 1978 by the MIT, demonstrating for the first time the real power of immersive experiences as a tool to virtually walk inside a real city.

Many other technologies came one by one after that, such as: Vcass by Thoms Furness in 1982, Vived by Nasa Ames Research Center in 1984 and then the Virtual Windtunnel, Cave, Boom, etc. up to the first head mounted display as we know it today, that is the Oculus Rift, invented in 2010 by Palmer Luckey [4, 5].

From 2010 software and hardware improvement became extremely faster and this situation allowed to overcome all the limits that Extended Reality tools had had since the very beginning of the 2000s, especially in terms of graphics.

Today not only a headset is more affordable than before, but there are many free softwares that allow anyone to even create his own virtual reality executable.

1.2 VR for Heritage: State of the Art

Of course, the one described in this paper is not the first attempt to use this VR in order to enhance the cultural field. International studies have already evaluated the efficiency of Extended Reality applications, that are considered a very efficient way to transmit knowledge [6]. The benefits of the use of both VR and AR have been highlighted through specific experiments referred especially to historical heritage and museums [7, 8].

The first experiments were conducted between the end of '90 s and the early 2000s and they were associated to very uncomfortable situations, in which people had to move around with a heavy equipment.

An example is Archeoguide [9], a project developed in 2001, whose aim was to enhance the archaeological site of Olympia through the use of a VR/AR app. Tourists had the possibility to use three different devices—a laptop, a pen tablet or a palmtop—to see virtual reconstructions of monuments, artifacts and life on top of the existing ruins and landscapes. The most complete experience was offered by the laptop unit, which was composed of a laptop—which had to be carried in a backpack by the user—a USB web camera, a digital compass, and a head mounted display.

Fortunately, as we already said, this technology has been far improved and research on the theme of its connection with educational purposes have been carried on, because of the promising results obtained in the past.

A study carried out by the Department of Leisure and Recreation Management of the Ming Chuan University, in Taiwan, [1] has shown how the use of VR is an effective way to encourage people of every age and social background to visit places that aren't popular destinations among tourists. Furthermore, the study confirms that VR simulations help to understand if and how the chosen places can be equipped to encourage sustainable tourism.

At the same time Virtual Reality can increase people's capability of perceiving and understanding places. Another study conducted by a rich group of cultural centers and universities—among which the Raymond Lemaire International Centre for Conservation in Belgium, the Carleton Immersive Media Studio in Canada and the Assiut University in Egypt [10]—has indeed shown that VR experiences allow users to easily recognize materials, features and details of the things they see as well as dimensions and state of conservation. Moreover, Virtual Reality raises the awareness of users about cultural heritage and helps experts in their research.

A study conducted by the Polytechnic of Milan in 2019 [11] has demonstrated that VR and AR reconstructions of monuments—in this case a digital twin of the Basilica di S. Ambrogio—in which also an interactive session is included, enrich the experience of casual users and experts, who can deepen their knowledge depending on their interests and purposes. Above all, Virtual Reality sessions open visitors' mind, giving them different perspectives on cultural heritage, and at the same time give monuments the chance of reaching a wider number of users.

This happens because Virtual Reality applications are conceived to enhance the users' curiosity, through an amusing and intriguing setting. This is the basis of the so-called *gamification* [12, 13], that brings to the production of *serious games*. Serious games can be effectively considered video games with goals and objectives that must be achieved, making the global experience stimulating and more effective [14, 15].

Of course, lots of other Extended Reality studies and experiments have been conducted in the latest years, especially because of the extreme development of technology, which has gone far beyond the limits found in many scientific papers from the early 2000s [16].

Since then and especially since 2010, when the first prototype of Oculus Rift was built, experts from different knowledge fields—architecture, engineering, medicine, archeology, design and computer graphics, etc.—kept doing research that underlines the importance of immersive experience in order to enhance cultural heritage and events [17], also looking for new ways of connecting virtual and physical world. Another experiment made at the University of Geneva, in Switzerland, has been carried out trying to develop the virtual experience with digital characters acting a storytelling drama on sites like Pompei [18].

Therefore, in a difficult historical moment such as the Covid-19 pandemic, which prevented people from being able to move freely, the use of VR became a strategic way both to take part in the Exhibition held in Venice and to think about important contemporary problems, for example the topic of virtual representations and digital twins as a way to enhance cultural heritage and the desire of phisically visiting it.

La Biennale di Venezia Foundation has clearly taken sides against the idea of creating a simple replica of the exhibition due to the risk that this operation could lead the exhibition itself to lose value. For this reason, it is necessary to specify that we focused only on one part of the Italian Pavilion, the Giardino delle Vergini, where it was possible to visit not only *Porzione d'Infinito*, but also the pavilions designed by Zaha Hadid Architects and Tecno, by Orizzontale, by David Turnbull (together with other designers), by Gianni Pettena and by Pongratz Perbellini Architects (together with Dustin White and Dario Pedrabissi). Furthermore, it is necessary to specify that an exact replication of reality has not been carried out: even if the aim was to reproduce the context of the garden and the installations as faithfully as possible, a simplification has sometimes been carried out in order to provide the general idea of the project linked to an alternative user experience.

2 The Method

The VR executable was made using the free software Unreal Engine 4, while for the 3D modeling and the texture mapping Rhinoceros and 3D Studio Max were chosen. The process consisted basically in three phases: the reconstruction of the context, the optimization and placement of pavilions, the interactivity implementation. Part of this work, especially the context modeling, has been done with the help a group of three more people—Maria Lucia Valentina Alemanno, Alessandro De Bellis and Isabella Giordano—in order to reach the best result possible.

As a matter of fact, the virtual reconstruction of a specific environment or architecture is a complex process that requires many different abilities linked to no less different fields of knowledge such as computer graphics, architecture, visualization, programming and 3D modeling.

2.1 The Context

The first step was the digital reconstruction of the Giardino delle Vergini, in which the models of the various pavilions had to be placed.

The garden is part of the Arsenale della Biennale, a building of the pre-industrial era. The Arsenale consists of a series of construction sites, where Venice fleet was built. It hosts part of the Exhibitions organized by La Biennale Foundation since 1980 and the site has been enhanced since 1999 thanks to a valorization program. The Giardino delle Vergini, which is accessible since 2009, hosts part of the Italian Pavilion. The garden is made of a wide green area surrounded by historical brick buildings [19].

From a practical point of view, the approach to the situation was not easy. As a matter of fact, in normal conditions working on an existing place would require a long session of inspections with the help of cameras and drones in order to carry out surveys that help to have an accurate three-dimensional clone of the intervention area. This is necessary to make all the subsequent operations more coherent and precise and to give the future users a sense of total immersion that makes them feel part of the virtual scene and that guarantees the VR executable not to be "reduced" to the same perception of a video game. This is not an easy operation because of the technical limitations of viewing via headset and the deleting of many physical characteristics of the human body. At the same time this issue is fundamental: the more an individual is involved, the more it will be easy to forget the outside concrete world, leading the user to the best Virtual Reality experience.

In this case it was impossible to personally visit Venice due to the pandemic, so it was firstly necessary to resort to Google Earth: the free software made available by Google allowed the extraction of a three-dimensional reproduction of the land and a high quality orthophoto. Moreover, several screenshots were taken through the Street View tool. These data, together with some photographs of the site and a CAD planimetry sent us by the curator of the 17th International Architecture Exhibition of Venice Dario Pedrabissi, were fundamental both for the three-dimensional modeling and for the rendering of the place.

It is important to focus on the optimization strategies adopted during this phase, since the context was the most complex element in terms of polygons account and materials to render and a bad optimization would have inevitably compromised the virtual experience.

The model extracted from Google Earth was processed using Sketchup and was useful to evaluate terrain heights. It was possible to notice that there weren't significant depressions or elevations, therefore terrain was approximated to a plane, in order to have the lowest amount of polygons and to simplify the scene. It was split in different portions according to the different soil materials (grass, concrete and pebbles) with the help of the ortophoto. The single planar parts were mapped and then imported in Unreal Engine, where proper materials were applied.

The Google Earth model was useful also to obtain the correct dimensions of the buildings surrounding the garden. The volumes of these buildings were transformed into more accurate models using Rhinoceros.

A special focus should be done on the arched building, which was the nearest one to the area where installations were placed and that required a good compromise in terms of optimization and realism. This building was treated as a modular one, therefore a single arch was modeled and repeated. The measure of a single module was obtained dividing the total length of the building for the number of arches. Of course, it is important to underline that the purpose of this work wasn't to obtain an accurate survey of the site, but to have a light 3D model that globally appeared as similar as possible to it.

Even if the dimension and the global geometry of the module is always the same, it is possible to distinguish four variations looking at the photos: a module with a simple wall, a module with a round arched window, a module with a pointed arched window and a module with both a pointed arched window and a door (Fig. 1).

In order to have a good optimization, it was impossible to have a different model for each variation of the arched element, therefore the differentiation was made using custom textures that could be applied to the same 3D object (Fig. 2). When strictly

Fig. 1 The different variations of the arches. Photo taken from Google Earth's Street View

necessary, a few 3D objects were added to increase the realism (for example sills or jutting brick arches).

Furniture elements as lamps and trash cans were modeled and placed referring to the photos in order to improve the realism of the scene.

For the lighting a HDRI sky with low sun was chosen and some parameters were modified in order to increase the contrasts of the scene.

Vegetation choice and placement required some time and various trials, since it is what influenced the performances of the VR application the most. Some lowpoly tree models similar to the real trees of the garden were chosen and placed with the help of the photos and the planimetry. We tried to scatter different types of 3D grass, but the FPS always dropped dramatically, so we decided to use a grass texture. We modified some parameters of the grass map included in the Unreal Engine starter content, in order to increase variation and reduce the perception of a texture repeated all over the place.

Finally, some decals were used to simulate stains and imperfections.

2.2 The Pavilions

After the setting of the context, we proceeded with the optimization of the pavilions. Most of the designers gave us high resolution models, so we had to obtain simpler models, deleting details and reducing the number of polygons as much as possible.

Only Gianni Pettena's and Zaha Hadid Architects' ones were modeled from scratch because the authors provided only 2D drawings.

As we already said in the introduction, not all the installations were rendered to look exactly like the real ones. For example, in the case of Christian Pongratz's

Fig. 2 The textures used for the variations of the module

pavilion, the designer himself requested the model displayed in VR not to be reproduced neither with its real materials nor in a photorealistic way, but it had to be a white painted version of his project that could allow a deeper focus on its geometry.

We treated similarly David Turnbull's pavilion, too (Figs. 3, 4, 5 and 6).

Fig. 3 A comparison between the digital reproduction of the garden and the photos taken from Google Earth

Fig. 4 A screenshot of the VR application environment. From the left: *Flux* (by Giuseppe Fallacara and Maurizio Barberio), *Watershed* (by David Turnbull with Fred Avitaia, Lorenzo Bertolotto, Marc DiDomenico, Saul Golden), *Archipensiero* (by Gianni Pettena) and *Crispr-Locus* (by Pongratz Perbellini Architects with Dustin White and Dario Pedrabissi)

Fig. 5 A screenshot of the VR application environment. From the left: *High-Performing Urban Ecologies* (by Zaha Hadid Architects with Tecno), *Prossima Apertura* (by Orizzontale) and *Flux* (by Giuseppe Fallacara and Maurizio Barberio)

2.3 Interactivity

Interactivity is the basis for a good VR experience, since the possibility to explore, interact with objects and acquire information is what distinguish an app from a simple 360° image.

Given the extent of the garden, it was impossible to explore the area just walking around in a room with the headset on, so a teleport system was set up. This way the user can move with the help of a luminescent indicator, which shows the destination of the teleport (Fig. 7).

Future developments of this research may be linked to tests carried on expert and casual users in order to evaluate their feedback and use them to improve the experience.

We tested a navigation system based on the use of joysticks, too, but at the end it was rejected because it caused more problems of motion sickness than the teleport system.

Then we decided to further improve the VR experience providing some information about each pavilion. To do so we added the official descriptive panels of the exposition, which appear when the user goes near the installations (Fig. 8). In order to indicate the precise area where the panel appears, we put in the scene some luminescent info signs.

In the case of the Pettena's pavilion, which is an anamorphosis, a luminescent signal was placed on the floor in the exact point where it would have been possible to perceive the false perspective generated. This is a way to make the user immediately grasp the meaning of the artefact itself, without any other explanation (Fig. 9).

Virtual Reality Application for the 17th International Architecture …

Fig. 6 A comparison between the real Christian Pongratz's Paviliom (photo By Pongraz Perbellini Architects) and the digital one

3 Ethical Matters

Creating a digital twin of a real place or architecture is not a simple operation, not only because of the required technological skills, but also because of what we may call an ethical problem. As a matter of fact, the digital world is something that can be completely controlled by its creators: situations, events, sounds, everything is guided by just one hand. This means that this world can be shaped and guided in

Fig. 7 A screenshot of the teleport indicator

Fig. 8 A screenshot of one of the information panels

one direction, maybe choosing to give a user just a specific view in order to evoke specific emotions, feelings and thoughts.

This is an important matter that must not be underestimated, especially considering the topic of the Metaverse, that's spreading today especially because of advertising actions made by big companies like Facebook.

Lorenzo Scaraggi is a famous italian storyteller and videomaker who has experienced the use of virtual reality both as a user and as a creator of contents like immersive movies and this experience gave him the chance of thinking about the moral matter behind the extended reality experiences.

Fig. 9 A screenshot which shows the signal for the correct perception of Pettena's anamorphosis

[…] The risk is that the new range of emotions coming from the use of this technology is conveyed in the worst way: in a consumerist dictatorship where those emotions become a coercive vehicle for opinions, moods, political orientations, easy consent» he wrote.

I think of a world of isolated people wearing headsets and whose emotions become invisible threads maneuvered from above; something far more powerful than what happens with social networks; something that gives us the power of talking to the heart of the large masses through an algorithm, creating a place where we decide what they want, what they buy and what they choose. So, what we need to do as a responsible society is to shift the focus of virtual production onto culture and onto a representation of reality as clear as possible […] [20].

These quotes can help us also explain the reason why it is important not to build an exact replica of the real world, as already said. The real advantage that immersive visualization brings is strictly linked to a well balanced relationship between virtual and physical so that they do not harm each other. Also, it is important for what it may concern the topic of emotions and feelings, which has been already discussed.

Virtual places, also the fictitious ones made for videogames, are catalysts of emotions just like books and movies. In this regard, during an interview published on IoArch Magazine n. 93, the architect Eric de Broche des Combes from Luxigon Studio said: «I personally have played World of Warcraft for ten years (although I should maybe not have), and for me these places really exist. I remember them; I discover new ones exactly like visiting Rome, or Milan. For me they do exist. The interesting part of the story is that, psychologically speaking, there is an emotional involvement with these places, which is similar to the one you commonly experience in reality. They are cognitively and (to a certain extent) physiologically real. Of course, you do not experience pain, death and things like that, but most of what you normally experience like empathy, love, aesthetic appreciation, do exist and are reflected in the environment, even though it is virtual» [21].

This is important to reflect on the possibilities given by Virtual Reality for cultural purpose, since interactive experiences, and in particular the immersive ones, produce a much greater involvement of the user than other traditional educational tools, as written panels, oral explanations, etc.

4 Conclusions: Potentials and Limitations

In conclusion, Virtual Reality is an instrument with a high potential for the enhancement of cultural heritage, and to increase the collective participation in events all around the globe.

On one hand VR applications can be used to improve the experience of a real tour in a museum or in a cultural site giving the user additional information. On the other hand VR executables can be sent everywhere and can become a means to break down barriers and to keep people all over the world able to visit a specific place and to interact with it. Especially after the Covid-19 pandemic, when limitations prevented people to travel, virtual tours have increased their popularity and it is important to reflect on the possibilities offered by immersive experiences as new means of knowledge acquirement and cultural entertainment.

Of course, it is necessary to underline that Virtual Reality cannot and must not be used as a mere substitute for reality, but rather as a way to build augmented and alternative visions of the reality itself in order to allow to grasp different facets and to acquire additional information in an uncommon way.

Moreover, this kind of parallel experiences can encourage people to visit the real places and discover their concrete side, bringing benefits to the tourism field, which almost completely stopped because of the Covid-19 pandemic.

In this paper we have described the entire process for the creation of a VR application of a cultural site. Of course, even if the technology has great potentials, it still shows some limits. Firstly, this kind of executable requires time and specific skills to be correctly realized; moreover, complex Virtual Reality executable still need specific tools to be used. Even though Augmented Reality is more and more used through smartphones, virtual reality is still mostly linked to headsets and workstations, which can be expensive and difficult to move. In order to obtain the best result in terms of education, it must be sent to schools and organizations that already have specific laboratories. In our case, with the help of prof. Christian Pongratz, we were able to share the app with the New York Institute of Technology, which has labs equipped with various Oculus headsets.

References

1. Lin, L.P.L., Huang, S.C.L., Ho, Y.C.: Could virtual reality effectively market slow travel in a heritage destination? Tour. Manage. **78**, 1–11 (2020)

2. Steuer, J.: Defining Virtual Reality: Dimensions determining Telepresence. In: Biocca, f., Levy, M. (eds.): Communication in the Age of Virtual Reality, pp. 33–56, Lawrence Erlbaum, Hillsdale (1995).
3. Greenbaum, P.: The lawnmower man. Film and video 9(3), 58–62 (1992)
4. Mazuryk, T., Gervautz, M.: Virtual Reality History, Applications, Technology and Future, https://www.cg.tuwien.ac.at/research/publications/1996/mazuryk-1996-VRH/TR-186-2-96-06Paper.pdf (1996), last accessed 2022/04/29.
5. Marr, B.: The Fascinating History And Evolution Of Extended Reality (XR)—Covering AR, VR And MR. Forbes (17 May 2021), https://www.forbes.com/sites/bernardmarr/2021/05/17/the-fascinating-history-and-evolution-of-extended-reality-xr--covering-ar-vr-and-mr/, last accessed 2022/12/21.
6. Ibañez-Etxeberria, A., Gómez-Carrasco, C.J., Fontal, O., García-Ceballos, S.: Virtual Environments and Augmented Reality Applied to Heritage Education. An Evaluative Study. Applied Sciences 10(7), 1–20 (2020)
7. Tsai, S.: Augmented reality enhancing place satisfaction for heritage tourism marketing. Curr. Issue Tour. 20(9), 1078–1082 (2019)
8. Trunfio, M., Della Lucia, M., Campana, S., Magnelli, A.: Innovating the cultural heritage museum service model through virtual reality and augmented reality: the effects on the overall visitor experience and satisfaction. J. Herit. Tour. 17(1), 1–19 (2021)
9. Vlahakis, V., Karigiannis, J., Tsotros, M., Gounaris, M., Almeida, L., Stricker, D., Gleue, T., Christou, I. T., Carlucci, R. and Ioannidis, N.: Archeoguide: First results of an Augmented Reality, Mobile Computing System in Cultural Heritage Sites. In: Vast '01—Proceedings of the 2001 Conference on Virtual Reality, Archeology and Cultural Heritage, pp. 131–140, Association for Computing Machinery, New York (2001).
10. Paladini, A., Dhanda, A., Reina Ortiz, M., Weigert, A., Nofal, E., Min, A., Gyi, M., Su, S., Van Balen, K., Santana Quintero, M.: Impact of Virtual Reality Experience on Accessibility of Cultural Heritage. In: International Archives of the Photogrammetry, Remote Sensing and Spatial Information Sciences, vol. XLII-2/W11, pp. 929–936, Copernicus Publications, Hannover (2019).
11. Banfi, F., Brumana, R., Stanga, C.: Extended Reality and Informative Models for the Architectural Heritage: From Scan-To-Bim Process to Virtual and Augmented Reality. Virtual Archaeology Review 10(21), 14–30 (2019)
12. Swacha, J.: State of Research on Gamification in Education: A Bibliometric Survey. Education Sciences 11(2), 1–15 (2021)
13. Mazur-Stommen, S., Farley, K.: Games for Grownups: The Role of Gamification in Climate Change and Sustainability, Indicia Consulting LLC (2016).
14. Mariotti, S.: The Use of Serious Games as an Educational and Dissemination Tool for Archaeological Heritage. Potential and Challenges for the Future. Magazén 2 (1), 119–138 (2021).
15. Ye, L., Wang, R., Zhao, J.: Enhancing Learning Performance and Motivation of Cultural Heritage Using Serious Games. Journal of Educational Computing Research 59(2), 287–317 (2021)
16. Paranandi, M., Sarawgi, T.: Virtual Reality in Architecture: Enabling Possibilities. In: Ahmad Rafi, M.E., Chee W.K., Mai, N., Ken, T.-K. N. and Sharifah Nur, A.S.A. (eds.) CAADRIA 2002, Proceedings of the 7th International Conference on Computer Aided Architectural Design Research in Asia, Prentice Hall, Petaling Jaya, pp. 309–316 (2002).
17. Jacobson, J., Vadnal, J.: The Virtual Pompeii Project. In: G. Richards (ed.), Proceedings of E-Learn 2005—World Conference on E-Learning in Corporate, Government, Healthcare, and Higher Education, pp. 1644–1649, Association for the Advancement of Computing in Education (AACE), Vancouver, Canada (2005).
18. Papagiannakis, G., Schertenleib, S., O'Kennedy, B., Arevalo-Poizat, M., Magnenat-Thalmann, N., Stoddart, A., Thalmann, D.: Mixing virtual and real scenes in the site of ancient Pompeii. In: Computer Animation and Virtual Worlds, 16 (1), 11–24 (2005).

19. La Biennale di Venezia website, https://www.labiennale.org/en/venues/arsenale, last accessed 2022/04/29.
20. Scaraggi, L.: La creazione dei video immersivi: cultura della verità o dittatura tecnologica? In: Costantino, D., Cavaliere, I.: Virtual Architecture. L'architettura al tempo della Realtà Estesa: compendio di esperienze 2019–2021, pp. 12–17, Amazon Books (2021).
21. de Broche des Combes, E.: SpaziFantasma. IoArch 93, 88–92 (2021)

Towards a Digital Shift in Museum Visiting Experience. Drafting the Research Agenda Between Academic Research and Practice of Museum Management

Giuseppe Resta and **Fabiana Dicuonzo**

Abstract This chapter reviews the state of the art in digital applications for museums and exhibitions, with a particular focus on the visiting experience. The authors have measured the gap between academic research and the current practice of museum management through a mixed-methodology approach. On one hand, the text presents the result of a systematic literature review of articles on museum digitalization that have been published since 2000. On the other hand, it includes the results of an interview with a group of experts, directors, and curators of Italian museums to understand the degree to which digitalization is currently adopted in those cultural institutions. COVID-19 is an additional factor that has been considered in terms of its impact on scientific production and museums' strategies. Such cultural institutions, having ticketing and similar forms of revenue related to physical visitors at the core of their model of economic sustainability, suddenly realized the need for a different approach to promoting art, namely forms of engagement from a distance. Within the frame of industry 4.0, it has become evident the crucial role experts play in the field of digitalization and implementation of virtual environments for the art sector. This text aims to draft a research agenda on museum digitalization for the near future, looking at trending topics, academic networks, and research geographies. The qualitative survey with experts' opinions discussed whether regular employment of digital platforms and virtual tours can engage new visitors in the long term, and established the current status of their day-to-day activities.

Keywords Survey · Literature Review · Digital Shift · Museum Sector · Digitalization

---

G. Resta (✉)
Faculdade de Arquitectura, Universidade do Porto, Porto 4150-564, Portugal
e-mail: giusepperesta.arch@gmail.com

F. Dicuonzo
CITCEM, Universidade do Porto, Porto 4150-564, Portugal
e-mail: fabiana.dicuonzo@gmail.com

© The Author(s), under exclusive license to Springer Nature Switzerland AG 2024
M. Barberio et al. (eds.), *Architecture and Design for Industry 4.0*, Lecture Notes in Mechanical Engineering, https://doi.org/10.1007/978-3-031-36922-3_34

United Nations' Sustainable Development Goals 9. Build resilient infrastructure, promote inclusive and sustainable industrialization and foster innovation · 10. Reduce inequality within and among countries · 17. Strengthen the means of implementation and revitalize the global partnership for sustainable development

1 Introduction

1.1 Outline

In the last twenty years, the digital shift in the field of art and architecture has been forming a new body of theory that encompasses: a revision of the design phase [1]; new manufacturing processes and robotic fabrication [2]; the adoption of intersectoral educational models [3]; and the introduction of virtual experiences in connection with or in replacement of existing spaces and heritage sites. The latter will be at the center of this analysis, with the intention to draft the research agenda for the museum sector facing such new paradigms as: digital twins, data-driven strategies, Virtual Reality and Augmented Reality, real-time digital representation, and visitor-computer interaction.

This chapter intends to review the status of digitalization in museums, with a particular focus on the visiting experience. In this regard, we have decided to measure the gap between academic research and the current practice of museum management. On one hand, the text presents the result of an extensive literature review of articles that have been published since the year 2000. On the other hand, the authors have conducted a semi-structured interview with a group of experts, directors, and curators of public museums to understand the degree to which digitalization is currently adopted in museums.

The digital shift in the museum visiting experience has been occurring for many decades. It started with the concept of museum computing at the end of the 1960s [4], firstly integrating archives and records and then affecting the visiting experience with the evolution of audio–video guides [5]. This text intends to lay out the updated state of the art for the issue of museum digitalization by studying the current research panorama together with trends of future strategic implementations.

Among the many impacts of COVID-19 restrictions in the last two years, the digitalization of the museum experience has secured the attention of many researchers [6, 7]. Cultural institutions devised multiple communication strategies and virtual environments to target visitors' engagement during this period. In this chapter, we will also examine how this affected the direction of academic research considering the output published since 2020, when restrictions were enforced. Simultaneously, if we are to draft a research agenda for the near future, we should be able to contextualize the pandemic event as a prominent but circumscribed occurrence. Hence, it is vital to step back to a more comprehensive vantage point from which it is possible to trace the whole trajectory of museum digitalization in academia and practice.

1.2 Theoretical Framework: Interaction and visitor's Experience

Interaction in visitors' experience is usually addressed as synonymous with digital environments. Though, as pointed out by Levent and Pascual-Leone [8], sensory engagement and immersive experience can be obtained by triggering senses of smell, touch, sound, space, and memory in exhibition spaces. In this regard, Classen [9] has discussed how museums are essentially focused on the visual experience, while many masterpieces and historical artworks are intertwined with the overall bodily experience that the subject perceives. Hence, this aspect is essential to the visitor's feeling of being present in the venue and will be addressed throughout the expert interviews presented in this study. Additionally, the act of art appraisal by visitors is mediated through a behavioral code established by the institution (i.e., museum, gallery, collector), and it is not always clear how the author intended their work to be experienced in the first place. For example, if touch and manipulation are allowed, and to what degree [10]. In archaeological museums, such impasse has been solved with partial or integral 3D printed replicas of the original that would satisfy the necessity of object handling as an exploratory phase of the visit [11].

Interaction aims then at increasing the level of engagement with the subject, mainly to produce long-term involvement [12]. In this sense, engagement as "the willingness to have emotions, affect and thoughts directed towards and aroused by the mediated activity in order to achieve a specific objective" [13] would be the ultimate goal of professionals working in the field of culture.

Societies have always constructed alternative worlds to engage an audience of visitors, religious believers, gamers, etc., projecting to another environment activities or representations that the physical world couldn't afford [14]. In museums, the traditional visit can be augmented with a narrative structure (storytelling), additional content (multimedia), and immersive experience (virtual reconstructions). Bekele and Champion [15] compared virtual reality technologies in virtual heritage, examining the most used interaction interfaces: Augmented Reality (AR), Virtual Reality (VR), Augmented Virtuality (AV), and Mixed Reality (MxR). The latter is seen as the most viable option for heritage sites and museums to establish a relationship between users, virtuality, and reality, without losing the social dimension of cultural learning. In fact, the educational value of museums is seen as a primary form of interaction [16], both within and outside 3D virtual environments [17].

This bond between century-old institutions and digital interactive tools opens another issue we will address through the expert interviews: the digital preparedness of museums. Hanussek discussed the supposed enhancement of the visiting experience through ad-hoc smartphone applications, pointing out that "museum apps have not brought the impact so often promised to visiting audiences" because "professional expertise in information technology and data analysis seem still to pose a huge challenge for museums, as evidenced by the technical issues of the discussed apps and the lack of proper assessment of their user experiences" [18]. Hence, we have asked the interviewed experts to describe the consistency of the information technology personnel in their institutions, if any.

Engagement also has an online phase that is conducted on social media channels. In relation to the visit, digital content can help build the construct of the visitor's motivation before the visit, or complement the acquired information after the visit [6]. We have addressed this issue in the bibliographic review and with specific questions to the experts. It should be noted that COVID-19 restrictions have impeded physical visits, offering for a certain period a unique opportunity to measure the delivery of cultural content through online platforms only [7]. This raised the question of the degree of replaceability of online experience in opposition to onsite presence. The use of digitization and social media also has profound political implications because it is being directed by choices that imply a selection ex-ante, and received on devices that are subject to digital divide disparities [19]. Hence, authority and curation of content are not secondary to the impact on visitors' experience through social media. In turn, different platforms have different audiences, making the overall assessment fragmented by definition: while Twitter has stronger involvement with political and social issues [20], Instagram's feed is predominantly visual and has more aesthetic connections with the experience of an exhibition [21]. Contents on Facebook create virtual communities of users interested in a specific topic: it allows interaction in both directions, but at the same time users expect the cultural institution to be consistently responsive to maintain an effective engagement [22]. The creation of content is then tailored to the specific platform if museums intend to gain maximum engagement, requiring an effort in communication strategies that is constant and with a long-term perspective.

Online communication covers a broad spectrum of channels, from institutional websites to chatbots. The former is unidirectional and aimed at a generic prospective visitor; the latter is "a computer program designed to simulate conversation with human users" [23] with one-to-one interaction.

Finally, the visitor's experience can be considered interactive when the museum activates participatory projects of co-creation. This social aspect has an extensive literature and is widely studied among practitioners and researchers [24–27]. We will address it several times in the bibliographic collection and with the questionnaire only in relation to the digitalization of the visitor's experience.

This study builds upon an article on the impact of virtual tours on museum exhibitions that we have recently published [6]. We have decided to define the perimeter of the investigation through the following parameters:

- Definition of a specific setting: museum. Art galleries, fairs, temporary exhibitions, and art parks are not taken into consideration.
- Definition of a specific subject: visit. Laboratories, archives, museum libraries, happenings, and talks are not taken into consideration.

- Definition of a specific aspect of the visitor's experience: digitalization.

Other literature reviews partially cover these three elements, but none is updated, with a systematic approach, or contains all the aforementioned components. Xu et al. [28] analyzed research results published between 2011 and 2021 on the impact of technology applications on museum learning outcomes retrieved from the core Web of Science collection. Ayala et al. [29] examined research articles on audience development in museums and heritage organizations, combining results from three different databases. Serravalle et al. [30] focused their attention on research items on augmented reality in the museum with reference to the pool of stakeholders.

2 Literature Review

2.1 Methodology

This bibliometric analysis is structured in two components that will be addressed separately in the result section. One concerns descriptive metrics in the domain of museum digitalization in terms of overall scientific production and its yearly evolution. The second looks at knowledge structures across the pool of articles considered for this research.

To collect a reliable and representative number of articles, papers were retrieved from the core Web of Science collection, containing journals in the Science Citation Index Expanded and Social Sciences Citation Index.

Bibliometrix R-package and Microsoft Excel were used for analysis. Bibliometrix is a science mapping open-source tool programmed in R that elaborates research distribution, subjects, and citations [31]. The following objectives have guided this quantitative analysis:

- Establish ground for comparative evaluation with expert interviews
- Identify research trends and specific geographies interested in the topic of museum digitalization
- Visualize the collaborative network that shares an interest in the topic of museum digitalization
- Study the use of keywords in scientific production
- Identify the most cited articles, journals, and authors.

Articles had to contain the three components that form the construct of this research: keywords "visitor", or "visit", or "engagement"; and keywords "digital" or starting with "digitali"; and keywords "museum", or "exhibition". With the Boolean operators "AND NOT" we have excluded those articles that contain the keyword "archive" as it is within the domain of museum studies but beyond the scope of our study on the visitor's experience; and the keywords "machine learning", "deep learning", or "artificial intelligence", that characterize articles beyond the scope of our study. The analysis was conducted in March–April 2022, and the records are

Table 1 WoS search query

Search syntax
Visitor OR visit OR engagement (All Fields)   AND digitali* OR digital (All Fields)   AND museum OR museums OR exhibition OR exhibitions (All Fields)   AND NOT "machine learning" OR "deep learning" OR "artificial intelligence" (All Fields)   AND NOT archive (All Fields)   AND 2022 or 2021 or 2020 or 2019 or 2018 or 2017 or 2016 or 2015 or 2014 or 2013 or 2012 or 2011 or 2010 or 2009 or 2008 or 2007 or 2006 or 2005 or 2004 or 2003 or 2002 or 2001 or 2000 (Publication Years)

updated to April 22nd. We have included all articles published in the last 22 years, considering that 2022 is represented only for the first four months of the year and will have limited relevance in certain aspects of the result section. A total of 1257 results were obtained with the syntax shown in Table 1.

After removing duplicates, 1240 articles were left. Then a close reading of titles and abstracts reduced the number to 1109, considering articles whose content is not covering any issue related to museums, visitor engagement, or virtual reality. We reported that some research published in journals of environmental sciences, ecology, and zoology, contain the same key terms but address very different research fields. Finally, after discarding reviews, editorials, data papers, and meeting abstracts, the pool of items reached the final number of 1082.

3 Results and Discussion

Overview. Articles are spread across 675 sources (books, journals, proceedings) with an average number of citations per document of 4.43, and 0.72 average citations per year per document. Items are mainly journal articles (54%) and conference papers (42%), and only 4% are published as book chapters. The total of authors involved is 2886, meaning 0.38 documents per author and 2.67 authors per document. Multi-authored items are 850 (79%), with 3.2 co-authors per document and a Collaboration Index (CI) of 3.13. The latter measures the mean number of authors of multi-authored papers per joint paper [32, 33], while co-authors per document measures authors' appearances per total number of documents. This suggests that the research team is generally formed of three authors. In terms of annual scientific production (Fig. 1), starting from 2016, publications constantly total 100 or more. The graph shows a considerable jump in 2018, maybe because the hardware for immersive reality started to become affordable and adopted by major entertainment companies [34]. Another spike is positioned between 2020 and 2021, when COVID-19 restrictions have amplified the debate on digitalization of cultural institutions. Compound Annual Growth Rate returns a constant rate of 14.59% over the examined period. Considering the average article citation per year (Fig. 2), articles that collect the highest number of yearly citations were published in 2000 and 2008.

It should be noted that we will differentiate between global citations, those that are provided by WoS metrics gauging the impact of an article in the whole database and across all disciplines, and local citations, those that are received from documents that are present in the analyzed collection as is formed through the search query in Table 1. Hence, the latter measure the impact in the field of museum digitalization.

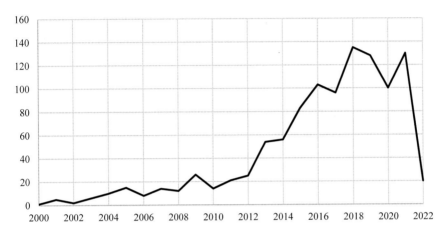

Fig. 1 Annual Scientific production. Y: Articles, X: year

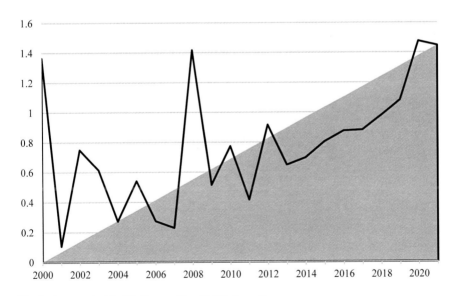

Fig. 2 Average Article Citation per Year. Y: Citations, X: year

Analytics and graphs. The analysis covers statistics on sources, authors, and documents. First, we examine the relationship between topics and geographies by looking at the keywords of the academic works.

Examination of keywords (Fig. 3) shows that the authors' countries are mainly the United Kingdom (237), Italy (229), Greece (101), Spain (101), and the USA (90). The keyword "cultural heritage" is mostly used by Italian authors. British scholars prevail in the use of "digital heritage". Keywords "virtual reality" and "augmented reality" have a similar distribution; the former is used by most of the Austrian and Chinese authors that were considered in this research; British authors mostly address "engagement"; "social media" and "virtual museum" are frequently cited by Italian authors; "education" is the second most used keyword by Spanish scholars. If we look at the authors' affiliations (Fig. 4), "cultural heritage" is mainly used by authors from Sheffield Hallam University and Università Politecnica delle Marche. The former prevails in the use of "virtual reality", the second in the use of "augmented reality". Some keywords are almost exclusively linked to one university: "social media" to Politecnico di Milano, "engagement" to King's College London, and "heritage" to Newcastle University. Vice versa, certain universities are very much focused on specific topics: the University of Peloponnese on "cultural heritage" and the University of Patras more generally on "museums", which is part of the search query, and is not linked to any of the top 20 keywords. The University of the Aegean and the University of Nottingham distribute their contributions in most of the top 20 keywords.

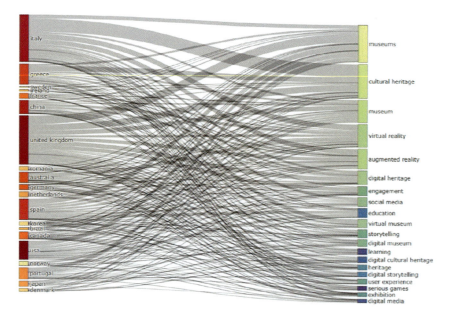

Fig. 3 Fields plot elaborated by Bibliometrix. Left column: author's country, right column: keywords

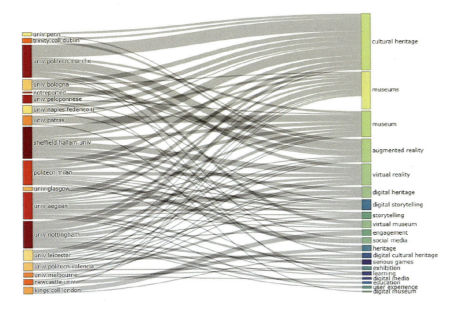

Fig. 4 Fields plot elaborated by Bibliometrix. Left column: author's affiliation, right column: keywords

It should be considered that the keywords mentioned above are the author's keywords. Although many publications suggest a preference for using Keywords Plus [35, 36], which are index terms generated by an algorithm that scans the titles of an article's bibliography [37], they are usually more generic and linked to methodological aspects [38]. For this reason, we will employ Keywords Plus to better understand the structure of scientific production on museum digitalization, while author keywords are considered better descriptors of the content of the articles [20, 38].

The journal Museum Management and Curatorship (Humanities, AHCI), providing 31 documents, is the most relevant source in terms of published articles. Second is the ACM Journal on Computing and Cultural Heritage (Computer Science and Humanities, SCIE, and AHCI), with 25 articles. The journal of Museum Education (Education & Educational Research, ESCI) with 14 articles, and Curator – The Museum Journal (Humanities, AHCI) follow. Particularly relevant are the two volumes of the 2015 Digital Heritage International Congress, with a total of 34 contributions pertinent to the subject. The congress held in Granada, Spain, is then the venue where museum digitalization has been more organically discussed in the last 22 years. The following 2018 Digital Heritage International Congress held in San Francisco, California, is well represented with 12 papers. Source clustering through Bradford's law [39, 40] shows that the core area that represents the nucleus of journals that cover the examined issue is quite broad. Bradford's model suggests that the core literature is scattered across 41 periodicals, confirming that the issue of museum digitalization is highly interdisciplinary and is present in sources of different scientific fields.

To evaluate sources' impact, we have considered the *g-index* developed by Egghe, which is the "unique largest number such that the top g articles received (together) at least g^2 citations" [41]. It has been demonstrated that this index, compared to the *h-index*, is not influenced by the total number of publications [42]. Hence, in our case is preferable because the initial publication year varies considerably. Museum Management and Curatorship has the highest *g-index* (17), and ACM Journal on Computing and Cultural Heritage is second with 12. Considering the number of examined articles, those published in Digital Creativity ($g = 8$) and in Visitor Studies ($g = 6$) have had a significant impact. Of the 12 top journals with a *g-index* of 5 or above, 5 are published in England, 3 in the USA, 1 in Greece, 1 in Italy, 1 in Poland, and 1 in the Netherlands. The category of humanities is the most represented with 6 periodicals, then computer science and archaeology with 3, art with 1, tourism with 1, and social sciences with 1. Most are indexed in AHCI collection (8), SCIE (3), ESCI (2), and SSCI (1). Source dynamics performed on these journals (Fig. 5) shows that periodicals concerned with museum studies have constantly investigated museum digitalization starting from the period 2004–2008, while periodicals more centered on computer science have considerably increased their interest only in the last years.

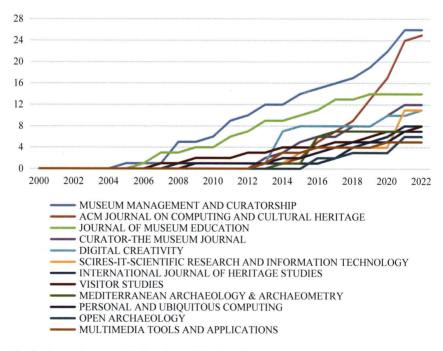

Fig. 5 Source dynamics. Y: Cumulate publications, X: year

Focusing data analysis on authors, we observe that the top-ten most relevant authors per fractionalized number of documents [43] range from 0.8 to 4.4 documents. Top-ten authors per fully counted documents, ranging from 6 to 11 papers, are the same as the fractionalized count with slight differences in terms of rank. Benford has the most extended production on the subject over time, while Antoniou has the most protracted timeline if we consider active authors that have already published an article in 2022. Petrelli has constantly published every year 1 to 4 articles from 2016 to 2020, and Lepouras has continuously published 1 to 4 articles from 2016 to 2019. COVID-19 is having a remarkable impact on authors' production on museum digitalization: the two most cited articles were published only in 2020 and 2021 by Arnaboldi and Agostino (joined by different co-authors) with the titles "New development: COVID-19 as an accelerator of digital transformation in public service delivery" [44] and "Italian state museums during the COVID-19 crisis: from onsite closure to online openness" [7]. Both look at how Italian state museums implemented strategies of engagement during the lockdown. This element confirms that researchers have shifted their focus in the last three years. Another interesting aspect is that most of top authors' timelines start in 2013–14 and end in 2020, suggesting that 2021 imposed a halt in terms of production. Top authors published the majority of their articles in 2017–18.

Frequency distribution of scientific productivity studied with Lotka's Law [45] shows that 88% of items are authored by occasional contributors, while core authors have published at least 5 articles on the topic. The 0.2%, 19 researchers, can be considered core contributors in the field. Rounding this number to the 20 top authors, their *g-index* ranges from 4 to 9, with Petrelli and Pierdicca that record the best local impact.

The most relevant affiliations per number of articles are Sheffield Hallam University (25), University of Nottingham (21), Università Politecnica delle Marche (20), University of Peloponnese (16), University of the Aegean (13), Politecnico di Milano (12).

In Fig. 6, we can see the corresponding author's geographical distribution. Almost the same number of articles have Italian or UK corresponding authors, followed by USA, Chinese, and Spanish researchers. The total is then split into Single Country Publications (SCP), which are co-authored by researchers of the same country, and Multiple Countries Publications (MCP), with at least one co-author from a different country. Hence, the MCP ratio measures the intensity of international collaboration of a country. In this regard, the Netherlands (44%), Sweden (33%), and Denmark (29%), have the best ratio of international collaboration. Low international collaboration is measured with Brazilian, Romanian, Japanese, and French authors.

Counting instead the number of documents per country (Fig. 7), namely the affiliation countries' frequency distribution, the USA is represented in 518 documents, the UK in 366, Italy in 285, and China in 144. Large parts of Africa and central Asia are not present in any affiliation. Though, in terms of total citations per country, UK authors collect a total of 936, prevailing on the USA with 759, Italy with 672, and after that is a considerable gap to the fourth, China, with 347 citations.

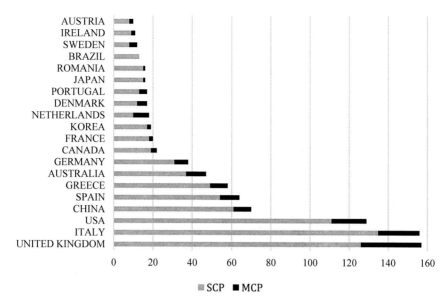

Fig. 6 Corresponding Author's Country. Y: number of publications (SCP = Single Country Publications; MCP = Multiple Countries Publications), X: country

Fig. 7 Country scientific production. Scale: white (unrepresented countries) to dark grey (USA = 518)

The last aspect of our analysis is related to the 1082 items retrieved with the search syntax mentioned above. We will call "documents" all items that are included in the bibliographic collection; "references" all articles that are cited in the bibliography of each document; "cited documents" all articles that are included in the bibliographic collection and at the same time cited as references.

Within the examined bibliographic collection, the most locally cited source is the book series Lecture Notes in Computer Science, published by Springer with 340 citations. Curator (247), Museum Management and Curatorship (234), and Journal of Cultural Heritage (215) score a similar number of citations. Thesis works are also quite present, with 185 citations. The most cited author in museum digitalization is Petrelli (25 local citations), which is not surprising as her works are all centered on the relationship between museums and digital platforms. Marty has 18 local citations, Not has 13 local citations (she co-authored three works with Petrelli), Agostino and Arnaboldi both have 12 local citations, having also co-authored three works together.

Most globally cited documents are published in computer science journals: "Using augmented reality and knowledge-building scaffolds to improve learning in a science museum" [46] 108 citations, "Effects of the inquiry-based mobile learning model on the cognitive load and learning achievement of students" [47] 78 citations, "Leveraging explicitly disclosed location information to understand tourist dynamics: a case study" [48] 75 citations. All three address museum issues only partially. This is demonstrated by the fact that top cited local documents are instead all published by Museum Management and Curatorship and are centered on museum issues: "Museum websites and museum visitors: digital museum resources and their use" [49] has 16 local citations, "The presence of Web 2.0 tools on museum websites: a comparative study between England, France, Spain, Italy, and the USA" [50] has 9 local citations, "Heritage in lockdown: digital provision of memory institutions in the UK and US of America during the COVID-19 pandemic" has 7 local citations. Their local/global citation ratio is 25% to 28%, meaning that more than one-fourth of their citations fall into the examined topic-specific bibliographic collection.

Concerning the most locally cited references, it is interesting to observe that although the majority of sources are journal articles, among the first 8 documents, only two are articles. The most cited source is Nina Simon's "The Participatory Museum" (53 local citations), which tackles the issue of community engagement through the design and practice of participatory projects. In fact, the author looks at the institution of the museum from a social point of view, examining the hiatus that the audience feels in terms of authority and relevance to their life [25]. John Howard Falk's books are second ("Learning from museums: visitor experiences and the making of meaning", 43 local citations), fourth ("Identity and the Museum Visitor Experience", 33 local citations), and eighth ("The Museum Experience", 20 local citations). The former [16] interprets museums as learning environments proposing a model underpinned by theories from psychology, education, anthropology, and neuroscience. The second [51] has a similar approach, focusing on the construct of visitors' motivations influenced by their identity. And suggests that some of these motivations occur even before a visitor enters the museum. The third [52] can be considered as the starting point of Falks' research, in collaboration with Lynn Diane Dierking, where the framework of the interactive experience is studied in its physical, personal, and social dimensions. All three books combine accessible language with broad multidisciplinary contributions. Similarly, the book "Learning in the Museum" [53], published by George E. Hein in 1998 (23 local citations), has a foundational role in laying out how the educational theories of John Dewey, Jean Piaget, and

Lev Vygotsky can be adapted to museum contexts. Tallon and Walker's edited book "Digital technologies and the museum experience: handheld guides and other media" [5] is the most cited document (25) that explicitly addresses the digital in its title.

The most cited articles are "Beyond virtual museums: Experiencing immersive virtual reality in real museums" (35 local citations), which examines the positive and negative aspects of immersive VR [54], and "Virtual museums, a survey and some issues for consideration" (26 local citations), on preservation and dissemination of cultural heritage through Web3D, VR, AR, MR, haptics and handheld devices, in a virtual museum environment [55]. Both are published in the Journal of Cultural Heritage.

In terms of the year of publication, references range from 1709 to 2022. Reference Year Publication Spectroscopy (RYPS) is a quantitative method that identifies the temporal roots of research fields, and is based on the analysis of the distribution of frequencies with which references are cited [56]. The RYPS of the studied bibliographic collection (Fig. 8) shows that the historical papers relevant to the field are quite recent, mainly published in 2012–14. The deviation curve shows only one distinct peak in 2010, when Simon [25] and Parry (ed.) [57] published their books, while Carrozzino [54] and Bruno [58] published their articles on virtual reality in the Journal of Cultural Heritage. Particularly relevant is 2012–13, when the personalization of visitor's experience has been widely discussed for the first time in separate articles by Ardissono [59], Lombardo [60], Capriotti [61], Charitonos [62], and Fletcher [63]. Also, Petrelli [64] and Coenen [65] discussed tools and applications for interactive visits. Additionally, the proceedings of the SIGCHI Conference on Human Factors in Computing Systems (2013) and the updated version of Falk and Dierking's book [66] contributed significantly. Other relevant historical references are published in 2004–5 and 2000–1.

The most relevant word in the bibliographic collection (Fig. 9), after having excluded the words used in the search query, is "heritage" among Keywords Plus occurrences (44) and "cultural heritage" among author's keywords (96). Both with

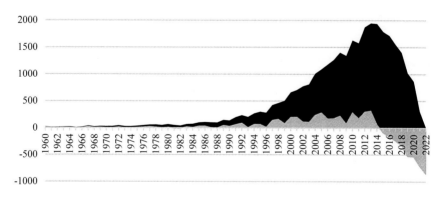

Fig. 8 Reference Year Publication Spectroscopy. Black: Number of cited references per year, Grey: Deviation from the 5-year median

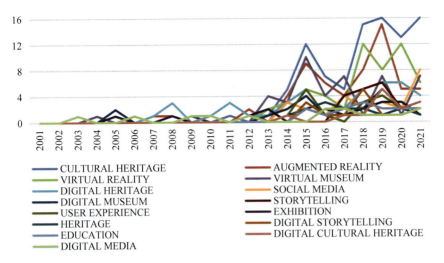

Fig. 9 Word dynamics. Y: Annual occurrence of author's keyword, X: year

a considerable gap on the second most used keyword. The frequency of "design" and "model" in Keyword Plus suggests recurrent works on methodological aspects. Author's keywords are very much referred to immersive reality ("augmented reality" and "virtual reality"). Abstract's words confirm the use of bigrams "cultural heritage" (297), "social media" (134), and "augmented reality" (126).

Over the years, starting from 2015, "cultural heritage" has been the most used keyword by authors (Fig. 9). Words that contain "virtual" ("virtual reality, "augmented reality", "virtual museum") also started to be used consistently in 2015. While words like "digital media" are being used from 2000, "storytelling" in association with the museum has been used only from 2017. The term "social media" shows the highest growth in 2021.

If we group frequencies of n words year by year, it is possible to know how trend topics vary within the examined collection. Searching the $n = 5$ most frequent author's keywords, "new media" and "website" were used until 2012. Then in 2012–16 gamification was introduced in several articles ("game-based learning", "3D modeling" and "usability") together with the concepts of "virtual heritage" and "participation". In 2016–19, "virtual reality" and "augmented reality" are the most studied topics together with the concept of "digital heritage". In 2019, there is a meaningful shift towards "storytelling" and "social media" until 2020–21, which shows another significant linguistic shift in terms of processes ("digitization", "digital culture" and "digital transformation"), tools ("3D printing"), social engagement ("museum education"), and events ("Covid-19"). The word "guide", which records frequent usage in many articles until 2014, is not frequently employed after 2016.

The words used in abstracts have similar dynamics, with an evident prevalence in the last three years of the words "eco museum", "HBIM", and "olfactory".

Trends essentially confirm a growing interest in visitor engagement in both social and technological acceptation, together with studies on social media. COVID-19 dramatically impacted titles and keywords, though it should be considered what will be the long-lasting effect of this event over time. The increasing use of "touch" and "olfactory" suggests that the visitor's experience is being studied beyond its visual dimension.

Structure of knowledge. The analysis outlines a conceptual, intellectual, and social structure of the research field. The visualization of this knowledge domain is expected to reveal the main themes and trends of the bibliographic collection, how certain authors influence the overall scientific production, and the geography of the research network [67].

Cluster map by documents coupling divides the items into subsets that are internally homogeneous and externally homogeneous. Figure 10 represents the five clusters positioned according to their impact and centrality (relevance to the field). This cluster analysis selects the top 250 documents with a minimum of 10% cluster frequency. Coupling is measured by references and the articles' impact through local citation score. The figure shows 5 clusters labeled with the main Keywords Plus terms: orange, blue, red, purple, and green.

The largest cluster (purple) is in the upper-right quadrant, with impactful and relevant documents. It has an impact of 2.62, a centrality of 0.42, and 82 documents. Marty [49] and Lopez [50] are the main contributors with research documents on museum websites and the use of web 2.0 tools. In general, articles in the purple cluster discuss how internet enhances the experience of a museum visit.

The green cluster is across the two right quadrants, with the highest centrality (0.43), average impact (2.28), and 57 documents. Smith [68] and King [69] are prominent authors in this cluster centered on social engagement with virtual environments and social media. Smith combines principles from participatory design with

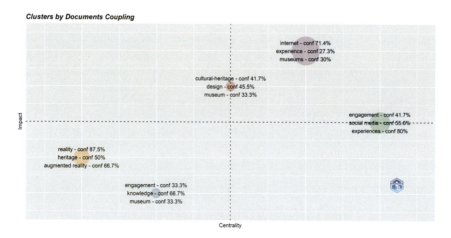

Fig. 10 Clusters by documents coupling positioned by impact and centrality. For color interpretation, refer to the text

themes of contemporary digital culture to create heritage innovation; King analyzes literature on digital engagement, interactivity, and participation in combination with a survey of heritage professionals.

The red cluster is across upper quadrants, with average centrality (0.41), relevant impact (2.45), and 33 documents. Most contributions are published in computer science journals and provide case studies of interactive exhibitions for cultural heritage. Among authors, Pierdicca [70] suggests the implementation of the Internet of Things framework to study visit patterns for a personalized museum experience, while Petrelli [71] explores the design, implementation, use, and evaluation of tangible data souvenirs for interactive museum exhibitions.

The lower-left quadrat has two clusters with lower impact and lower centrality, suggesting topics that might be emerging or ending in the context of museum digitalization. The orange cluster is mainly represented with the keywords "reality" and "augmented reality", suggesting a focus on interaction interfaces: Augmented Reality (AR), Virtual Reality (VR), Augmented Virtuality (AV), and Mixed Reality (MxR). Orange has a centrality of 0.31, an impact of 2.1, and 45 documents. Barsanti [72] discusses the optimization of 3D models of artifacts for virtual reality, Yoon [46] studies informal learning in a science museum using augmented reality, and Caggianese [73] analyzes interaction design focusing on a holographic projection system equipped with a gesture-based interface. The blue cluster has a similar focus on interaction interfaces but is more directed toward the learning impact rather than the design implications addressed in the orange cluster. Blue cluster has a centrality of 0.39, impact of 1.90, and 33 documents. One representative article is the study by Damala [74] with a qualitative and quantitative analysis of an augmented reality prototype to achieve an interactive learning experience in museums.

Moving to the visualization of the conceptual structure, Fig. 11 represents the co-occurrence network of authors' keywords. The network is based on simple similarities between words that are hierarchically grouped in clusters. After removing the words of the search query, we can see the core cluster, in red, which is formed around the concept of digital heritage and the use of VR and AR. Associated with these, we see other tools such as 3D printing and mobile applications, forms of visit augmentation such as storytelling and gamification, and hybrid approaches such as mixed reality.

Most of the terms are strongly connected with the center of the blue cluster that revolves around the virtual museum as a setting for the exhibition. Satellite words refer to 3D reconstruction, virtual heritage, and the issue of digitization itself.

The green cluster is isolated but internally coherent with the topic of social media. The terms "communication", "participation", "education", and "digital culture" complete the cluster together with "covid-19". The latter is also the only connection of this cluster with "virtual museum". This result confirms that the impact of the pandemic has been primarily studied in connection with the social media activity of museums.

The purple cluster is centered on the user experience and has stronger ties with the red and blue clusters. The words refer to the visitor's perspective, and especially to interaction design and personalization. Finally, the isolated yellow cluster suggests an interest in informal learning through games.

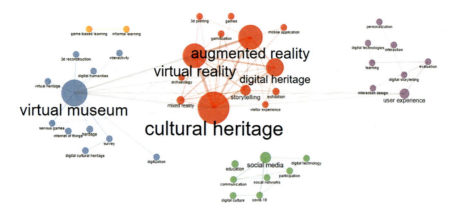

Fig. 11 Co-occurrence network of author's keywords. Elaborated by Bibliometrix

The thematic map of authors' keywords (Fig. 12) visualizes four types of themes based on two dimensions: centrality (importance of the topic in the given research field) and density (level of development of the theme). The motor themes of the discipline are based on aspects of communication, education, and visitor experience. Basic and transversal themes relate to cultural heritage on the one hand, and virtual museum (with augmented reality) on the other. Blue cluster on social media and COVID-19 is being consistently developed together with another cluster that contains digital humanities and technology. Two clusters are in the quadrant of niche themes, namely highly developed and isolated topics. One is virtual archaeology; the other refers to informal learning through gaming applications.

Two clusters collect themes that are less developed. The orange cluster with "digital storytelling" and learning scenarios is also in the field of basic themes. Instead, the purple cluster containing "survey", "co-design", and "community engagement" is peripheral to the research field, suggesting that it is possibly emerging or declining.

When the thematic map is evaluated over time, it draws a trajectory of the evolution of the topics, and how they are developed and connected together. In order to set the time span of each period, Fig. 1 shows that 2012 and 2017 are two crucial turning points in scientific production. Hence, time slices are set accordingly, weighting occurrences of 250 words year by year.

Figure 13 shows that in the first sub-period, museum digitalization is dominated by discussions around the virtual museum and digital media, mainly supported by research on augmented reality and interaction design. Starting in 2013, the virtual museum concept grows and assimilates issues related to digital media and digital heritage. In the second sub-period, topics are much more specialized and introduce a social aspect in the field: social media, education, children, and storytelling. The last period is short but characterized by massive scientific production. Most of the technological issues investigated in 2013–17 (augmented reality, 3D printing, gamification, user experience), converge to redefine a new understanding of cultural heritage. Social media also collects various research strands, especially those related

Towards a Digital Shift in Museum Visiting Experience. Drafting …

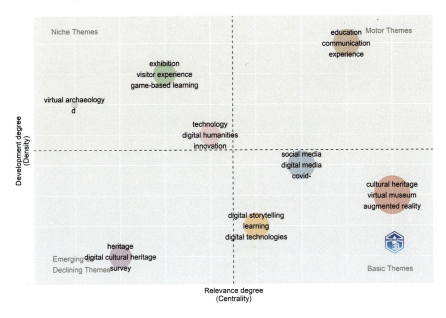

Fig. 12 Thematic map of author's keywords. Elaborated by Bibliometrix

to younger generations. In turn, learning and education are now underpinned by social media and augmented reality. Digital storytelling started as a personalized visit experience to become now a separate research issue.

The intellectual structure is based on a co-citation network of articles that are both cited in another article. In other words, it is "the degree of relationship or association between papers as perceived by the population of citing authors" [75]. The co-citation network visualizes 50 papers on museum digitalization clustered

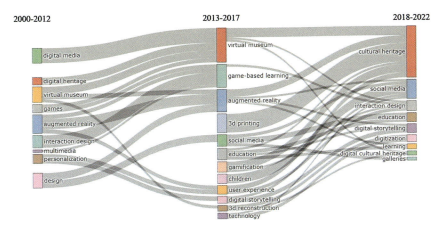

Fig. 13 Thematic evolution map of author's keywords. Elaborated by Bibliometrix

with the Louvain algorithm (Fig. 14). The analysis confirms the existence of 4 main streams of literature: the dimension represents the normalized number of citations received by the paper, and the thickness is the strength of co-citation bonds. Their position indicates centrality in the research field, and their proximity shows the density of the stream of literature.

The red cluster contains the core publications of the bibliographic collection and overlaps with the documents with the most locally cited references we have previously discussed. It is not surprising that red nodes are grouped in the center of gravity of the network. These citations can be summarized with "learning in/at the museum" and are used to build the theoretical framework that aims at educational goals through technologies, experiments, and social engagement. Some are co-cited only internally in the cluster, such as Hein [53], Tallon [5], Capriotti [61], and Parry [57], while others have strong connections with different clusters, such as Simon [25] and Falk [16]. The red sub-set of articles is generally transversal and very well connected with

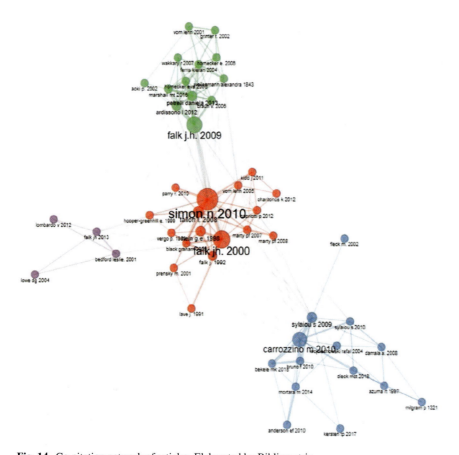

Fig. 14 Co-citation network of articles. Elaborated by Bibliometrix

the green cluster. The latter discusses how to integrate virtual and physical experiences. Falk [51], Petrelli [64], Ardissono [59], and Ferris [76] are the main nodes of this stream of literature that can be labeled as "personalization". The blue cluster is densely populated and more peripheral to the center. We label this stream of literature that covers all issues of immersive reality applied to cultural heritage as "virtual museum", also referring to one of the most used keywords analyzed in this study. Carrozzino [54], Bekele [15], Sylaiou [55], Damala [74], and Mortara [77], represent the main co-cited articles of the cluster.

The small purple cluster that is close to the center of gravity contains the stream of literature that falls under "storytelling", based on Bedford [78], Falk [66], and Lombardo [60]. It confirms the existence of a residual and possibly growing interest in this direction, as shown in our previous analyses.

Finally, we analyze the social structure of the research field by looking at the collaboration network among authors, institutions, and countries.

Co-authorship network identifies research groups working in the same sub-field in order to cluster groups of regular authors and the most influential figures within the analyzed research field [79]. Museum digitalization shows a fragmented collaborative network (Fig. 15).

Most are biunivocal collaborations, such as the three publications in which Benford and Bedwell have worked on ways to augment museum visits with visual markers, hidden objects, or card games. Among three-author research groups, Nisi-Cesario-Coelho shows a robust collaboration around the relationship between museums and teenagers through games and interactive stories. Vayanou-Katifori-Ioannidis have collaborated in 5 publications on personalized storytelling and human-led hybrid guides. A four-author group is composed of Petrelli-Ciolfi-Marshall-Not with significant contributions by the first author and separated collaboration with other authors. These collaborations are generally positioned on the relationship between museum and information, spanning from the Internet of Things to advanced storytelling techniques. Web of Science categorizes their contribution under "computer science". Another solid research group is formed at Università Politecnica delle Marche with Pierdicca-Malinverni-Frontoni-Angeloni-Clini. Their work on digital platforms is especially aimed at archaeological sites. Antoniou-Lepouras-Wallace-Vassilakis-Poulopoulos form the most consistent research group, and the first author also has strong connections with Vayanou-Katifori-Ioannidis, resulting the focus in the wider collaborative network in museum digitalization. Antoniou et al. work on games, guides, and social media engagement for museum visits.

Figure 16 shows the collaboration network of institutions. One populated cluster is formed by Northern American universities and American national academies, having the University of Pennsylvania as the most contributing affiliation with publications concentrated in the period 2015–18. Another populated cluster is led by Greek universities such as the University of Peloponnese and the University of the Aegean, but extended to Universidade de Vigo, Università di Napoli Federico II, University of Glasgow, and the University of York. Among small clusters, Università Politecnica delle Marche has strong collaborations with the Italian National Research Council.

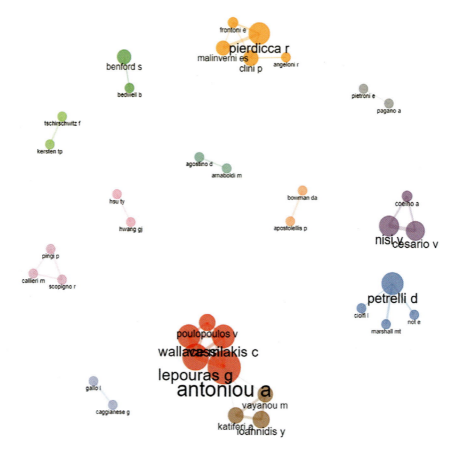

Fig. 15 Co-author network. Elaborated by Bibliometrix

Country-wise, the collaboration network confirms that the main clusters are led by the UK, Italy, and the USA (Fig. 17). The latter is mainly related to Canada and eastern countries; the UK collaborates with all European countries and has the largest reaching network, Italy has a smaller cluster but many collaborations with European countries and American countries. Interestingly, while the UK has very strong collaborations with both USA and Italy, ties are relatively weak between USA and Italy. The clusters mentioned above are very much polarized towards one country; however, a fourth collaborative cluster (purple) is formed by Greece, France, Germany, Spain, and Austria with multiple connections among nodes (distributed network) and central to the analyzed topic of the bibliographic collection. This purple cluster is at the intersection of the other three main clusters. Biunivocal relations are observed between Brazil and Portugal, and Germany and Turkey. It should be mentioned that China is usually among the top contributors in bibliographic analyses [80–82], but in museum digitalization is still not a major contributor and is relatively isolated. Russia, central Asia, and Africa have residual or null impact on the collaboration network.

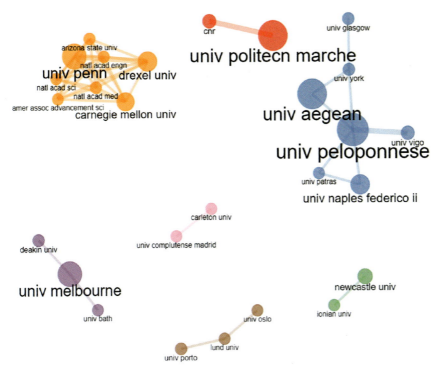

Fig. 16 Collaboration network of institutions. Elaborated by Bibliometrix

3.1 Limitations

This literature review has some limitations. First, the WoS database is one of the main databases and is generally regarded as the source with the highest quality of entries [82]. However, other databases, such as Scopus, might have partially different entries according to the typology of the document [83]. Hence, articles not indexed in WoS have not been analyzed. Second, publications whose abstract language differs from English have not been included as well. Some essential publications in French, Spanish, and Chinese have been excluded. Third, some analyses imply the use of mathematical models that simplify a large amount of data to allow interpretations and visualizations. This process may omit perspectives that are relevant to the study.

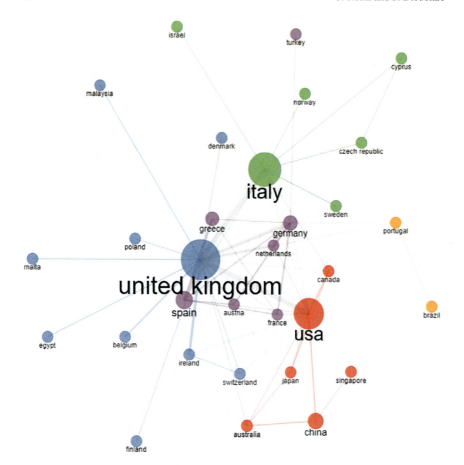

Fig. 17 Collaboration network of countries. Elaborated by Bibliometrix

4 Survey

4.1 Methodology

The survey aims at completing the descriptive quantitative analysis performed in the literature review with a qualitative evaluation of the issue of museum digitalization through interviews with museum professionals. This critical review with a group of experts will highlight similarities and differences with the research strands that we have identified in the literature review. Additionally, we decided to source participants among curators and directors of museums to measure the distance between academic publishing and practice. This will allow us to see if emerging topics in literature align with professionals' opinions.

To achieve maximum exploration of the topic, the questionnaire is composed of open-ended questions so that interviewees can introduce new concepts on museum digitalization. The survey follows a qualitative design through in-depth, semi-structured interviews with ten professionals (Table 2). This group includes seven directors, one curator, one expert in historical heritage conservation, and one expert in digital storytelling. All experts are well-known and affiliated with one of the institutions listed in Table 2. Answers have not been associated with the corresponding institution to guarantee their anonymity.

Table 2 Interviewee affiliation

Museum	Location	Nr of inhabitants	typology	Nr of visitors (2018)*	Nr of visitors (2019)*
Museo Sigismondo Castromediano	Lecce	795.134	Provincial	8000	4000
Museo dell'Ara Pacis	Roma	2.848.084	Civic	216,806	203,586
Museo Archeologico Nazionale di Taranto—MArTA	Taranto	576.756	National-autonomous	73,237	71,032
Museo Archeologico Regionale Paolo Orsi	Siracusa	399.224	Regional	63,239	42,290
Museo internazionale delle marionette Antonio Pasqualino	Palermo	1.253.000	Private	40,000	29,374
Museo archeologico nazionale di Napoli—MANN	Napoli	3.085.000	National	616,878	670,594
Museo Egizio	Torino	2.260.000	National	848,923	853,320
Civico Museo Archeologico	Milano	3.250.000	Civic	70,200	44,930
Museo di Storia Naturale di Venezia Giancarlo Ligabue	Venezia	853.338	Civic	79,870	70,660
Museo Archeologico e d'Arte della Maremma	Grosseto	221.629	Civic	15,033	16,030

Sources *(microdati Istat, Visitatori nei musei del Sistema Musei Civici, Annuario Statistico Roma Capitale, MIBACT, www.museodellemarionette.it, Annuario del Turismo—Città di Venezia, Rapporto Musei 2019 e 2020 Regione Toscana)

To get more homogeneous answers, we have sourced only institutions from one country. Hence, these museums operate under the same regulatory framework. Italy has been chosen for the following reasons:

- As demonstrated in the literature review, Italy is the second most frequent country of origin of authors
- As demonstrated in the literature review, Italy is the second most frequent country of authors' affiliation
- As demonstrated in the literature review, Italy is one of the core clusters of scientific production
- As demonstrated in the literature review, Italian museums have been widely studied with reference to COVID-19 impact [7, 44, 84, 85].

Museums are spread over nine different regions, from north to south. Their typology and size have been differentiated into four civic museums, three national museums, two regional museums, and one private museum. All are positioned in cities of different sizes and administrative statuses: Lecce, Roma, Taranto, Siracusa, Palermo, Napoli, Torino, Milano, Venezia, and Grosseto. Certain centers are more subject to tourism; others have local relevance.

Interviews took place through individual online meetings in April–May 2022, recorded and transcribed by the authors. All participants have been contacted by email or phone and asked to participate in the study. They have been provided with a privacy statement signed by the authors and had the chance to request and review the recorded meetings. The duration of the interview was 45 min up to 60 min.

With the transcripts, we first analyzed the text using the software Voyant to find recurrent words and concepts. Then we performed a qualitative assessment of the answers. Finally, we compared these answers and key concepts with those that emerged in the literature review.

The survey of museum professionals is structured on four key areas that have been highlighted in literature: (1) Digitalization in museums; (2) Engagement; (3) Interaction; (4) Virtual Environments. Each key area is explored with four open-ended questions.

Digitalization in museums. This set of questions generally enquires about the expert's view on the topic. Question 1a asks about the main challenges for a museum in relation to digitalization. Question 2a asks how experts feel about the migration of museums to online platforms (i.e., websites, virtual tours, web galleries, Instagram, etc.) as new forms of engagement. Question 1c asks about the existence and consistency, in their museum, of a department dedicated to the development of digital content and platforms. Question 1d asks about initiatives adopted during COVID-19 restrictions and whether such strategies were further developed after the re-opening to visitors.

Engagement. This set of questions enquires about engagement through digital platforms. Question 2a asks how important is the image and presence of their institution on social media. Question 2b asks whether the target audience of digital programs is the same audience as their in-person programs. Question 2c asks if they have evidence that social media presence increases the museum's engagement with the public. Question 2d asks what target audience they would like to attract more in the future.

Interaction. This set of questions enquires about interactive experiences during the museum visit. Question 3a asks to list the interactive platforms/systems adopted. Question 3b asks what part of these systems is digital and when they were designed. Question 3c asks to elaborate on the weaknesses of interactive platforms/systems. Question 3d asks if interaction is an essential component of the visitor's experience.

Virtual Environments. This set of questions enquires particularly about interactive interfaces and their development besides the physical visit to the museum. Question 4a asks if they think that regular employment of virtual tours can engage new visitors in the long term. Question 4b asks what experts think about the contribution of virtual tours to the visitor experience in museums and if the virtual tour can replace the physical visit. Question 4c asks if they provide a virtual tour of the museum. Question 4d asks about their strategy to implement digital content from a distance other than the virtual tour.

5 Results and Discussion

The cities where the ten museums are located range from 220.000 to 3.200.000 inhabitants, while the number of visitors measured as an average of the two-year period before COVID-19 restrictions (2018–19) is between 6000 and 850,000 (Table 2). Most used words during the interview were "audience" and "social", 36 times. The term "virtual" has been mentioned 32 times, and "communication" 27 times. It should be noted that the word "game", although never introduced by the interviewer in any question, has been used 14 times (Table 3). This confirms the rising attention on gamification of the visitor's experience that is evidenced in literature.

Moving to the first key area of the questionnaire, digitalization, for question 1a on main challenges, interviewees affirm the following in order of relevance and process:

Table 3 List of 10 most cited keywords

Word	Frequency (number of times)
Audience	36
Social	36
Virtual	32
Communication	27
Contents	25
Tour	21
Experience	17
Heritage	16
Video	15
Game	14

(1) re-organization of the collection in new catalogues and displays to create a more rational digitalization workflow (30%). (2) make the digitalization functional to: (a) research, (b) communication to the public, and (c) conservation of perishable material (40%). (3) make digitalization more inclusive and accessible to the different target and social strata (40%). While smaller museums express the primary need to digitalize their collection, large museums have already achieved this step and are already focused on the next stage, that of content accessibility. One expert said:

"The challenge is to reach a point of balance between the materiality of the objects and immateriality of digital data". Another expert focused on the inclusivity issue:

> To guarantee the utmost inclusion, access should be oriented to all targets and designed through research and digitalization programs. At the same time, differentiated communication channels will adequately reach various types of visitors.

Answers to question 1b on online platforms favor migration to such realms in as much as the message and content of a cultural institution can be communicated to multiple audiences. A specific point of view comes from one expert who considers the transition unnecessary if it is just for the sake of doing it. The online transition becomes substantial when technological innovation goes together with social innovation. One respondent said:

> We are working on some hypotheses to be present in the metaverse. I consider the metaverse a chance to fulfill the dream of perfect worlds. These solutions should be supplementary and not alternatives for those with limited time or who visit the museum in groups with a predetermined schedule.

Question 1c on the presence of dedicated digital departments collected 7 "no". Two of the remaining respondents are part of a network of museums in which the communication/digital department is centralized within a broader institutional framework. One expert said:

"The team is transversal. Archeologists work with external experts (videogame, digital, anthropologists, and sociologists). The digital product creates bonds with certain audiences (kids, elderly people, but also visitors from Eastern countries)". They all agree that the job of a digital expert must be continuously coordinated with specialized consultants such as archeologists, historians, art historians, and other professionals. Technicians are not expected to give their contribution independently. Additionally, these services are often performed by external companies because of the lack of specialized staff, making the integration with museum strategies very problematic.

Regarding question 1d on the long-term effect of Covid-restriction, in most cases the work during the pandemic was an implementation phase of strategies that were already active. Initiatives: intensive use of the website and social media; development of digital content and ad hoc virtual programs, detailed studies of specific artworks, 3D videos, conferences, contests, games, online laboratories, live performance in streaming, and digital classrooms. Most of the activities are still operational on digital platforms. Instead, COVID-19 helped to strengthen their presence on social media. Most of the traditional cultural activities (conferences, seminars, performances, and guided tours) resumed their in-person format. All other activities are still offered online. Only one expert said that they are aiming to return to pre-Covid strategies. Another expert said:

> This forced closure allowed us to become very resilient to changes and to strategize a new way of communicating with the visitors. This communication goes on and must evolve; we should never look back.

In the key area of engagement, question 2a asks about museums' curation of social media image (Table 5). It is seen as a fundamental component. All experts agree on its relevance, although specifying the following caveats: have a strong visual identity, coordinate communication across all different activities, and customize their social media presence on different social media. Facebook is considered by one expert a mere repository of information. For another expert, it's essential to be friendly in order to attract people and create a community. One expert said:

> We use a program called 'travel appeal', created for hotels, that allows us to understand visitor's appreciation. It monitors the digital reputation of the museum. This generates a ranking of satisfaction. It is monitored carefully but retains some problems since performances are measured with keywords related to the hotel sector.

Question 2b on the possible overlapping of targets of digital platforms and physical visit received various responses. Two experts agreed on this correspondence. Five experts did not agree, especially considering age differences and respective education to the use of interactive devices. Serious games are mainly used by adults (40–50 years old). In terms of nationality, non-Italian visitors interact more on digital platforms. Two experts affirm that it's not possible to make a difference between the physical and digital public: they are all part of the same community of the museum, and considered at the same level. One expert didn't elaborate specifically on the issue. Noticeable feedbacks are:

> No. [touristic city] is a reality unto itself in the sense that the museum attracts a variable percentage of tourists that pass by and visit the exhibition. Therefore, it is very popular among children and grandparents or families.
>
> No. There is a group of users who follow the web and social pages with interest but do not necessarily become physical visitors. It might happen. The community is transformed into a physical community if there is social awareness. The museum must regain possession of its social centrality; it must be an expression of a community to have a consistent audience.
>
> We must not consider the digital visit inferior and secondary to the physical visit. We must consider all as a single community.

To question 2c on proof of public engagement with digital platforms, two experts answered that it is not possible to give an evaluation on this question because of the COVID-19 restrictions. Eight experts confirm that they have proof that digital platforms increase public engagement. They need to be complementary to the rest of the visit and functional in order to be coherent with the identity of the museum. Two of them said that the number of visits or likes reveals the engagement, but it is not proof of the real impact of the content that has been delivered. One expert pointed out:

Communication with visitors is much easier on social media, but the goal is also to create a physical agora (in gardens, bar, restaurant) to build new spaces for social engagement.

Question 2d on the typology of target that they believe their museum should aim at in the future, over 60 and young people (15–30) are indicated as the preferred targets to be engaged in the next strategies. Two experts mentioned the local community because with them they have the potential to build a long-term relationship. Two experts refer to accessibility as one of the main factors to consider: people with different kinds of disabilities, people who cannot afford technological devices (social disparity), and non-native digital people (older adults). One said:

> We will continue to involve the community as much as possible. The 'threshold' effect still blocks the visitor at the entrance. Young people are very hard to attract outside school environments (school visits). We are trying with a dedicated language in communication and with games.

In the key area of interaction, Tabb. 4–5 show a classification of the systems used in museums. Question 3a, asking about interactive activities to be experienced during the visit, received the following list of items, ordered from the most mentioned to the least mentioned:

1. Museum guide
2. App
3. Audioguide
4. Video
5. Touch screen or tablet with database
6. Interactive games
7. Immersive experience for people with visual or hearing impairment
8. Interactive tour with museum guides or performer

One expert said:

> The museum is equipped with a visit system focused on artificial intelligence. We have digitized the visit path, and we have a google-centered ecosystem that can be programmed according to the tastes and position of the visitor. According to the target, the age, and the subject of interest, the AI adapts to whoever uses it. This system, however, needs many inputs; the pandemic did not allow its use and therefore the prototype in this period is being tested further.

Question 3b asked whether such interactive experiences are digital and when they were designed. Almost all are based on digital platforms. The guide is considered the first interactive method to create empathy and interest in the visitor. Dramas performed in the visited venue are regarded as an innovative device for one expert. Collateral activities such as performances and interactive visits through a virtual tour can be added to analog interactive methods. The oldest digital platforms date back to 2006, primarily to create digital twins of the collection. Others were designed in 2010 or 2015–2016. Most of them used funds from the European Commission, and the digitalization process in general is still ongoing. One expert mentioned that digitalization started in 2016 with European ERDF funds, "When the museum became autonomous in 2015, it was possible to create a strategic plan by choosing long-term orientations and guidelines. These kinds of projects need long-term planning".

Question 3c on weak spots of digital platforms was answered primarily with economic sustainability in the long term: maintenance and hiring of external technical experts because of the lack of existing personnel that can guarantee updates and curation of devices/technologies. This is connected to one of the most common weak points, the obsolescence due to continuous updates or shifts in operating systems (i.e., Android, Apple, Windows). Another negative factor is the lack of real interactivity, which can cause lower interest in the physical path in the museum, especially if digital systems are used during the visit. The accessibility of interactive scientific content needs to be improved and made available to people of different backgrounds and education. Concerning virtual tours, the quality of the digital product needs to be very high. Still, they usually present problems that are impossible to overcome, such as light refraction on displays. The alternation of exhibition layouts or display arrangements during the months or the years can be another weak point, especially for museums that cannot afford a different virtual tour every time the visit layout changes. One expert said:

> The visitor should be prepared before the visit. Usually, we assume minimum competence in this regard. In recent times, the level of reading concentration has decreased while the percentage of those who lack schooling is still high. So, we adapt to a zero degree of content to be experienced on digital devices. While the reconstruction of a statue in a temple is an arduous task, the technology that makes this content available to the public should be elementary.

All experts agreed on question 3d that interaction is essential for a visitor. Two experts specified that it is only a component of the visit, especially with digital interfaces. It needs to be customized to the individual experience (personalization) and can be extended before and after the visit.

In the key area of the virtual tour, question 4a asks whether this technology can create long-term engagement. One expert disagreed, six agreed with this statement, two agreed with reservations, and one couldn't say. Most experts confirmed that the regular use of virtual tours could attract more visitors in the long term if it enriches the visit in a process that extends "before, during, and after" the visit. Two of them considered essential the virtual tour only in a few cases in as much as the elements or the space represented no longer exists (virtual reconstruction of an archeological area, for example) or the virtual tour of the museum offers only a partial preview of the visiting experience in presence. One expert said:

> There are initial peaks of great interest, but then the use of the virtual tour reaches a stable level of a small number of users. There is no doubt, however, that in the case of archaeological sites, where what is no longer visible or difficult to reconstruct can be represented by the virtual tour, the virtual visit can be a standalone experience separated from the on-site visit.

On question 4b, on the degree of replaceability of the physical visit with the virtual tour, they all agreed that the virtual tour cannot replace the physical visit except for some cases related to the impossibility of reaching the museum (distance, political issues, disability, etc.) or for objects/areas that no longer exist. Two of them added that the virtual tour can support the visit as a preview and invitation to join the physical visit. It is crucial to find a balance between the physical and the virtual visit. One expert said:

> If the virtual tour is designed to provide a wide range of additional processed content, which can be transformed into an experience that integrates the material and immaterial, then it makes sense to undertake this path from the point of view of a museum.

Question 4c received eight affirmative responses on the presence of a virtual tour of their museum. Final question 4d, on the perspective of adopting digital experiences different from the virtual tour, received the following answers:

- No (10%)
- gaming (20%)
- digital platform dedicated to cultural heritage (20%)
- film, short film (20%)
- virtual reconstruction of the context in which the collection was based (not existing geographical contest of the past) or the museum is based (exterior and interior) (20%)
- augmented reality
- webGIS: interactive map
- digital storytelling
- 3D reconstruction of the cultural heritage
- podcast
- TikTok.

One expert explained:

> The aspect that is missing in an archaeological museum such as [museum name] is the context: the historical context, therefore the temporal distance that separates us from the time in which the objects in front of us were made, and the environmental context in which they were used. The digital environment can certainly help us to reconstruct both of these contexts.

5.1 Limitations

The interview was conducted with a qualitative methodology on a restricted number of experts. A structured interview with a larger sample of experts, as well as parallel research on other countries, would improve the spectrum of the issue of digitalization from the experts' point of view.

6 Conclusions

The interview offered a starting point to verify keywords, concepts, and directions that emerged within the bibliographic collection. The experts portrayed the state of the art in museum digitalization from the viewpoints of directors and curators that manage cultural institutions on a daily basis. This mixed methodology shows that academic production and practice intersect in most of the topics that emerged through the bibliographic analysis. The stream of literature on "learning in/at the museum" is also present in the interview, but educational goals are not mentioned as primary motivations for the process of museum digitalization. The research topic of "personalization," which investigates the integration of virtual and physical experiences, has a very strong parallel with the topics that emerged from the experts' responses. This is often considered their main concern when adopting digital platforms. Another topic that parallels academic research covers issues of immersive reality applied to cultural heritage, as is tackled in the stream of literature "virtual museum." Both academic researchers and directors have minor but rising interest in digital "storytelling." Experts also confirmed a significant interest in gamification as a way to engage specific targets.

We have verified two primary domains of interest: knowledge and reality. The first involves teachers and students and gives a social aspect to the interaction with the museum. The second examines the different degrees of reality attached to the visitor's experience: the analysis spans from the development and optimization of the devices to the theoretical discussion on the relevance of digital twins for cultural heritage. This implies a question, often present in our analysis, around the existence of a virtual museum.

The educational aspect of museum digitalization shows a more independent development than studies interactive interfaces. This is due to the fact that museums can be the right venue to deliver informal education through immersive experiences or hands-on activities. A thematic museum can expose school-age students to important societal issues, both outside and in collaboration with schools [86]. For instance, such a hybrid cognitive model can be shaped as a mobile learning environment, allowing students to access physical and virtual resources [47]. Alternatively, mobile location-based systems used in the museum are being experimented on teenagers to perform serious games and create new learning scenarios [87].

Issues related to COVID-19 are researched and discussed in relation to social media presence and social media engagement as lessons learned from the 2020–21 period.

Finally, the ultimate goal of museum experts seems to be the personalization of the visit. A customized experience renders strong visitation motives and possible long-term affiliation. This effect can be established with digital and analog interactivity, social engagement, and maximum accessibility.

Appendix

See Tables 4 and 5.

Table 4 Interactive systems used during the physical museum visit

Museum	Digital	Analog
Museo Sigismondo Castromediano	Artificial intelligence	Drama in the museum space, contemporary art show, performances
Museo dell'Ara Pacis	Audioguide, videoguide, video, Augmented Reality, app Sistema Musei Roma Capitale	Guided visit
Museo Archeologico Nazionale di Taranto—MArTA	App Past for Future (videogame), Augmented Reality, Artificial Intelligence	Guided visit
Museo Archeologico Regionale Paolo Orsi	Virtual tour	Guided visit
Museo internazionale delle marionette Antonio Pasqualino	Virtual tour	Interactive guided virtual visit
Museo archeologico nazionale di Napoli—MANN	Extramann app	Interactive virtual tour mediated by the guide
Museo Egizio	Audioguide	Guided visit, guided visit with the director
Civico Museo Archeologico	Tablet with data sheets	Guided visit
Museo di Storia Naturale di Venezia Giancarlo Ligabue	Touch screen, audioguide, interactive rooms, immersive game, audioguide	Tactile visit, tactile visit for the visually impaired, guided visit
Museo Archeologico e d'Arte della Maremma	Video, immersive itinerary for hearing impaired people, immersive itinerary for visually impaired, audioguide	Tactile 3D reconstructions, tactile tables, guided visit

Table 5 Digital systems/platforms disconnected from the physical visit

Museum	Online digital system	platform
Museo Sigismondo Castromediano	Social network	YouTube, Facebook, Instagram
Museo dell'Ara Pacis	Google Arts & Culture	artsandculture.google.com
	Virtual tour	website
	Video-story telling for kids	website
	Video-story telling	website
	App MiC Roma Musei	playstore, app store
	Video on temporary shows	website
	Social network	YouTube, Facebook, Instagram, Twitter
Museo Archeologico Nazionale di Taranto MArTA	Virtual tour 3D	website
	Artsupp	Artsupp website
	Google Arts & Culture (ongoing)	artsandculture.google.com
	MArTA Lab—e-learning lab	website
	Gaming: Past for Future app	playstore, app store
	Social network	YouTube, Facebook, Instagram, TikTok, Twitter
Museo Archeologico Regionale Paolo Orsi	Virtual tour	website
	Podcast	.izi travel
	Google Arts & Culture	artsandculture.google.com
	Social	Facebook, Instagram, YouTube
Museo internazionale delle marionette Antonio Pasqualino	Podcast	.izi travel
	Visual and sound archive	website
	The human library	website
	Pupi archive	website
	Video virtual tour Italian/ English/sign language	YouTube
	Social network	Facebook, Instagram, Twitter, YouTube
Museo archeologico nazionale di Napoli	Extramann app	playstore, app store
	Video	website
	Ppodcast	Ohmyguide tours
	Gaming: Father and Son1	playstore, app store
	Gaming: Father and Son2	playstore, app store
	Gaming: manncrafts	Minecraft for PC Java or mobile (Android e iOs)
	Google Arts & Culture	artsandculture.google.com

(continued)

Table 5 (continued)

Museum	Online digital system	platform
	Open-data	website
	Social network	Facebook, Instagram, YouTube, Twitter
Museo Egizio	Virtual tour	website
	Virtual tour for kids	website
	Google Arts & Culture	artsandculture.google.com
	Social network	Facebook, Instagram, YouTube, Twitter, LinkedIn
Civico Museo Archeologico	Virtual tour	website
	Didactic datasheet	website
	Archive of collection	website
	Informative material	website
	Social network	Facebook, Instagram, YouTube
Museo di Storia Naturale di Venezia Giancarlo Ligabue	Google Arts & Culture	artsandculture.google.com
	Database	website
	Thematic datasheet	website
	Virtual tour	artsandculture.google.com
	Social network	Facebook, Instagram, YouTube, Twitter, LinkedIn
Museo Archeologico e d'Arte della Maremma	Didactic laboratories online	website
	Podcast	.izi travel
	Google Arts & Culture	artsandculture.google.com
	Social network	Facebook, Instagram, YouTube, Twitter

References

1. Barberio M, Colella M.: Architettura 4.0. Santarcangelo di Romagna: Maggioli Editore (2020)
2. Figliola A, Battisti A.: Post-industrial Robotics: Exploring Informed Architecture. Singapore: Springer Singapore (2021)
3. Figliola, A.: The role of didactics in the post-digital age. AGATHÓN International Journal of Architecture, Art and Design. **3**, 29–36 (2018). https://doi.org/10.19229/2464-9309/342018
4. Parry, R.: Recoding the Museum: Digital Heritage and the Technologies of Change. Routledge, London (2007)
5. Tallon Lc, Walker K, editors. Digital technologies and the museum experience: handheld guides and other media. Plymouth: AltaMira Press (2008)
6. Resta, G., Dicuonzo, F., Karacan, E., Pastore, D.: The impact of virtual tours on museum exhibitions after the onset of COVID-19 restrictions: visitor engagement and long-term perspectives. SCIRES-IT. **11**(1), 151–166 (2021). https://doi.org/10.2423/i22394303v11n1p151
7. Agostino, D., Arnaboldi, M., Lampis, A.: Italian state museums during the COVID-19 crisis: from onsite closure to online openness. Museum Management and Curatorship. **35**(4), 362–372 (2020). https://doi.org/10.1080/09647775.2020.1790029

8. Levent, N., Pascual-Leone, A. (eds.): The Multisensory Museum: Cross-disciplinary Perspectives on Touch, Sound, Smell, Memory, and Space. Rowman & Littlefield, Lanham (2014)
9. Classen C.: The Museum of the Senses: Experiencing Art and Collections. London-New York: Bloomsbury (2017)
10. Bacci, F., Pavani, F.: First hand, not 'first eye' knowledge: bodily experience in museums. In: Levent, N., Pascual-Leone, A. (eds.) The Multisensory Museum: Cross-Disciplinary Perspectives on Touch, Sound, Smell, Memory, and Space, pp. 17–28. Rowman & Littlefield, Lanham (2014)
11. Cooper, C.: You Can Handle It: 3D Printing for Museums. Adv. Archaeol. Pract. **7**(4), 443–447 (2019). https://doi.org/10.1017/aap.2019.39
12. Schubert T, Friedmann F, Regenbrecht H. The Experience of Presence: Factor Analytic Insights. Presence: Teleoperators and Virtual Environments. 2001;10(3):266–81. https://doi.org/10.1162/105474601300343603
13. Bouvier, P., Lavoué, E., Sehaba, K.: Defining Engagement and Characterizing Engaged-Behaviors in Digital Gaming. Simul. Gaming **45**(4–5), 491–507 (2014). https://doi.org/10.1177/1046878114553571
14. Resta G, Dicuonzo F.: Isn't metaverse just a secular version of paradise? The visual experience of possible realities. In: Weidinger A, Müller-Nittel F, Reindl M, editors. Metaspace – Visions of Space from the Middle Ages to the Digital Age. Berlin: Distanz (2022)
15. Bekele MK, Champion E.: A Comparison of Immersive Realities and Interaction Methods: Cultural Learning in Virtual Heritage. Frontiers in Robotics and AI. 2019;6. https://doi.org/10.3389/frobt.2019.00091
16. Falk, J. H.: Learning from museums: visitor experiences and the making of meaning. In: Dierking LD, editor. Walnut Creek, CA: AltaMira Press (2000)
17. Dalgarno, B., Lee, M.J.W.: What are the learning affordances of 3-D virtual environments? Br. J. Edu. Technol. **41**(1), 10–32 (2010). https://doi.org/10.1111/j.1467-8535.2009.01038.x
18. Hanussek, B.: Enhanced Exhibitions? Discussing Museum Apps after a Decade of Development. Adv. Archaeol. Pract. **8**(2), 206–212 (2020). https://doi.org/10.1017/aap.2020.10
19. Taylor, J., Gibson, L.K.: Digitisation, digital interaction and social media: embedded barriers to democratic heritage. Int. J. Herit. Stud. **23**(5), 408–420 (2017). https://doi.org/10.1080/13527258.2016.1171245
20. Yu J, Muñoz-Justicia J. A Bibliometric Overview of Twitter-Related Studies Indexed in Web of Science. Future Internet. 2020;12(5). https://doi.org/10.3390/fi12050091
21. Suess A, Barton G. Instagram and the museum experience: theorising the connection through aesthetics, space and sharing. Museum Management and Curatorship. 2022:1–16. https://doi.org/10.1080/09647775.2022.2073563
22. Kelpšienė, I.: Exploring Archaeological Organizations' Communication on Facebook: A Review of MOLA's Facebook Page. Adv. Archaeol. Pract. **7**(2), 203–214 (2019). https://doi.org/10.1017/aap.2019.9
23. Tzouganatou, A.: Can Heritage Bots Thrive? Toward Future Engagement in Cultural Heritage. Adv. Archaeol. Pract. **6**(4), 377–383 (2018). https://doi.org/10.1017/aap.2018.32
24. Watson, S.: Museums and their Communities. Routledge, London (2007)
25. Simon N. The Participatory Museum. Santa Cruz, CA: Museum 2.0 (2010)
26. Bishop C, editor. Participation. Documents of Contemporary Art. Cambridge, MA: MIT press; (2006)
27. Feng X. Curating and Exhibiting for the Pandemic: Participatory Virtual Art Practices During the COVID-19 Outbreak in China. Social Media + Society. 2020;6(3):1–6. https://doi.org/10.1177/2056305120948232
28. Xu, W., Dai, T.-T., Shen, Z.-Y., Yao, Y.-J., Effects of technology application on museum learning: a meta-analysis of 42 studies published between,: and 2021. Interact. Learn. Environ. **2021**, 1–16 (2011). https://doi.org/10.1080/10494820.2021.1976803
29. Ayala, I., Cuenca-Amigo, M., Cuenca, J.: Examining the state of the art of audience development in museums and heritage organisations: a Systematic Literature review. Museum Management and Curatorship. **35**(3), 306–327 (2020). https://doi.org/10.1080/09647775.2019.1698312

30. Serravalle, F., Ferraris, A., Vrontis, D., Thrassou, A., Christofi, M.: Augmented reality in the tourism industry: A multi-stakeholder analysis of museums. Tourism Management Perspectives. **32**, 100549 (2019). https://doi.org/10.1016/j.tmp.2019.07.002
31. Aria, M., Cuccurullo, C.: bibliometrix: An R-tool for comprehensive science mapping analysis. J. Informet. **11**(4), 959–975 (2017). https://doi.org/10.1016/j.joi.2017.08.007
32. Elango, B., Rajendran, P.: Authorship Trends and Collaboration Pattern in the Marine Sciences Literature : A Scientometric Study. Int. J. Inf. Dissem. Technol. **2**(3), 166–169 (2012)
33. Koseoglu, M.A.: Growth and Structure of Authorship and Co-Authorship Network in the Strategic Management Realm: Evidence from the Strategic Management Journal. BRQ Bus. Res. Q. **19**(3), 153–170 (2016). https://doi.org/10.1016/j.brq.2016.02.001
34. Sag A. Virtual Reality In 2017: A Year In Review. Forbes (2018)
35. Garfield, E., Sher, I.H.: KeyWords Plus™—algorithmic derivative indexing. Journal of the American Society for Information Science. **44**(5), 298–299 (1993). https://doi.org/10.1002/(SICI)1097-4571(199306)44:5%3c298::AID-ASI5%3e3.0.CO;2-A
36. Mao, N., Wang, M.-H., Ho, Y.-S.: A Bibliometric Study of the Trend in Articles Related to Risk Assessment Published in Science Citation Index. Hum. Ecol. Risk Assess. Int. J. **16**(4), 801–824 (2010). https://doi.org/10.1080/10807039.2010.501248
37. Clarivate Analytics: Web of Science Core Collection Help. https://images-webofknowledge-com.lproxy.yeditepe.edu.tr/images/help/WOS/hp_full_record.html (2020). Accessed 14/05/2022.
38. Zhang, J., Yu, Q., Zheng, F., Long, C., Lu, Z., Duan, Z.: Comparing keywords plus of WOS and author keywords: A case study of patient adherence research. J. Am. Soc. Inf. Sci. **67**(4), 967–972 (2016). https://doi.org/10.1002/asi.23437
39. Bradford, S.C.: Documentation. Crosby Lockwood, London (1948)
40. Vickery, B.C.: Bradford's Law of Scattering. Journal of Documentation. **4**(3), 198–203 (1948). https://doi.org/10.1108/eb026133
41. Egghe, L.: Theory and practise of the g-index. Scientometrics **69**(1), 131–152 (2006). https://doi.org/10.1007/s11192-006-0144-7
42. Costas, R., Bordons, M.: Is g-index better than h-index? An exploratory study at the individual level. Scientometrics **77**(2), 267–288 (2008). https://doi.org/10.1007/s11192-007-1997-0
43. Perianes-Rodriguez, A., Waltman, L., van Eck, N.J.: Constructing bibliometric networks: A comparison between full and fractional counting. J. Informet. **10**(4), 1178–1195 (2016). https://doi.org/10.1016/j.joi.2016.10.006
44. Agostino, D., Arnaboldi, M., Lema, M.D.: New development: COVID-19 as an accelerator of digital transformation in public service delivery. Public Money & Management. **41**(1), 69–72 (2021). https://doi.org/10.1080/09540962.2020.1764206
45. Lotka, A.J.: The frequency distribution of scientific productivity. J. Wash. Acad. Sci. **16**(12), 317–323 (1926)
46. Yoon, S.A., Elinich, K., Wang, J., Steinmeier, C., Tucker, S.: Using augmented reality and knowledge-building scaffolds to improve learning in a science museum. Int. J. Comput.-Support. Collab. Learn. **7**(4), 519–541 (2012). https://doi.org/10.1007/s11412-012-9156-x
47. Hwang, G.J., Wu, P.H., Zhuang, Y.Y., Huang, Y.M.: Effects of the inquiry-based mobile learning model on the cognitive load and learning achievement of students. Interact. Learn. Environ. **21**(4), 338–354 (2013). https://doi.org/10.1080/10494820.2011.575789
48. Girardin, F., Fiore, F.D., Ratti, C., Blat, J.: Leveraging explicitly disclosed location information to understand tourist dynamics: a case study. Journal of Location Based Services. **2**(1), 41–56 (2008). https://doi.org/10.1080/17489720802261138
49. Marty, P.F.: Museum websites and museum visitors: digital museum resources and their use. Museum Management and Curatorship. **23**(1), 81–99 (2008). https://doi.org/10.1080/09647770701865410
50. López X, Margapoti I, Maragliano R, Bove G. The presence of Web 2.0 tools on museum websites: a comparative study between England, France, Spain, Italy, and the USA. Museum Management and Curatorship. 2010;25(2):235–49. https://doi.org/10.1080/09647771003737356

51. Falk, J.H.: Identity and the museum visitor experience. Left Coast Press, Walnut Creek, CA (2009)
52. Falk JH. The museum experience. In: Dierking LD, editor. Washington, D.C.: Whalesback Books; (1992)
53. Hein, G.E.: Learning in the Museum. Routledge, Museum and Heritage Studies. London (1998)
54. Carrozzino, M., Bergamasco, M.: Beyond virtual museums: Experiencing immersive virtual reality in real museums. J. Cult. Herit. **11**(4), 452–458 (2010). https://doi.org/10.1016/j.culher.2010.04.001
55. Styliani, S., Fotis, L., Kostas, K., Petros, P.: Virtual museums, a survey and some issues for consideration. J. Cult. Herit. **10**(4), 520–528 (2009). https://doi.org/10.1016/j.culher.2009.03.003
56. Marx, W., Bornmann, L., Barth, A., Leydesdorff, L.: Detecting the historical roots of research fields by reference publication year spectroscopy (RPYS). J. Am. Soc. Inf. Sci. **65**(4), 751–764 (2014). https://doi.org/10.1002/asi.23089
57. Parry R, editor. Museums in a Digital Age. Leicester Readers in Museum Studies. London & New York: Routledge (2010)
58. Bruno, F., Bruno, S., De Sensi, G., Luchi, M.-L., Mancuso, S., Muzzupappa, M.: From 3D reconstruction to virtual reality: A complete methodology for digital archaeological exhibition. J. Cult. Herit. **11**(1), 42–49 (2010). https://doi.org/10.1016/j.culher.2009.02.006
59. Ardissono, L., Kuflik, T., Petrelli, D.: Personalization in cultural heritage: the road travelled and the one ahead. User Model. User-Adap. Inter. **22**(1), 73–99 (2012). https://doi.org/10.1007/s11257-011-9104-x
60. Lombardo, V., Damiano, R.: Storytelling on mobile devices for cultural heritage. New Review of Hypermedia and Multimedia. **18**(1–2), 11–35 (2012). https://doi.org/10.1080/13614568.2012.617846
61. Capriotti, P., Pardo, K.H.: Assessing dialogic communication through the Internet in Spanish museums. Public Relations Review. **38**(4), 619–626 (2012). https://doi.org/10.1016/j.pubrev.2012.05.005
62. Charitonos, K., Blake, C., Scanlon, E., Jones, A.: Museum learning via social and mobile technologies: (How) can online interactions enhance the visitor experience? Br. J. Edu. Technol. **43**(5), 802–819 (2012). https://doi.org/10.1111/j.1467-8535.2012.01360.x
63. Fletcher, A., Lee, M.J.: Current social media uses and evaluations in American museums. Museum Management and Curatorship. **27**(5), 505–521 (2012). https://doi.org/10.1080/09647775.2012.738136
64. Petrelli D, Ciolfi L, Dijk Dv, Hornecker E, Not E, Schmidt A. Integrating material and digital: a new way for cultural heritage. interactions. 2013;20(4):58–63. https://doi.org/10.1145/2486227.2486239
65. Coenen T, Mostmans L, Naessens K. MuseUs: Case study of a pervasive cultural heritage serious game. J Comput Cult Herit. 2013;6(2):Article 8. https://doi.org/10.1145/2460376.2460379.
66. Falk, J.H., Dierking, L.D.: The Museum Experience Revisited. Routledge, London & New York (2013)
67. Börner, K., Chen, C., Boyack, K.W.: Visualizing knowledge domains. Ann. Rev. Inf. Sci. Technol. **37**(1), 179–255 (2003). https://doi.org/10.1002/aris.1440370106
68. Smith, R.C., Iversen, O.S.: Participatory heritage innovation: designing dialogic sites of engagement. Digital Creativity. **25**(3), 255–268 (2014). https://doi.org/10.1080/14626268.2014.904796
69. King, L., Stark, J.F., Cooke, P.: Experiencing the Digital World: The Cultural Value of Digital Engagement with Heritage. Heritage & Society. **9**(1), 76–101 (2016). https://doi.org/10.1080/2159032X.2016.1246156
70. Pierdicca R, Marques-Pita M, Paolanti M, Malinverni ES. IoT and Engagement in the Ubiquitous Museum. Sensors. 2019;19(6). https://doi.org/10.3390/s19061387
71. Petrelli, D., Marshall, M.T., O'Brien, S., McEntaggart, P., Gwilt, I.: Tangible data souvenirs as a bridge between a physical museum visit and online digital experience. Pers. Ubiquit. Comput. **21**(2), 281–295 (2017). https://doi.org/10.1007/s00779-016-0993-x

72. Gonizzi Barsanti S, Caruso G, Micoli LL, Covarrubias Rodriguez M, Guidi G. 3D Visualization of Cultural Heritage Artefacts with Virtual Reality devices. Int Arch Photogramm Remote Sens Spatial Inf Sci. 2015;XL-5/W7:165–72. https://doi.org/10.5194/isprsarchives-XL-5-W7-165-2015
73. Caggianese, G., Gallo, L., Neroni, P.: Evaluation of spatial interaction techniques for virtual heritage applications: A case study of an interactive holographic projection. Futur. Gener. Comput. Syst. **81**, 516–527 (2018). https://doi.org/10.1016/j.future.2017.07.047
74. Damala, A., Hornecker, E., van der Vaart, M., van Dijk, D., Ruthven, I.: The Loupe: Tangible augmented reality for learning to look at ancient Greek art. Mediter. Archaeol. Archaeom. **16**(5), 73–85 (2016). https://doi.org/10.5281/zenodo.204970
75. Small, H.: Co-citation in the scientific literature: A new measure of the relationship between two documents. Journal of the American Society for Information Science. **24**(4), 265–269 (1973). https://doi.org/10.1002/asi.4630240406
76. Ferris K, Bannon L, Ciolfi L, Gallagher P, Hall T, Lennon M. Shaping experiences in the hunt museum: a design case study. Proceedings of the 5th conference on Designing interactive systems: processes, practices, methods, and techniques. Cambridge, MA, USA: Association for Computing Machinery; 2004. p. 205–14
77. Mortara, M., Catalano, C.E., Bellotti, F., Fiucci, G., Houry-Panchetti, M., Petridis, P.: Learning cultural heritage by serious games. J. Cult. Herit. **15**(3), 318–325 (2014). https://doi.org/10.1016/j.culher.2013.04.004
78. Bedford L. Storytelling: The Real Work of Museums. Curator: The Museum Journal. 2001;44(1):27–34. https://doi.org/10.1111/j.2151-6952.2001.tb00027.x
79. Peters, H.P.F., Van Raan, A.F.J.: Structuring scientific activities by co-author analysis. Scientometrics **20**(1), 235–255 (1991). https://doi.org/10.1007/BF02018157
80. Azad, A.K., Parvin, S.: Bibliometric analysis of photovoltaic thermal (PV/T) system: From citation mapping to research agenda. Energy Rep. **8**, 2699–2711 (2022). https://doi.org/10.1016/j.egyr.2022.01.182
81. Mumu, J.R., Saona, P., Russell, H.I., Azad, M.A.K.: Corporate governance and remuneration: a bibliometric analysis. Journal of Asian Business and Economic Studies. **28**(4), 242–262 (2021). https://doi.org/10.1108/JABES-03-2021-0025
82. Aria, M., Misuraca, M., Spano, M.: Mapping the Evolution of Social Research and Data Science on 30 Years of Social Indicators Research. Soc. Indic. Res. **149**(3), 803–831 (2020). https://doi.org/10.1007/s11205-020-02281-3
83. Visser, M., van Eck, N.J., Waltman, L.: Large-scale comparison of bibliographic data sources: Scopus, Web of Science, Dimensions, Crossref, and Microsoft Academic. Quantitative Science Studies. **2**(1), 20–41 (2021). https://doi.org/10.1162/qss_a_00112
84. Mason MC, Riviezzo A, Zamparo G, Napolitano MR. It is worth a visit! Website quality and visitors' intentions in the context of corporate museums: a multimethod approach. Current Issues in Tourism. 2021:1–15. https://doi.org/10.1080/13683500.2021.1978947
85. Magliacani M, Sorrentino D. Reinterpreting museums' intended experience during the COVID-19 pandemic: insights from Italian University Museums. Museum Management and Curatorship. 2021:1–15. https://doi.org/10.1080/09647775.2021.1954984
86. Mujtaba, T., Lawrence, M., Oliver, M., Reiss, M.J.: Learning and engagement through natural history museums. Stud. Sci. Educ. **54**(1), 41–67 (2018). https://doi.org/10.1080/03057267.2018.1442820
87. Rubino I, Barberis C, Xhembulla J, Malnati G. Integrating a Location-Based Mobile Game in the Museum Visit: Evaluating Visitors' Behaviour and Learning. J Comput Cult Herit. 2015;8(3):Article 15. https://doi.org/10.1145/2724723

Practice

The Humanistic Basis of Digital Self-productions in Every-Day Architecture Practice

Marco Verde

Abstract The intersection of robotics and architecture supports the search for new spatial, structural and construction models useful to support the innovation in conception and making of spaces towards a more sustainable production, and to refine the "Industry 4.0" paradigm from a humanistic perspective to meet the needs of the socio-ecological transition. One of the major emerging challenges of technological innovation, in fact, is to accelerate the realization of a high quality architecture that is responsive and sensitive toward the environmental and social context within which is designed and implemented. This requires a holistic and transdisciplinary approach during the whole process, from conception to construction. In this regard, the self-production in every-day architecture practice (starting from small scale projects) represents an important field for theoretical and empirical investigations that calls for a smarter use of traditional and non-traditional building materials, and innovative computational ways of dealing with craftsmanship for more sustainable manufacturing methods along the whole factory life-cycle, suggesting a greater insight into the humanistic basis of architecture. This chapter will frame the design-research in contemporary strategies and processes for architecture undertaken at ALO, architecture and design studio based in the south Sardinia (Italy), and will showcase some of the computational design and robotic fabrication research carried out within the daily practice.

Keywords R&D and entrepreneurship · Digital theory · Design thinking and human–computer interaction · Computational and parametric design · Performance-based architecture and design · Digital twin

United Nations' Sustainable Development Goals 9. Build resilient infrastructure, promote inclusive and sustainable industrialization and foster innovation · 11. Make cities and human settlements inclusive, safe, resilient and sustainable · 12. Ensure sustainable consumption and production patterns

---

M. Verde (✉)
Founder and Director at ALO S.R.L, 09123 Cagliari, Italy
e-mail: m.verde@alo-architettura.com

1 Introduction

By reconnecting the design process to that of construction through a holistic and material driven approach, sustainability can emerge as a synthesis of different knowledge. Moreover, thanks to the contribution of a new computational thinking and processes to both moments, a more sustainable production can become actualised in a new tectonic and spatial conception of architecture that goes beyond utilitarian, decorative and mannerist paradigms and aims. The divergence from the conventional fragmented practice is becoming increasingly necessary.

However, recent developments show that this innovation tends to be relegated to elitist projects, landmarks, exotic canopies or sculptural pavilions [1]. The sustainable transfer of innovation in design and construction must aim at tailor-made strategies starting with small-scale interventions that, by sheer numbers, will have a greater impact on people's daily lives and the quality of urban development already from the micro-scale, while being oriented towards criteria of scalability and habitability.

At ALO we tackle this problem through a holistic applied research that focuses on investigating a novel conception and making of space by exploring innovative uses of traditional materials, like wood and concrete, and researching non-serial prefabrication as critical for a more sustainable production.

Our work is material-driven and supported by advanced computational design and fabrication strategies that help us to transfer into practice smart design criteria borrowed from nature.

To provide an overview of our design and fabrication research, the chapter is organized as follow: Sect. 1 introduces the research territory discussing then need for a humanistic approach to technology, some pivotal design criteria and key topics such as performative revamping, digital craft and the search for lightness.

The following sections present the structure of the ALO laboratories and provide a sample of three current holistic research strands unfolding through our practice: Innovate use of traditional materials (Sect. 2.1); Collaborative robotics for advanced building components (Sect. 2.2) and Computation in design and making for augmented human experience (Sect. 2.3). Finally, in conclusions, a trajectory for future developments will be introduced and a novel paradigm for collaborative robotics will be proposed.

1.1 The Humanistic Turn of Industry 4.0

By the middle of the first 20 years of the digital turn in architecture [2] the need for closer relationships among the processes of design and making, as well as computation and matter emerged. The exploration of novel possible futures was especially fed by theorists and scholars [3–9] who provided the ground for a novel culture in architecture more focused on the "how" rather than the "why".

The renewed approach to architectural design, already passed through several, micro-phases, especially supported by the "democratization" of digital fabrication technologies, which starting from 2000s is bringing designers back in touch with the making. Fablabs [10], Arduino microcontroller or open source communities are only some of the expressions of a new culture bringing machines out from factories and closer to people. The concept of 3D printing, for example, is already mainstream. The vision for a procedural practice in architecture suggests the move from arbitrary top-down choices and empirically reconnects with matter through bottom-up strategies. Matter itself, and subsequently digital fabrication, started being regarded as active agents within the process of architecture itself, becoming generative, genetic and evolutionary.

To a part of the architectural domain already, design and making are conceived as a single non-linear data flow. This new paradigm is pushing architects and engineers to rethink their processes and strategies through computation. We are now bridging the gap that has emerged between the design and construction industries due to the mechanistic and fragmented view of complexity that has prevailed over the past two centuries.

Meanwhile, robotics landed inside architecture schools; and some research-oriented firms (among which ALO) are already prototyping their custom end-effectors to self-produce their own building components directly inside their proprietary laboratories, making research in matter and fabrication an actual instrument for design within their daily practice.

However, a technocratic race towards novelty seems also emerging driven by a fanatical attitude towards technology. Similarly this is also happening within the Industry 4.0 sphere. But technology per se doesn't suffice to generate content and to impact everyday practice. We need a humanistic theoretical framework within which to understand why emerging technologies may be valuable for the next generation of architectures, and how to develop such a strong thinking to overcome even the possible unavailability or non-immediate accessibility of certain technologies, especially those for digital fabrication.

In "Mechanization takes command" Giedion provides a transversal analysis of the influence of the total phenomenon of mechanization correlating among several symptoms the *"decaying sense of materials"* not so much to mechanization but to *"the manner in which mechanization was employed"* [11].

Giedion work is still surprisingly relevant. In this regard, possibly one of the biggest weaknesses of the Industry 4.0 project lies in the way it has been proposed and received by industry, i.e. mainly from a technocratic point of view. In Italy, it often turned into a digital upgrade of obsolete machinery and procedures, as obtaining important tax incentives was not directly linked to the innovation of the products themselves. In 2021, up to 95%[1] of the cost of new machinery could be converted into a tax credit without, however, the benefit being tied, as a necessary condition, to the innovation of the products themselves. At the same time actions in favour of

---

[1] https://www.mise.gov.it/index.php/it/incentivi/impresa.

R&D towards design innovation offered tax credits up to only 10% of the investments.[2] The diversity of the two incentive policies somehow shows that, the main concern is to boost the market by stimulating the blind acquisition of new machines, of course connected to the Internet, equipped with monitoring sensors and able to share production data. Returning to Giedion, is there a sufficient interest in what is being done (or could be done) with these machines such as robots, especially in the architecture and construction sectors?

This open question highlight one of the major missed opportunity to trigger a tangible effort toward a new approach to industrial production in the pursuit of a sustainable industrialization through: (1) product differentiation, (2) a true collaborative and inclusive ecosystem among humans and machines, (3) the extension of product life through upgradability of components, (4) the overcoming of the constraints of mass production leading to overproduction, and (5) the pursuit of lightness [12]. We should indeed be especially concerned about reducing the weight of simple and complex products, as it is directly related to the amount of raw materials required for their production, transport, use and disposal or recycling.

As for architecture, perhaps the slowest sector to embrace technological innovation, the utilitarian I4.0 mindset may need a major revision if our goal is to become seriously environmentally responsible. Despite the ease of access to advanced fabrication techniques, more and more architectures are appearing that promise disruptive innovations but often turn out to be almost theatrical stage sets, illusory shining architectures heedless of the cost to keep them up.

For these reasons, both Industry 4.0 and computation in architecture shall aim to expand human capabilities by designing collaborative processes in which human and machine efforts are ideally deployed simultaneously (and not sequentially) to make innovative products and spaces that are consistent with the objectives listed above and that otherwise, without the intersection of the contributions of the two parties, could not be produced.

1.2 Tight-Fit and Performative Revamping

Invoking the concept of sustainability is already a cliché, a mandatory requirement, but in the end it is scarcely implemented in ordinary urban development. Deep down, nothing has changed in the way new residential buildings or shopping malls are designed and constructed. Sustainability has mostly turned into a list of qualities to be met and ingredients to be mixed, but at the scale of everyday life in general, interest in ordinary design and construction that does not disrupt established practices still seems to prevail. Just as sweeping the dust under the carpet is not a real solution, building greenish façades wrapping obsolete design concepts in the name of a certain biophilic luxury mood is also not a real answer.

---

[2] https://www.mise.gov.it/index.php/it/incentivi/impresa/credito-d-imposta-r-s.

The construction industry in large part, especially in small local communities, is not yet ready to dismiss older building systems, and this prevents the diffusion at daily practice of novel solutions for new buildings. For this reason, for a small research-oriented firm, bringing innovative concepts to the daily buildings practice can be still tricky. However, being equipped with a proprietary laboratory for the digital fabrication of complex building components offers great opportunities to explore an alternative path towards sustainability in architecture focusing on the advanced, high-quality revamping of existing, exhausted architectures.

In this context, small and old historical buildings, which often have complex and irregular spaces, are one of the most suggestive places to take an innovative approach in which technology becomes an instrument for finding new design opportunities that cannot be foreseen in advance, rather a tool for problem solving. In this sense, light-weight, tight-fit and performative interventions can be now carried out with a meticulous control of the liminal space between the old and the new.

Especially the search for light structural and construction models based on redundancy and differentiation becomes critical to preserve older buildings integrity and identity while exploring alternative models for a contemporary living through non-mimetic work.

Lightness, in a strict physical sense, is one of the greatest challenges for the contemporary culture that is widely educated to the idea that heavier is better while the lightest is ephemeral. Saving weight can have an exponential impact on the entire life cycle of a building, from its structural conception down to the furnishing and all the processes relating to its existence, including the industry that orbits around it.

Thomson discusses the concept of form in nature as the expression of the forces acting on a body [13]. Learning from nature, lightness is achieved by shape, by the differentiation of building components, by distributing forces across redundant adaptive systems made of small entities. In 1904 A.G.M. Michell presents a model for the minimization of material in frame-structures showing how their weight could be drastically reduced shifting from continuous elements to networks of multiple small building components [14]. His diagrams resemble the porous, organization pattern of the upper human femur bone tissue, which first exact mathematical analysis was originally published by John C. Kock [15]. In nature, lightness is a survival strategy indeed.

Next to these criteria of differentiation and distribution of forces, the integration of multiple functions into single building components, surpassing the segregation of functions into mono-purpose components, becomes as well critical towards the minimization of weight and matter consumption. Nature provides an infinite number of examples among which the eggshell, which performs as a protective envelope and a porous membrane that regulates the moisture level by allowing only water particles of a certain size to pass through, or sea sponges such as the Venus Basket sponge, whose structure allows the organism to resist the forces of sea currents but also to create low-speed interstitial micro-vortices that promote feeding and reproduction [16].

If we transfer the concepts of structural differentiation, lightness and polyfunctionality of building components to architecture, space becomes a differentiated body, whose components, from skin to furniture via structure, are no longer segregated to perform only single functions. Hereby, the figure of the architect, the engineer and the builder, as currently conceived, blur and embrace a new transdisciplinary approach, supported an active adoption of computational design and advanced manufacturing strategies. Hereby, the design process is no longer linear, but becomes performative as it produces knowledge by transgressing the boundaries of the disciplines through computation.

In this process, we need to be concerned about building a new responsible relationship with local artisans: it is necessary to breathe new life into a sector that suffers as much from the standardization and cost-cutting processes put in place by large retailers (offering increasingly low product quality well hidden behind more accurate finishes) as from the growing shortage of skilled labour. There is perhaps a need to refine the conception of digital craftsmanship, which has emerged over the past decade, recalibrating the concept not so much as an evolved expression of the individual, but as a collaborative virtuous process across disciplines.

1.3 Holistic Research-Driven Practice at ALO

ALO's roots back to 2005 and the firm was founded in 2012 in Cagliari, in the south of Sardinia (Italy). The context of a small island imposes many constraints that risk slowing down innovation processes. Architecture needs time to think and make mistakes, and making architecture the main activity of a small company could force one to accept too many compromises in order to survive. Moreover, in Italy, too many figures overlap and clash within a sector that is actually poorly regulated in terms of hierarchies and competences.

Despite our small company size, we push ourselves to be engaged in a significant production. We manage to keep architecture in a safe place; we feed the studio through complementary design-research and R&D services, and deeply dedicate ourselves to architecture only when there is the change to do something truly significant with open-minded clients.

Research has a pivotal role within our daily work. Every built, unbuilt or architectural competition project is taken as an opportunity to push further our agenda.

In order to make actual our thoughts through everyday studio life, we are developing an in-house laboratory for digital fabrication, and we are already able to strategically combine several techniques from laser cutting, to 3d printing and robotic processing within our projects. Exploring a strategic combination of different digital technologies and defining an ethical and meaningful role for robots in architecture is indeed one of the biggest questions we are addressing.

In this search, the idea that digital fabrication has become a broader transversal concept. It is a holistic synthesis of thought, design and realisation; a transformative entity that triggers new possible futures and fosters our agenda.

The hybrid studio setup, blurring from studio to advanced industry grade workshop, supports a cost-effective research towards the cross-pollination of architecture, engineering and a performative manipulation of traditional materials.

At ALO we took such a challenge, and we work to merge advanced design and fabrication protocols with traditional craftsmanship so that all parties within the process contribute to the achievement of unique final result that are an expression of the intersection of the best of their abilities.

Having embarked on a path of constant innovation and research has helped us to trigger a virtuous activity of high-quality sartorial and performative architectures.

1.4 The Project of Villa Vi. A Study Case on the Innovate Use of Traditional Materials

Villa Vi is a villa located on the coast of Golfo Degli Angeli in Quartu Sant'Elena (Cagliari). It was built in the'60 s and the aim was to transform the building into an experiential, fine guest house. Villa Vi project embodies some of our thoughts about the intersection of contemporary living and hospitality with matter, digital fabrication and the aim of providing an immersive architectural experience for guests. For this purpose, architecture and matter have a pivotal role; the search for innovative applications of traditional materials like wood and concrete was central (Fig. 1a, b).

The entire work was triggered by the original traits of the existing structure of the villa featuring uncommon triangular arches surrounding two porches on the main facades. Their polygonal topology became the generative seed for the entire design which aim was to blur the boundaries between exterior and interior spaces across scales. Beside the full renovation of the façades, interiors spaces and exteriors, including the design of the performative landscape surrounding the villa, we designed and robotically fabricated a collection of bespoke wooden, faceted furniture, POLYHEDR.a, the new concrete main entrance portal POLYHEDR.a/r and several parts and jigs to facilitate an high-precision and fast construction (Fig. 2).

1.4.1 Innovation of Traditional Materials Through Computation in Design and Making

POLIHEDR.a are irregular polyhedral pieces of furniture that, by shape, integrate several functional accessories typical of a guest room. A desk, a bookcase, and the luggage rack find their place in a compact but articulated hybrid wall system that wraps around a stand-alone shower space. These are the main elements organizing the layout of Villa Vi guest rooms. From the very beginning, the pieces were intended for self-production. Therefore, we developed a sophisticated parametric system that served both to digitally adapt the topology of the layout of each guest room to the

Fig. 1 a, b View of Villa Vi's main garden and building. A computational environmental analysis informed the differential organization of the landscape so to provide a beneficial microclimate for outdoor living during the hot summer season

different dimensions of the floor surface and to achieve state-of-the-art finishing qualities as a unique expression of the synthesis of design and fabrication.

From the very beginning, we decided to use birch plywood as building material. This material is characterized by exceptional structural performance and CNC machinability. This was important to achieve well-refined, self-supporting structures

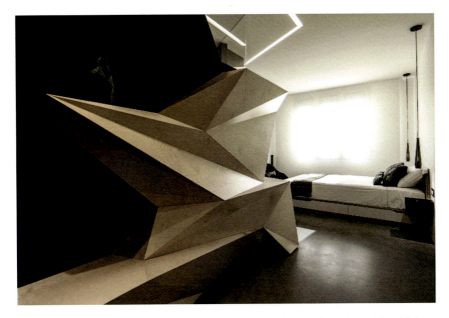

Fig. 2 Digital wood: the architectural furniture is made of self-supporting sub-modules. All facets are designed through a computational procedure that generates all geometrical features necessary for the six-axis circular-saw robotic-cut carried out directly at ALO laboratory

even while using panels of reduced thickness. However, processing birch plywood requires special attention. This is because, for example, the colour one perceives can change from honey to chocolate depending on his/her point of observation. In this regard, our goal was to achieve a final monolithic appearance, with continuous grain flowing in one sole direction as well as a homogeneous colour between the faces while moving around the piece.

From this initial brief, we decided to start the design-research looking at the "Quartabuono" joint to assembly all faces without showing the thickness of the panels. This type of joint is used in carpentry to join wooden elements that must form a right angle with the section of the material not visible to give continuity to the grain.

However, when it comes to irregular polyhedral, both concave and convex, with thick faces and more than three faces converging to one vertex, exactly the collision of the thicknesses of the facets at the vertexes would require an empirical trial and error assembly procedure making impossible a precise fabrication of the parts by traditional means. Therefore, due to the complexity of the parts and over all assembly, the design-research aimed to develop e fully digital design and fabrication protocol to avoid any kind of manual intervention and achieve a high quality finishing.

Given the amount of aspects to intersect as a whole, the project required clearly a computational design approach (Fig. 3). A first software module allows to work on a simplified parametric figure, a "skin without thickness", in order to facilitate the adaptation of each piece to rooms size while minimizing calculation time and the real time verification of the compliance with all manufacturing constraints, among which, for example, the minimum acceptable values for the angles between adjacent faces (set to 20°). Then, once the morphology is set up, a second module independently processes the digital twin by transforming all simple faces into 3D panels with a given thickness, calculating the bisector of the shared faces and eliminating excess material in the joints, the number and position of connecting dowels, part numbering, and all the data needed to drive the six-axis circular saw cutting and multi-tool robotic manufacturing as a whole. The entire system is parametric, so changes can still be applied and reflected in new digital manufacturing layouts.

POLYHEDR.a parametrics delivers further sophisticated design features. It recognizes the spatial orientation of each face of the polyhedral figure with respect to a given observation vector and then, getting ready for their digital fabrication, it unfolds all parts flat on a plane and carry out the nesting negotiating among (1) the initial reference vector for grain, (2) the direction of the grain of the raw board to be machined (3) the minimization of residual waste material. As a result, once facets are assembled back in real world, the grain wraps the figure as a continuous flow, enhancing the global monolithic appearance of the furniture still reducing the amount of raw matter necessary for the construction (Fig. 4).

In addition to this, in order to facilitate the assembly, every part is numbered according to a specific protocol that provides for each face (1) a unique tag identifying the part itself, and for every edge, (2) the tag of the adjacent part. Artisans taking

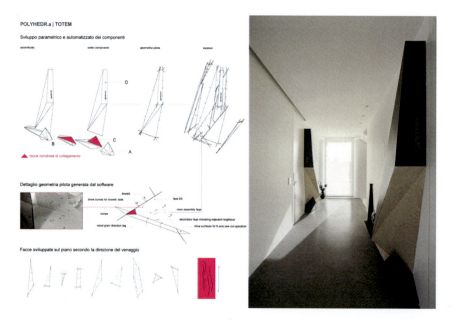

Fig. 3 The diagram illustrates the phases of recursive computation for the generation of the executive 3D digital twin of the guestrooms' totems

Fig. 4 Detail of the grain flowing along the body of the furniture along one consistent direction

care of the assembly could then assembly all parts just following the tagging system and a few 3D representations of the final piece.

Starting from this research we derived the POLYHEDR.a/r formwork system to robotically fabricate the scaffoldings system for the production of the concrete structural components of the entrance (Fig. 5).

Fig. 5 Digital Concrete: a view of the FRC portal

This reverse version of the parametric system, as the name suggests, builds the thickness of the scaffolding and all production drive data calculating the thickness of the material outside the initial 3D reference geometry of the part.

In addition, POLYHEDR.a/r parametrically generates all components of the inner steel reinforcements as well.

1.5 The RCC Project. A Study Case on Collaborative Robotics for Advanced Building Components

The concept of collaborative robotics is spreading across industrial production; however, it is largely presented from a reductive perspective of problem-solving.

Differently, human–robot collaboration should go beyond the idea of simply replacing human labour in repetitive or heavy tasks in order to become instrumental to deep product innovation.

To tackle such utilitarian approach, in 2018 ALO initiated the research project originally named "Robotic Collaborative Construction" (RCC) focusing on designing products and processes where humans would work in unison with machines summing both best skills to achieve complex productions not otherwise possible.

Our agenda is to effectively transfer the research on robotic fabrication in architecture into daily architecture practice, with a focus on the scalability of the results to small-scale projects with limited budgets too. For this purpose, RCC addresses a further architectural elaboration of some of the criteria enunciated by Pearce (1980) and, specifically, that of *"Minimum Inventory/Maximum Diversity"* [17] from a holistic and computational perspective to the advantage of a greater flexibility with respect to the possible architectural conception and making of space.

In this sense, RCC explores the idea of a non-standardised digital prefabrication that is efficient in terms of cost, processing energy consumption and the architectural quality of the results in terms of their completeness, structural integrity and refinement.

Imagining such a scenario, perhaps starting from the architectural domain, we could enunciate a new model of industry, the *"Industry X"*, where "X" is not a letter but a symbol representing the symbiotic interaction between man and robot.

1.5.1 RCC Hardware and Software Design-Research

The project stems from the idea of combining to our advantage, on the one hand, the capability of robots as spatial positioners for building components and, on the other hand, the ability of a skilled worker, such as a welder, to complete complex assembly operations that need human awareness ore solve unexpected faults in real time in a viable manner.

For this purpose we designed a double-curved membrane consisting of steel rings to by assemble without the use of any kind of scaffolding or jigs. The rings had to be welded in place by an operator while the robot was employed to recursively load and position each component, piece by piece, directly according to a dynamically linked 3D parametric model (Fig. 6). Some might argue that welding could also be automated. But in the event that an artificial-intelligence robotic system could be adapted for this application to handle unforeseen complexities in real time by autonomously adjusting processing paths, what would be the cost of such a total automation development in terms of research, necessary infrastructure, and long-term flexibility? Would such total automation be a reasonable option to make the process sustainable even for a small firm or for application to small-scale projects? Even with an access to infinite resources, we believe these questions should be given more consideration.

For this reason, the RCC project departs from some of the totalitarian and technocratic positions that are emerging in the field that seem to increasingly distance research from viability in everyday practice in the name of novelty itself. On the contrary, RCC aims to enhance the skill, versatility and intuition of human operators towards greater architectural quality, flexibility and agility in a lean digital production.

The overall objective of the basic research was to develop a set of proprietary hardware and software tools to prove the concept, evaluate potential issues to be addressed and possible future research directions and application in daily practice.

On the hardware side, we designed and 3D printed a custom self-centring gripper (Fig. 7) to perform parametrically controlled pick-and-place-and-weld operations driving the robot directly from the design interface of the digital-twin.

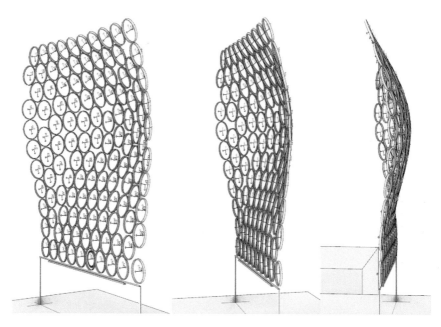

Fig. 6 Details of the 3D double-curved membrane parametrically generated as to preserve the tangency between building components and manufacturing tolerances

The gripper is electrically isolated to allow welding operations while the robot still holds the part in place. In common practice, the part to be welded is fixed on an external positioner (1 or 2 axis) while the robot performs the welding; we inverted the roles, living the welding to a skilled worker and using the freedom of six-axis positioning to facilitate complex assemblies (Fig. 7). Moreover, the gripper has a certain level of autonomous intelligence. Through embedded sensors, it can detect when the part has been picked and then control the force applied to hold it firmly without overloading motors. It has also an embedded LED lighting system to communicate its state with the operator. All building parts and mechanisms have been designed to be 3d in-house printed in the perspective of a sustainable research economy and further developments.

On the software side, we have expanded the control and simulation capabilities of existing methods based on the Kuka PRC plug-in [18], the parametric programming environment for robots that since 2016 has made robot programming easily accessible to designers. We developed a series of customized codes to facilitate simulation control, to trigger specific actions along the process based on factors of time, distance and Boolean operations. We also developed a system to check the state of the end-effector and provide feedback consents during the process. The control engineering of the gripper was realized in-house too connecting the parametric digital-twin to our kuka robot via Arduino micro controller to carry out online simulations and off-line programming too.

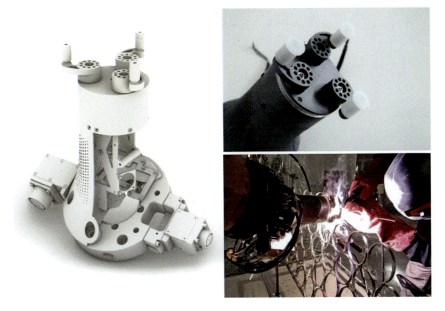

Fig. 7 Digital assembly of the self-centering gripper (left). In-house 3d printed functional end-effector (top-right). Testing the collaborative construction protocol (bottom-right) with the robot positioning building components and the welder fixing parts in place

Finally, we developed a general parametric digital environment that accurately simulates complex end-effector actions to be verified for both safety and assembly accuracy purposes. In the perspective of future architectural applications, the system has proved suggestive for hypothesizing complex assemblies (even of non-standardized building components) to more accurately fit complex design contexts with clean prefabricated solutions, potentially useful in the case of an intervention in a historic building.

The research work was supported by the "Microincentivi per l'Innovazione" grant program awarded to ALO by Sardegna Ricerche (2019) (Fig. 7).

1.6 The APTICA Project. A Study Case of Computation in Design and Making for Augmented Human Experience

The need for accessibility of spaces and culture is finally spreading; architecture has first and foremost a responsibility towards people who, for whatever reason, have a disability. The accessibility of museums is one of the critical issues with respect to which this renewed attention is beginning to be prolific and translates into an innovation of methods and processes to facilitate the enjoyment of museum

Fig. 8 The figure shows the APTICA interfaces with the 3D miniature of the artwork. The dark top surfaces are the touch feelers, sensing surfaces that enable the multimedia contents navigation

content. In this context, starting from the small scale of museum devices and installations, new digital fabrication technologies and advanced multimedia strategies are gaining ground in the research and development of new methods to engage visitors in inclusive, immersive and rewarding experiences.

Our research work on this subject started in 2019, thanks to the collaboration with CRS4 (Sardinia's Centre for Research, Development and Advanced Studies) that commissioned ALO to design and prototype a functional device for the multimedia and tactile exploration of pictorial works. Our work stemmed from the initial studies developed by researchers at CRS4 on the transformation of pictorial works into scaled 3D representations, equipped with sensors to enable the multimedia content. From here, we developed a new complete and functional interface named APTICA.

Beyond the utilitarian aspects, which were particularly relevant to this project, the brief immediately showed the possibility of opening a new chapter within philosophical, historical and semiological discussions about the frame, an object that has so far assumed the status of a theoretical object as observed by Pinotti [19].

The frame, which appeared in a context that intended art as a reality detached from the one we live in, was a boundary between the representation and real space. Its purpose was intended to intensify the perception of the depth of the field or to project the movement outwards.

The historical evolution of its role went through various phases until its denial. However, the whole debate starts from the assumption of a visual fruition of the artworks; therefore for people with severe visual impairment, the frame in its traditional conception loses its value. This consideration raised a new question: Can we reformulate the role of the frame and envision a new type that encourages alternative cognitive processes and facilitates the tactile exploration of pictorial artworks? Indeed, the challenge of this project was to move from the concept of the frame as a boundary for the eye to that of an intensifier of the senses.

1.6.1 Accessibility by Design

A complex tactile experience requires training and preparation. A strategy is needed to see objects with the hand, and from the perspective of a designer there are specific criteria that must observed in order to achieve effective results [20]. APTICA is ultimately designed to prepare the hands for the tactile experience.

The device is a table-top physical object with part of its body that seems floating over the supporting surface with the aim of isolating it from the surrounding world and stimulating the perception of a suspended object on which to focus one's senses.

The first part that user comes into contact with is the perimeter of the frame: a fragmented body made up of a sequence of blades, all different and oriented like rays towards the centre of the subject. The fragmentation aims to provide an initial intense tactile transition from the surrounding solid objects to the sensorial space.

Then the hands encounter a second element, the skin: a smooth surface surrounding the tactile subject at its centre. The skin is a neutral transition zone to reset the touch. The concave shape of the skin then guides the hands towards its deepest part suggesting a tactile immersion in the sensory space of the 3D tactile miniature. To achieve a smooth and robust surface, the skin is made of glass fibre moulded on a mould made by robotic milling.

Finally, at the centre of the membrane is a platform to accommodate the interchangeable tactile cards that are automatically recognised by the system to enable the respective audio and video descriptions. The tactile surfaces on the top layer of the 3D miniatures feature a sophisticated design that takes advantage of a computational design strategy to fill the area with a single line pattern necessary to build conductive tracks with a minimum amount of electrical connections.

1.6.2 The Computational Design of Single Line Touch Feelers

The subjects of the paintings are simplified and transformed into small-scale three-dimensional figures, each with a different height according to the different degrees of depth in the artwork. The upper part of each figure has been developed as a tactile surface, the touch feeler, which perceives visitor's touch and allows multimedia content stored on the integrated PC to be activated.

Next to Aptica body, the development of the touch feelers was the second major research topic we carried out. We were asked, for technical and usability reasons, to design a new sensor working based on two electrical connections (positive and ground) that would work without bracelets or additional connections. Accordingly, the work focused on finding a computational strategy to generate the conductive traces while minimizing the number of electrical connections and ensuring in-house feasibility of the prototypes.

Various geometrical approaches were explored (Fig. 9), however, in the case of non-compact figures, many failed to produce a single curve occupation. But this was very important to avoid overly complex wiring and an excessive number of electrical connections. Moreover a non-directional pattern would have been beneficial to avoid confusing the tactile perception of the figures. Hence, we searched for a computational strategy to generate an occupation pattern consisting of a single curve capable of filling any kind geometric figure, compact or non-compact.

Various geometrical approaches were explored (Fig. 9), however, in the case of non-compact figures, many failed to produce a single curve occupation. However, this was very important to avoid overly complex wiring and an excessive number of electrical connections. Moreover a non-directional pattern would have been beneficial to avoid confusing the tactile perception of the figures.

Finally, hooking back to the studies of D. Hilbert or W.Sierpinski [21] on space-filling curves such as those, we developed a computational protocol that, starting from a given boundary condition (the edge of the figure) and a set of genotypic parameters and constraints, modulates and folds back the edge on itself up to fully occupy the figure with a single complex curve. The single curve folds on itself according to a set of guide parameters that define: (1) the distance of the folds from the boundary of the figure, (2) the distance between the end and start point, (3) the minimum interstitial space between folds, (4) the minimum radius of curvature of the folds. The complex figure generated becomes then a geometrical skeleton for the construction of the conductive traces of the Touch-Feeler (Fig. 10).

Through this procedure, which combines physical-computing methodologies and parametric-associative design strategies, we obtained a generative system that occupies the space of the figure with a single continuous trace, while still matching production and functional constraints. The graphite based touch feelers were then prototyped in our laboratory by fine-tuning a reliable fabrication protocol combining painting, cutting and laser engraving as to achieve repeatable and precise results.

For this project, various digital fabrication techniques were strategically combined to address both the performative needs and budget economy; from this intersection, we could successfully develop a new type of interface that, by its shape and materiality, offers new opportunities to answer the demand for the accessibility of pictorial artwork. A new design iteration has been already undertaken for the development of a still more stable and durable version of the touch feelers. Nevertheless, APTICA is an expression of a novel design culture where computation, matter and fabrication actively combine to make avant-garde innovation possible inside small scale design studios too.

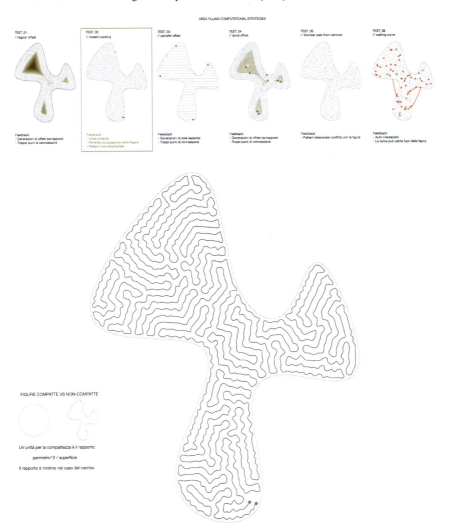

Fig. 9 The diagrams illustrate the different geometrical and computational approaches tested in order to find a single line filling procedure adaptable to both compact and non compact figures

2 Conclusions

The chapter outlined some of the criteria supporting ALO's design research in the field of contemporary strategies and processes for architecture. The projects sampled are the expression, at different scales, of an effort made in the course of everyday practice towards the formulation and acceptance of new concepts of space, structure and furniture, as well as of making strategies and human experience.

Fig. 10 Detail of the conductive graphite pattern of the touch feelers prototype

In this framework, digital fabrication and robotics become instrumental towards a novel material culture focusing on the achievement of a performative materiality. The search for new spatial repertoires and construction models according to the criteria illustrated aims to reduce the consumption of resources through a novel logic of differentiation and polyfunctionality. The concept of architectural composition is surpassed by the concept of formation; the design process is no longer linear and becomes performative as well; non-standardized prefabrication provides novel opportunities for a more responsible and higher quality production even for everyday small projects; matter and fabrication themselves become active agents within the conception and making of space and designs for human experience.

To date, dealing with small-scale interventions has certainly facilitated the implementation of some of the key aspects of our research. As also noted by Colarevich [22], certainly the scale factor is significant, especially since the behaviour of materials changes precisely in relation to the size of the parts. But we are aware that the performance of a material is as much an expression of its intrinsic capacity as of its form. We therefore believe that, although the shift to larger scale in general cannot be straightforward, the construction criteria highlighted and derived from a holistic and biomimetic approach to architecture pave the way for a new generation of smart and intrinsically more sustainable and smart architecture. Certainly, the smartness we pursue goes far beyond the widespread idea of smartness, which often focuses on the mere integration of digital technologies that can make architectures almost alive, capable of informing us about their state of operation and their relationship to their surroundings.

Rather, as a primary and unavoidable condition, smartness should be a matter of the materiality itself of architecture, both spatial and constructive. We should look for a type of intelligence which could provide the ability to passively contribute to the microclimatic well-being as much as to the programmatic and energy needs of both the occupants and the surrounding environment. The new architectures will possibly be interactive constituents of an augmented urban and territorial ecosystem.

2.1 Toward the Industry "X"

Like many other digital technologies that were initially confined to factories, robotics is undergoing a phase of democratisation. However, while the potential of robotic applications in production and human life is increasingly arousing enthusiasm, a sense of scepticism is also emerging because these technologies are often seen as a substitute for human labour. While it is true that machines are often used only to maximize production to meet the needs of mass production at the expense of certain labours, the digitization of production, if aimed at mass customization, can foster the specialization of existing figures and the emergence of new figures and markets. In fact, this is already happening.

In this context, we have the opportunity to embark on a virtuous path in which robotics is not a mere substitute for humans, but an instrument to enhance their skills and capabilities. By overcoming mistrust through a humanistic education in technology and developing critical thinking regarding certain fetishist approaches, we could develop a true deontology to elaborate strategies that privilege human–machine interaction and lead to collaborative productions that would not be possible without such interaction. Industry X, would be a new model of production based on a symbiotic and ethical interaction between man and robot; the X would just represent such a powerful convergence.

References

1. Saiu, V.: The Three Pitfalls of Sustainable City: A Conceptual Framework for Evaluating the Theory-Practice Gap. Sustainability **9**, 2311 (2017). https://doi.org/10.3390/su9122311,lastaccessed15/12/2022
2. Carpo, M.: The Digital Turn in Architecture 1992 - 2012. Wiley (2013)
3. Alexander, C.: A Pattern Language: Towns, Buildings. Oxford University Press, Construction (1977)
4. Cache, B.: Earth Moves: The Furnishing of Territories. MIT Press, Cambridge, MA, USA (1995)
5. Landa, MD.: A Thousand Years of Nonlinear History. Princeton University Press (2021)
6. Deleuze, G., Guattari, F., Massumi, B.: A thousand plateaus: capitalism and schizophrenia. Minnesota Press, Minneapolis (1987)
7. Leach, N.: Rethinking Architecture: A Reader in Cultural Theory. Psychology Press (1997).
8. Saggio, A. The IT Revolution in Architecture. Thoughts on a Paradigm Shift. Lulu.com (2008)

9. Weinstock M (2004) Morphogenesis and the Mathematics of Emergence. Archit Des 10–17
10. Gershenfeld, N.A.: Fab : the coming revolution on your desktop–from personal computers to personal fabrication. Basic Books, New York (2005)
11. Giedion, S.: Mechanization Takes Command: A Contribution to Anonymous History, Illustrated Univ Of Minnesota Press, Minneapolis (2014)
12. Beukers, A., Hinte, E.: Lightness: The Inevitable Renaissance of Minimum Energy Structures. 010 Publishers (2005)
13. Thompson, D.W.: On Growth and Form. Cambridge University Press, Cambridge (1992)
14. Michell, A.G.M.: LVIII. The limits of economy of material in frame-structures. Lond Edinb Dublin Philos Mag J Sci **8**, 589–597 (1904). https://doi.org/10.1080/14786440409463229,lastaccessed15/12/2022
15. Koch, J.C.: The laws of bone architecture. Am J Anat **21**, 177–298 (1917). https://doi.org/10.1002/aja.1000210202,lastaccessed15/12/2022
16. Falcucci, G., Amati, G., Fanelli, P., Krastev, V.K., Polverino, G., Porfiri, M., Succi, S.: Extreme flow simulations reveal skeletal adaptations of deep-sea sponges. Nature **595**, 537–541 (2021). https://doi.org/10.1038/s41586-021-03658-1,lastaccessed15/12/2022
17. Pearce, P.: Structure in Nature is a Strategy for Design. MIT Press (1980)
18. Stumm, S., Braumann, J., Brell-Cokcan, S.: Human-Machine Interaction for Intuitive Programming of Assembly Tasks in Construction. Procedia CIRP **44**, 269–274 (2016). https://doi.org/10.1016/j.procir.2016.02.108,lastaccessed15/12/2022
19. Ferrari, D., Pinotti, A.: La cornice: storie, teorie, testi. Johan & Levi, Monza (2018)
20. Levi, F., Rolli, R.: Disegnare per le mani. Manuale di disegno in rilievo. Silvio Zamorani Editore, Torino (1994)
21. Sagan, H.: Space-Filling Curves. Springer-Verlag, New York (1994)
22. Kolarevic, B.: Actualising (Overlooked) Material Capacities. Archit. Des. **85**, 128–133 (2015). https://doi.org/10.1002/ad.1965,lastaccessed15/12/2022

Digital Twins: Accelerating Digital Transformation in the Real Estate Industry

Mattia Santi

Abstract Digital twins have been introduced in 2002 at the University of Michigan by Michael Grieves. Digital twins are part of the Industry 4.0 revolution and are a strategic technology that, after being implemented in numerous industries such as aerospace and automotive, is now becoming important in the real estate sector. This chapter will introduce the concept of digital twins and their current applications, exploring different types of digital twins and their characteristics. The analysis will then focus on the existing issues facing the real estate sector, with an appreciation of how digital twins could significantly impact this industry. Digital twins will be explained covering the basic principles behind the technology. The chapter will continue analysing some contemporary applications of digital twins in the real estate industry developed by leading companies in the sector.

Keywords Digital twins · Building automation · AI and machine learning · BIM

United Nations' Sustainable Development Goals 9. Industry Innovation and Infrastructure · 11. Sustainable Cities and Communities

1 Introduction to Digital Twins

In recent years, digital twins have become increasingly popular and their use cases are spreading across all industries, from aviation to marketing, from automotive to real estate. Dr. Michael Grieves first introduced the concept of the digital twin in 2002 at the University of Michigan [1]. The concept presented was based on the idea of generating a digital clone of a physical entity and linking them together. Based on this concept, a digital twin can be defined as a digital clone of a physical system that covers its entire life cycle and is connected to it via real-time data exchange.

These data are analysed and visualized thanks to digital simulations and machine learning providing a real-time interconnection between the physical element and the

M. Santi (✉)
SASI Studio Ltd, London, N1 8JT, UK
e-mail: mattiasanti@sasi-studio.com

digital element. This real-time connection makes it possible to analyse and understand the behaviour of the physical asset and provides useful insights to enable a predictive maintenance strategy (Fig. 1).

Dr. Michael Grieves, in his original formulation of the digital twin concept, describes a digital twin as a system consisting of two elements, the physical asset and the digital replica. In the model proposed by Dr. Grieves, these two elements are connected during the entire lifecycle of the physical asset, from creation to production, operation, and disposal. According to this model, the physical asset and its digital clone exchange data in real time providing feedback to each other in a sort of mirroring effect. The name of this conceptual model was the "Mirrored Space Model" [1].

A digital twin can describe a physical asset at different scales. To achieve a digital twin, it is necessary to have a physical asset, a system of sensors connected to this asset that collects data, a system that analyses this data in real-time and a system that helps visualise it. Therefore, the concept of a digital twin goes beyond the traditional notion of a visual model, if we consider for example a BIM model, the primary distinction between a BIM model and a digital twin is the connection between the physical item and the digital model via a feedback loop. The main scope of a digital twin is to monitor and manage an asset during its lifecycle, supporting the decision-making process and predictive maintenance.

Digital twins can be applied at different scales, for instance, there are digital twins representing individual assets or digital twins representing systems of multiple assets, therefore the level of detail adopted for the representation of the physical reality can vary based on the purpose of the digital twin. Digital twins can be used for different types of reasons and this affects the level of detail that needs to be achieved and the quality of data required to make sure that the information collected and analysed is suitable for the scope of the project. Considering that digital twins can collect and process large amounts of data in real-time, it is essential to use the appropriate level of detail required to achieve efficient and reliable digital twins.

A digital twin is characterized by three main elements: the data exchanged with the physical asset, the model and the visual representation of the model. Considering that the central idea is to establish a real-time connection between the physical and

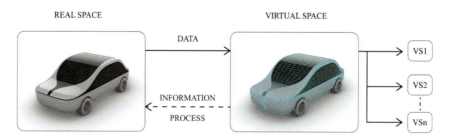

Fig. 1 Diagram reproducing the "Conceptual Ideal for PLM" introduced in 2002 by Dr. Michael Grieves

virtual realms, the data exchanged are defined as dynamic data, meaning data that can change asynchronously over time as soon as updated information is available.

Depending on the purpose of the digital twin, the model can be interpreted in a variety of ways. For instance, the models could be physics-based models to simulate the real context of the digital twin, agent-based models to simulate specific behaviours and interactions, or a combination of the two.

The other element to consider is the visualization of the digital twin model, the visualization strategy depends on the purpose of the digital twin but also the target users of the digital twin. For example, in some cases, a 2D visualisation may be sufficient, while in other cases a more immersive experience may be preferable. The correct visualization strategy is essential to allow the digital twin to absolve its main function of helping to make informed decisions.

Considering a wind turbine as an example, to build a digital twin to monitor this energy production infrastructure it is necessary to have sensors capable to measure in real-time the physical conditions of the turbine and its surrounding environment, such as temperature, wind speed, weather conditions, etc. In parallel, there should also be a digital model of the physical asset that contains the information needed to monitor the physical condition of the asset and its performance. In addition, there could be a 3D visualisation of the wind turbine that can help engineers observe its physical characteristics, its context and combine this information with the data received from the sensor to gain a multidimensional understanding of the asset and its performance, for example by measuring how much energy is produced and under what specific environmental conditions. More advanced models may also allow for simulation to test emergency scenarios and identify predictive maintenance strategies.

Digital twins should not be identified with simulation models. These are typically focused on simulating specific phenomena or processes. Although digital twins can also be used for simulation, digital twin models should be able, within their scope, to give a complete representation of an asset, allowing to run of multiple simulations in multiple scenarios. Another difference is that a digital twin establishes a two-way connection with a physical asset, allowing it to inform the simulation with real-time data collected by the interaction with the physical asset [2].

2 Origins of Digital Twins

Digital twins are a key technology in the context of Industry 4.0, although they are not a new technology. In fact, the idea of using digital twins was originally introduced by NASA in the 1960s and the Apollo 13 rescue mission in 1970 could be considered the first application of the idea of the digital twin, more than 30 years before the term "digital twin" was coined [2]. Each spacecraft was replicated in an earthbound version used to simulate and study the operation that the crew had to perform in space and to train the crew. The Apollo 13 simulators used the most advanced equipment in the entire space programme: only the crew and mission control consoles were real, while the rest was a simulation developed with advanced computers, complex calculations and expert engineers.

After the launch of the Apollo 13 spacecraft, there were technical problems and the mission changed from an exploration mission to a rescue mission. This became the first application of digital twins. NASA had a copy of Apollo 13 on the ground which enabled the engineers to carry out the necessary tests and make informed decisions to help the crew manoeuvre the damaged spacecraft to safety [3].

Contemporary digital twins are connected to physical assets through the use of the Internet and can exchange data in real time between the physical asset and the digital model. Apollo 13 did not have the "Internet of Things", but NASA was using state-of-the-art technology to achieve a near real-time connection with the crew 200,000 miles away. Considering this, Apollo 13 simulators could be considered early-stage instances of digital twins since they were connected to the actual asset in near real-time and could respond to changes in the physical asset [3] (Fig. 2).

This was the first application of digital twins before the term digital twin existed. Although Dr. Michael Grieves formally introduced the concept of a digital twin in 2002 under the name "Mirrored Space Model", it was actually in 1991 that the idea of a digital twin first appeared in the publication "Mirror Worlds: or the Day Software Puts the Universe in a Shoebox...How It Will Happen and What It Will Mean" by David Gelernter. The term "Digital Twin" (DT) first appeared in the draft version of NASA's technology roadmap in 2010. In this publication, NASA first introduces a formal definition of Digital Twin, describing it as "*an integrated multi-physics, multi-scale, probabilistic simulation of a vehicle or system that uses the best*

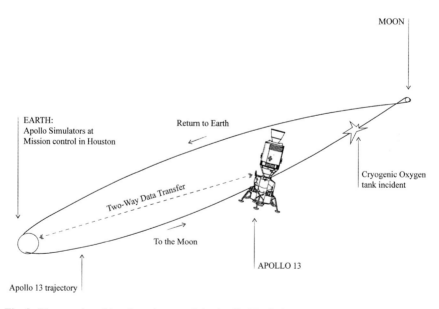

Fig. 2 Diagram describing the trajectory of the Apollo 13 mission

available physical models, sensor updates, fleet history, etc., to mirror the life of its flying twin" [4].

A few years later, leading industry media such as Gartner, Lockheed Martin and Forbes recognised digital twins as a strategic technology trend. Until a few years ago, limitations in processing power, data scarcity, storage constraints, bandwidth costs, and technological hurdles slowed company adoption of digital twins, but we are now seeing digital twins expand across different industries. Digital twins are now within reach of many companies, supported by the quick integration of IoT devices.

Digital twins became a key technology part of the Industry 4.0 revolution, a new evolution of industry characterized by revolutionary technologies such as Quantum Computing, Spatial Computing, the Internet of Things (IoT), Artificial Intelligence (AI), Genetic Engineering, 3D printing and more. Digital twins can play an essential role in the Industry 4.0 landscape, helping to bridge the gap between the physical and digital worlds. Digital twins, depending on their implementation, may enable a completely new way of interacting with physical reality thanks to spatial computing and could open the way to a completely new range of services and business applications.

3 Different Types of Digital Twins

In several sectors, digital twins have been widely applied, helping to digitise different types of products. According to the type and the scale of their application, digital twins can be categorized into 4 different types: Component Twins, Asset Twins, System Twins, and Process Twins [2].

Starting from the smallest units that define a system, Component Twins are digital twins used to mirror the basic components of a product. The component twins allow monitoring of these individual system components, considering their operational status and performance.

Combining multiple components, a more complex digital twin can be obtained to simulate the overall behaviour of an asset, considering the interactions between individual components. This digital twin is focused on analysing the system at the product level and can provide a considerable amount of data considering complex interactions happening between multiple components.

Multiple assets create a system that can be identified as a System Twin. System Twins are useful for representing complex elements characterised by the interaction of multiple asset twins. They can be helpful when it is necessary to control a group of assets and study their interactions at a system level.

Lastly, process twins are used to study the interaction between multiple system twins. These are large-scale digital twins that operate at the macro level and focus on understanding how multiple complex systems interact with each other. For example, they can represent large industrial plants or entire cities [2].

4 Current Applications of Digital Twins

Digital twins aim to provide a real-time copy of a physical asset and require a combination of IoT sensors and software development to build a reliable digital twin.

Therefore, digital twins still require a considerable investment that currently limits the range of applications of digital twins. Despite this, digital twins are currently applied in several industries, for example, in power-generating equipment, which includes large-scale engines such as jet engines, locomotive engines and turbines for power generation, where digital twins are applied in the management of maintenance requirements. Buildings, infrastructures, and their systems are other important examples to consider, as they can all benefit from digital twin technologies from the design to the operational stages, allowing for the management of these assets while considering their spatial characteristics and the way users interact with them. Digital twins play an essential role in manufacturing processes, and they have a natural application in product life cycle management. Digital twins are also used in the healthcare business to assist in the creation of digital profiles for patients receiving healthcare services. Another possible application in this industry is the employment of digital twins by pharmaceutical firms to model genomic codes, physiological parameters, and patient lifestyles [2] (Fig. 3).

As previously mentioned, digital twins also find applications in the automotive industry.

Cars are complex assets with several interdependent systems, therefore digital twins play a crucial role in their coordination, maintenance, and customer care.

In the automobile industry, digital twins are also utilized in product testing, where the digital twin of a product aids in evaluating its quality and performance by conducting digital experiments with different compounds and raw materials to improve the design and maximize the product's performance.

Another important application in this industry is the development of self-driving cars, which requires the use of several sensors to comprehend the interaction between the vehicle and its surrounding environment. Digital twins are necessary to test numerous scenarios and imitate the behaviour of these vehicles in the digital world [2].

Digital twins also find wide applications in the aerospace industry. As previously mentioned, one of the first applications of digital twins was in the aerospace field with Apollo 13 and all subsequent applications developed by NASA. Today, digital twins are essential for monitoring and simulating the behaviour of mechanical components, complex systems, entire vehicles or vehicle systems. They are very important when predictive maintenance is required and in complex projects that require complex simulations and monitoring. For example, they are used to monitor jet engines to check the level of degradation and simulate engine behaviour under different conditions, informing maintenance and design strategies.

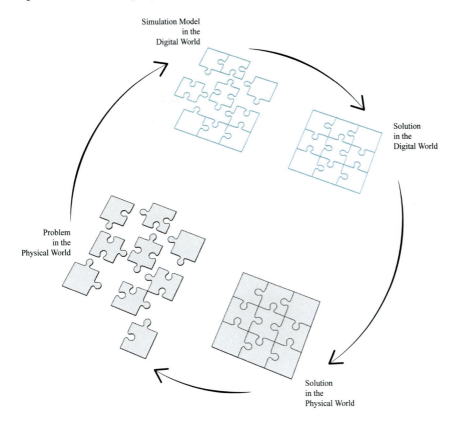

Fig. 3 Digital twins provide a risk-free environment to test new solutions

Digital twins can also play an important role in smart cities, where urban planning efforts can be guided by actual data to make better design decisions and simulate numerous scenarios to meet the needs of multiple stakeholders.

Digital twins are currently being applied at the urban level to simulate entire urban areas and to monitor traffic by extracting statistical information on public transport, pollution levels, traffic intensity, etc. Many of these projects are still in the initial stages and it is possible to imagine that the number of such applications will increase and that these projects will be further developed in the coming years [2].

4.1 Case Study: The Flying Catamarans

Digital twins are having an impact in several sectors and the sports industry is one of them. In 2012 the Emirates Team New Zealand (ETNZ) introduced new 72-foot hydrofoiling catamarans, which were subsequently adopted in the 36th America's Cup. In 2018, the America's Cup 75 Class Rule was published, defining the design rules for boats eligible to participate in the 36th America's Cup. No physical testing was authorized during the 36th America's Cup, therefore simulation-driven digital twins proved to be an essential design and testing tool. These racing yachts, rather than having a keel, have foil cant arms which can move outside or under the boat to provide stability and make these catamarans fly on top of the water [5] (Fig. 4).

These innovative racing yachts are 7.6-tonne boats with the crew, sailing at maximum speeds that can reach 50 knots, flying on the water, propelled by a double-sail skin mainsail, combined with a D-shaped mast to form a wing [5]. Since physical testing in wind tunnels or towing tanks was prohibited by AC75 class rules and teams could only build two racing boats, simulation-based digital twins became the solution for designing these racing yachts. The design team needed to create a comprehensive digital model that could be used by the entire team, from designers to boatbuilders and sailors, to create these high-performance products. To do this, they had to combine CAD tools with computational fluid dynamics (CFD), structural analysis and simulation, as well as product lifecycle management tools. The team's Velocity Prediction Program (VPP), which predicts boat speed under various conditions, is fed with CFD (Computational Fluid Dynamics) data from the simulation of many unique boat configurations characterised by different parameters, such as hull position or foil angles. This allows different design options to be evaluated against different scenarios and optimisation strategies to be defined. Another challenge in which digital simulation models have been important is the study of the interaction between aerodynamic and hydrodynamic forces operating on these vehicles. With the limitation of building only two physical models, having a digital twin of the vessel was essential to the entire design process, allowing for a high-fidelity copy of the physical asset on the one hand, and simulating the physical conditions in which the vehicle will operate on the other [6].

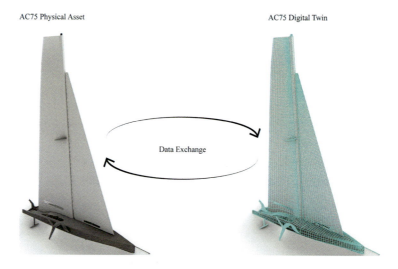

Fig. 4 Digital twins were adopted in the design of the AC75 racing yachts

4.2 Case Study: Digital Clones of the Earth

The function of digital twins as a predictive tool can be widely scaled, creating digital clones of complex systems like an entire planet. At the time of writing, there are multiple initiatives attempting to develop a digital twin of the Earth to monitor climate change using predictive models.

One of these projects is Nvidia's Earth-2 project, which aims to build a supercomputer to implement a digital twin of the Earth to study and predict climate change. The difficulty in predicting climate change is that, unlike weather forecasting, the period is too long and there are too many variables to consider. Designing the best strategies for reducing the effects of climate change and adapting to the changes requires climate models that can predict the climate in various places of the world over decades.

These models are "multidecade simulations" of multiple complex systems such as the atmosphere, oceans, land, human activities, etc. This level of simulation requires ultra-high-resolution climate modelling therefore this was not possible until a few years ago [7] (Fig. 5).

Another project that aims to create a digital twin of our planet is Destination Earth (DestinE), a project promoted by the European Commission in collaboration with partner organisations, which aims to develop an accurate digital model of the Earth that can collect real-time data and investigate how natural events and human behaviour interact.

The DestinE project is a complex system composed of three main elements, the "Core Service Platform", "the Data Lake", and the "Digital Twins". The "Core

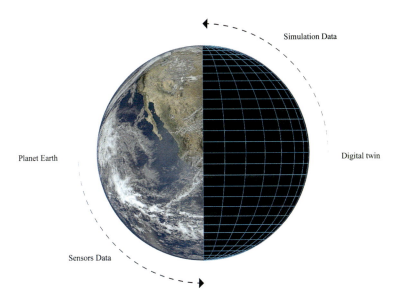

Fig. 5 The Destination Earth project intends to create a digital replica of the planet Earth

Service Platform" will be the platform and the interface for the users to access DestinE model. The "Data Lake" will be the pool of data collected by the data holdings provided by Copernicus the European Union's Earth observation programme, data holdings from ESA (European Space Agency), EUMETSAT (European Organisation for the Exploitation of Meteorological Satellites) and ECMWF (European Centre for Medium-Range Weather Forecasts) and additional data provided by various sources. Then the "Digital Twins" will be implemented to create digital clones of complex systems (e.g. oceans, lands, etc.) that will all be connected to the main model.

Through DestinE, users will have access to theme-related data, services, models, scenarios, simulations, forecasts, and visualisations. In order to produce accurate and useful scenario predictions, the underlying models and data will be regularly evaluated. DestinE initially will be accessible mainly by public authorities but in the future will be accessible also by a wider spectrum of users [8].

These case studies are some examples of how digital twin technology is currently being applied in different sectors. The implementation of this technology has advantages and disadvantages. Analysing the benefits first, the digital twins facilitate a more effective research and design process for product development and design. Indeed, because of the volume of data generated, they enable the prediction of an asset's performance and the potential of making essential changes prior to initiating the manufacturing process. This enables enterprises to obtain more accurate information and hence make more informed decisions more quickly. On the other hand, digital twins enable continuous monitoring of products and systems, resulting in a more effective product life cycle management process. Thus, digital twins can aid in addressing the difficulties of material resource scarcity, pollution, climate change consequences, and the transformation towards a net-zero greenhouse gas emissions economy.

The United Nations Sustainable Development Goals (SDG) serve as a framework for identifying objectives to address real-world problems that impact society and the environment. Regarding these goals, digital twins, especially when deployed as a large-scale digital twin network, can contribute to the achievement of several goals. For instance, regarding SDG 09 "Industry, Innovation and Infrastructure", digital twins can contribute to the development of robust systems to monitor infrastructure or production processes to maximize their resilience and efficiency. Regarding SDG 11 "Sustainable Cities and Communities", digital twins can help develop connected settlements with transparent access to information, facilitating participatory democracy through the inclusion of citizens in planning and policymaking, and promoting decarbonisation and resilience. Digital twins can help monitor and control energy use and its interaction with the distribution network in real-time, enabling the optimisation of resources, from individual products to buildings and cities.

On the other hand, deploying digital twins has some challenges, such as requiring the integration of several technologies, such as 3D models, sensors, data analysis, and machine learning, which need specialized knowledge and a significant initial financial investment. Several challenges are inexorably linked to technology: data privacy is one of them; sensors capture incredibly valuable data that must be managed and processed according to increasingly stringent cybersecurity standards. In addition, network partitioning and latency must be considered when designing these systems. Indeed, more processing power and bandwidth are required to maintain the service's real-time effectiveness as the system gathers more data and as more users interact with it. These are all aspects to be considered about this technology, regardless of the specific industrial applications.

The importance of digital twins in architecture and how they could change the real estate sector will be explored in the following paragraphs.

5 Current Challenges in Real Estate Industry

Digital twins can contribute to bringing innovation in Architecture providing solutions for some of the pain points currently affecting the real estate industry.

The digitisation of operations and data continues to be a significant challenge in the real estate industry. Existing buildings often have a history that spans several decades; thus, part of the information associated with these buildings is not available in digital format, and in other cases, is unavailable at all. Often, gathering data about buildings needs access to the facility's physical assets and human activities to extract data about the building. This makes it challenging to collect and maintain information on existing real estate assets [9].

The emergence of hybrid workspace models that blend remote and in-office employment is a prominent trend in the present real estate market. This necessitates the construction of more resilient working spaces that can adapt to a variety of usage circumstances and the provision of digital interaction with the physical area. For instance, it becomes crucial to provide a seamless experience that creates the same working conditions for remote and in-office workers and to ensure that all workers may access meetings even if they are not physically present in the office. On the other hand, it is essential to have digital and automated processes to maximize the use of space depending on user requirements, therefore creating spaces that can adapt to user demand [9].

Another challenge in the construction world is the need to innovate construction processes to reduce carbon emissions, improve health and safety, optimise the construction pipeline, and reduce costs and delays. The construction industry is currently characterised by the rising cost of raw materials, labour shortages and a decrease in the number of craftsmen available on the market [9].

Sustainability is a crucial challenge for the real estate industry; in 2021, 37% of global CO_2 emissions were due to the buildings and construction sector, which accounts for 34% of global energy demand [10]. Buildings are characterised by embodied carbon generated during the construction process and operational carbon emitted during the operational life of the building. It is difficult to monitor energy consumption, especially in existing assets, so it is important to incorporate IoT devices to monitor consumption and integrate circular models to reduce the carbon footprint of the building sector and develop optimisation strategies to improve building performance.

On the other hand, another challenge is the integration of IoT technologies in existing and new buildings. Implementing this technology would give the possibility to access building information remotely, an important element that becomes crucial when physical access to the asset is not possible. In this sense, Internet of Things (IoT) technologies, whether sensors or data systems, are essential to be able to gather information about the building in real time [9].

6 Digital Twins in Real Estate and Architecture, Engineering and Construction (AEC) Industry

Design, construction and management operations can be facilitated by the use of digital twins, as buildings are high-value assets with complex life cycles. Besides being physical assets, buildings are also environments where people live and work, building social relationships and forming communities. To further evaluate the possible uses of digital twins in the real estate sector, it is important to reflect on the profiles of users who may be interested in using this technology.

One of the first user profiles to be considered are professionals involved in the design of buildings, such as architects and engineers involved in the design process and in the decision-making. Architects and engineers could benefit from accessing data on the performance of the assets designed, to provide better ongoing support to clients and learn how to improve design services.

Social workers, for example, may be interested in determining the social impact of specific policies and may need to discover resources and solutions to give effective assistance to individuals. Asset owners are also user profiles to be considered. Digital twins could provide them with greater control over their assets and lower management expenses. Another profile is elected officials who determine and vote on how public funds are distributed and invested and who, thanks to the implementation of city-wide digital twins, can analyse and understand the real needs of citizens.

The policymaker profile should also be considered as they are in charge of recommending and establishing laws and norms. This user profile would unquestionably benefit from having access to structured facts to comprehend the impact of past actions and make educated decisions for the future. A crucial profile is that of an emergency service planner since they are responsible for gathering data and simulating the effects of emergencies. In addition, tourists can benefit from a connected ecosystem of digital twins, as they would be able to locate the information they want more easily thanks to the interaction and exchange of information with the built environment. Finally, numerous forms of digital twins might help small company owners to identify local patterns and make key choices about the services they supply.

Citizens, in general, could benefit from accessing digital twins of public spaces to learn more about their neighbourhoods and how they can contribute to improving their community and environment [11].

Reflecting now on the potential applications of digital twins in the real estate sector, one interesting application is to model and anticipate how tenants would use and interact with the property, gaining insights into users' comfort and productivity. In order to increase the user's comfort, a digital twin might monitor the environment and provide recommendations and adjustments. This ability to foresee allows more informed project choices and favours the development of successful management strategies. Digital twins, for instance, may be used to examine complex utilization scenarios for commercial and public buildings, enabling flexible day-to-day space and function configurations.

Another important area of application for the digital twin is the design process. In fact, with this technology, designers can create more than just a 3D model, they can create a sandbox environment for a project, an interactive game in which design ideas, products and user scenarios can be tested against multiple design iterations, different environmental conditions and environmental data. The construction process can be replicated and visualised using digital twins. Computational analysis is already widely used to improve the design and many organisations now have access to sophisticated methodologies that link different disciplines, but with digital twins, designers have more reliable information to test design ideas in different scenarios, using environmental data to make their studies context-specific (Fig. 6).

Considering that digital twins are able to store maintenance information and can enable the automation of part of the building maintenance process, they can also enable better maintenance strategies. Given the need to move towards a more sustainable use of energy, buildings will no longer rely solely on programmed responses but will be required to provide autonomous responses to different user behaviour and different environmental conditions to optimise system behaviour and the use of resources in real time.

Digital twins could become an important tool for studying strategies to improve the sustainability of buildings. It is possible to analyse a building's performance and calculate its carbon footprint in real time by collecting environmental data. This allows the implementation of methods to optimise a building's energy use and make it more energy efficient. Digital twins could be useful to identify waste streams and their potential as resources, increase operational efficiency and resource utilisation through waste reduction, quantify environmental expenditures to encourage a circular economy, and help achieve zero-emission targets.

Another interesting application of this technology is at the urban scale, where it is possible to create virtual clones of entire cities and neighbourhoods, allowing large amounts of data to be recorded and studied at the community level, providing crucial information for making informed decisions at the urban level. According to a study conducted by the C40 Cities Climate Leadership Group, urban policy decisions made before 2020 might affect up to one-third of the global carbon budget that has not yet been determined by past decisions [12]. Therefore, political decisions and public administrations play an important role in creating the conditions to achieve

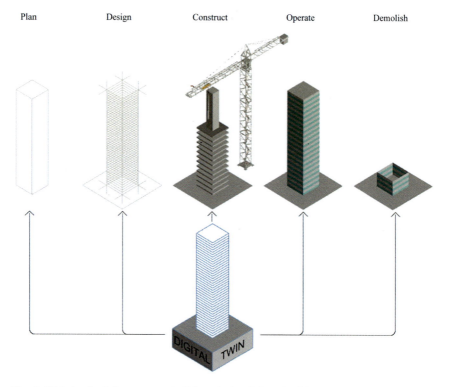

Fig. 6 Digital twins help to manage buildings during their entire lifecycle

a low-carbon economy. Digital twins can provide a simulation and monitoring tool, enabling the testing of policy decisions, identifying dependencies and enabling cooperation between policy sectors, all while improving the participation of citizens and communities.

7 Current Challenges in the Implementation of Digital Twins

To comprehend the present challenges in the adoption of digital twins in the Real Estate and Architecture, Engineering, and Construction (AEC) industries, it is necessary to examine the many types of user profiles that may be identified.

Tenants are one of the first user profiles to be considered because some of the data collected by the sensors may put their privacy at risk, but on the other hand, this data could be used to improve their living experience. Landlords, on the other hand, might benefit from having well-organised data on the digital assets they own and manage. Investors are another type of user profile who might be interested in learning more about the performance and management of a particular asset.

Vendors, technologists and start-ups are also important actors to consider, as they might be interested in receiving information on certain assets to advertise their services or adapt their business models to market demands. Nevertheless, it is vital to consider how local councils and governmental organizations might benefit from direct access to current information about assets and operations, enabling them to be more visible on the ground and providing citizens with better services. Research organizations and academic institutions are other interesting user profiles. They could use digital twins as virtual labs to collect field data for research and experimentation as well as to give students the chance to gain experience working with information that is typically challenging to obtain. Furthermore, government authorities, suppliers, insurance companies and similar businesses can benefit from having real-time data on the buildings for which they provide services [11].

Another important challenge to drive the adoption of this technology is the definition of standards. Standards are essential to establish definitions, concepts and processes such as data management and interchange, as well as technical requirements such as interoperability and data security.

In this sense several institutions are currently working to develop standards and frameworks to facilitate the implementation of Digital Twin technologies, an example is the National Digital Twin programme (NDTp) run by the Centre for Digital Built Britain, a partnership between the University of Cambridge and the Department for Business, Energy and Industrial Strategy [13]. Other examples are standards developed by the International Organization for Standardization (ISO) specific for the Digital Twin technology, such as ISO/IEC AWI 30173 "Digital Twin—Concepts And Terminology", ISO/IEC WD 30172 "Digital Twin—Use Cases" and ISO/TR 24464:2020 "Automation systems and integration—Industrial data—Visualization elements of digital twins". Also, important to consider are standards like ISO 10303 "Industrial automation systems and integration—Product data representation and exchange" and ISO 19650 "Organization and digitization of information about buildings and civil engineering works, including building information modelling (BIM)—Information management using building information modelling" [11].

Despite the increased use of digital twins in other industries over the past few years, which has made them more accessible, these technologies need a great deal of knowledge, and digital twins are challenging to implement in AEC processes, especially for small and medium-sized companies. Given the current state of technology, it is necessary to form multidisciplinary teams comprised of experts in fields such as software development, building information modelling (BIM), virtual reality, user interface design, interaction design, artificial intelligence and data science to connect IoT systems and create a system capable of streaming real-time data and allow users to interact with the system. Consequently, this technology requires considerable investment and requires the collaboration of multidisciplinary teams with different types of expertise. Therefore, it is important to consider the development of competency and skills frameworks to identify relevant roles within industry and organisations, as well as the key competencies needed to effectively implement these technologies. As a result, knowledge providers in this area can become important key players, because, without an adequate skill set, organisations risk employing staff with insufficient skills to develop their digital twin projects, with the risk of poorly designed results that do not work as intended [14].

8 Defining Building Ontologies

Real estate is one of the world's largest asset classes. Digital twins have the potential to significantly increase the value of real estate assets, from individual units to buildings and entire cities and facilitate the digitization of information relative to the physical assets. To achieve this and facilitate the implementation of digital twins it is necessary to define and set standards for digital twins [15].

The digital twin market is attracting tech giants like Microsoft, IBM, Siemens, Dassault Systems, Autodesk and others. These firms supply the digital infrastructure for IoT standards and protocols for data collection and analysis, enabling third-party companies to develop digital twins using the technology needed. The construction industry has been slower than other industries to embrace digital transformation. These companies are developing standards and integrations to support the acceleration of digital transformation in real estate. The software infrastructure that some of these companies are developing is important for the wider adoption of digital twins in the real estate sector (Fig. 7).

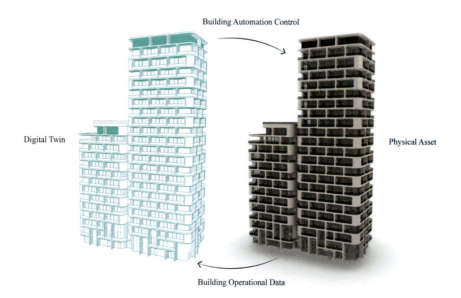

Fig. 7 The digital twin of a building allows for the digitisation of information about the physical asset through real-time data exchange

For example, if we consider a building element like a door, which is a common and essential item to have in a building, it involves several stakeholders to be specified, to be built and to be installed. Just to describe the characteristic of a basic item like a door it is required a considerable data structure, that nowadays is generally managed in BIM environment and requires specialized professionals and organizations to handle these building information models. To digitise real estate assets beyond the design and construction phases, it is important to simplify the way buildings can be described. This is why some of the tech giants interested in the digital twin market are developing new building ontologies that can be used to facilitate the implementation of digital twins.

An ontology in computer science is a formal representation of the concepts and interactions inside a certain domain. It is utilized to model and arrange the domain's knowledge in a machine-comprehensible format. Ontologies are frequently used in artificial intelligence, natural language processing, and the Semantic Web because they provide a standardized vocabulary for the comprehension and interpretation of data. An ontology consists of a collection of concepts, their attributes, and the links between them. It is used to describe knowledge so that it may be shared and reused across several systems and applications. In practical terms, a building ontology serves to conceptualize the type of data that are necessary to describe a building in all its constituent parts, describing the building elements, their properties and the relations between them. An ontology can be seen as a set of models for a specific domain. These models are developed through the use of a Digital Twins Definition Language (DTDL). A Digital Twin Definition Language (DTDL) is a standardized language or set of rules that can be used to describe the structure and behaviour of a digital twin. It enables developers to create a digital twin by defining the various components, their relationships, and the rules that govern their interactions [16]. For example, a digital twin definition language allows to define the various rooms in the building as components, specify the relationships between them (e.g., one room is adjacent to another), and define the rules that govern their behaviour (e.g., temperature control). The built environment is characterized by a high level of complexity, and a building ontology needs to express this reality in a manner that is simple for digital twin developers to use in order for this technology to achieve wide adoption.

By optimizing data categorization, integration, and accuracy, creators of digital twins may produce more accurate digital representations of buildings and their components. Therefore, the development of platforms that link data generated by IoT devices to the building topology is a necessary first step in setting standards for digital twins and facilitating the wider adoption of digital twins in the real estate sector [15].

9 Game Engines

Visual representation is an essential element that enables digital twins, allowing the user to understand the link between data and physical assets and to interact with them.

Game engines are playing an important role in facilitating the implementation of digital twins. Game engines offer technologies for handling the 3D data required to create a realistic representation of a physical asset. Most innovative manufacturers have started using game engines to simulate production processes and test their products with different usage scenarios before manufacturing them [17].

Game engines, such as Unity or Unreal Engine, include sophisticated 2D or 3D physics-based development environments that allow the development of photorealistic simulation while also providing an environment defined by physical forces and effects that can depict how the asset will interact with the real environment. This technology is important for the implementation of Digital Twins because, in addition to real-time data, the model may be simulated against physical circumstances that the user can adjust in real time. Game engines are used by innovative car manufacturers to simulate the creation of new cars and provide more immersive experiences by allowing users to try out a vehicle before it is built. Considering this example, digital twins are important for a wide range of people, including CEOs, marketing, sales teams and customers. Thanks to the visualisation and simulation capabilities provided by game engines, digital twins become more accessible: individuals no longer need a significant level of technical knowledge to understand or decode the simulation in front of them.

For example, non-technical people may experience how a new building will appear and work in a real-world scenario thanks to the strong visualisation capabilities of game engines [17].

10 Implementing a Digital Twin of a Building

The technology stack required to implement a digital twin of a building would depend on the specific requirements and goals of the project, but generally, it will include some essential components.

One of these components is the 3D modelling software, which is used to develop a virtual clone of the physical asset, including its geometry, spatial relationships, and materials.

The Internet of Things (IoT) system, which consists of sensors and devices, is another important element. It is used to gather information from the actual building, such as temperature, humidity, occupancy, and energy use. These sensors then feed the digital twin with the data they have acquired.

Another crucial element is the cloud infrastructure, which is utilized to store and handle the data gathered from IoT sensors and devices as well as any other data that is important to the digital twin.

Tools for analytics and data visualization are additional components required for creating a digital twin. These are used to analyse the data gathered from IoT sensors and devices and to present the findings in a form that is clear and useful to all stakeholders.

Other important elements are machine learning algorithms, which are used to identify patterns and trends in the data collected from the IoT sensors and devices and to make predictions about future behaviour or performance.

Application programming interfaces (APIs) are also important elements of a digital twin, these are used to connect the various components of the technology stack, and to enable integration with other systems or applications that may be relevant to the digital twin.

The user interface is another crucial element of a digital twin. This is the interface that stakeholders use to engage with the digital twin, and it may consist of web applications, mobile applications, or other types of user interfaces.

Developing the digital twin of a building requires a software architecture that can handle the various data sources, processing requirements, and user requirements involved in such a project. We can summarize the basic software architecture necessary for a digital twin in three main elements: the data management layer, the processing layer, and the user interface layer.

The data management layer is responsible for gathering and organizing the diverse data sources that will be utilized to create the digital twin. This data can include information about the building's physical characteristics, such as its 3D geometry and construction materials, as well as data on its systems and operations, such as energy usage and occupancy patterns. The information collected will be interpreted based on a building ontology specific to the type of building that thanks to the digital twin definition language (DTDL) will define the data model for the digital twin, which serves as a blueprint for how the data should be structured and organized. Once the data model has been defined using the digital twin definition language (DTDL), the data management layer can use this model to integrate data from a variety of sources and ensure that the data is organized and stored in a consistent and meaningful way [16]. This data can be collected by the data management layer from a variety of sources, including sensors, building management systems, and manual inputs. The data management layer should be able to obtain building-specific data updates in real time.

Another important component of the software architecture is the processing layer, which is responsible for analysing and interpreting the data collected by the data management layer. This can include tasks such as identifying trends and patterns in the data, predicting future behaviours and identifying potential issues or inefficiencies. The processing layer is not directly concerned with the structure or organization of the data but rather focuses on the meaning and significance of the data. The output generated by the processing layer can be used by users or other systems to make decisions or take actions. The processing layer should be able to handle large volumes of data in real time and should be able to scale up or down as needed to meet the needs of the digital twin.

Moreover, the software architecture should include a user interface layer, which allows users to interact with the digital twin and access the data and insights generated by the processing layer. This can include features such as dashboards, alerts, and visualization tools that help users understand and monitor the building's performance. The user interface should be intuitive and easy to use and should be accessible from a variety of devices, including desktop computers, tablets, and smartphones.

Overall, the software architecture for a digital twin of a building should be able to accommodate the project's complex and diversified data sources, processing requirements, and user expectations. The system of a digital twin must be designed with a scalable and reliable architecture to accommodate additional traffic and data volume.

11 Reference Projects

Having discussed the potential applications of digital twins and explored some of the challenges in implementing this technology, it is useful to discuss a couple of projects that provide a good example of how digital twins can be applied on the architectural and urban scale.

11.1 MX3D Bridge

The MX3D bridge, developed by the Joris Laarman Lab, is an innovative 12-m-long bridge 3D printed in stainless steel by the Dutch 3D printing company MX3D. The bridge is equipped with an innovative sensor system that will collect data about the bridge's structural behaviour and the environment. The data collected from the bridge is fed into a digital twin of the bridge, a virtual clone of the physical asset that analyses the condition of the bridge in real time, acquiring important information to understand how this innovative 3D-printed structure behaves from a structural point of view but also to understand how it interacts with its surroundings [18]. Funded by the Lloyd's Register Foundation, the project was developed in collaboration with several companies, including Arup, Imperial College London, Autodesk, University of Twente, Force Technology and the Alan Turing Institute. The sensors positioned on the bridge will measure real-time data like strain, displacement, vibration, and environmental elements like air quality and temperature acquired from the bridge while it is in use [19] (Fig. 8).

This data is used by research teams in the data-centric engineering program at the Alan Turing Institute to analyse material behaviour in diverse contexts and develop novel statistical methods to deepen understanding of advanced materials. The real

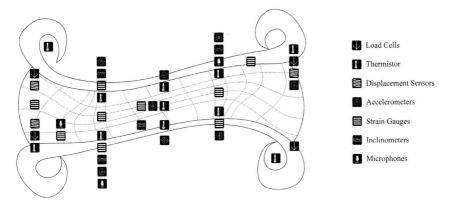

Fig. 8 Diagram representing IoT sensors embedded in and around the MX3D Bridge. (MX3D 2021)

bridge's performance is compared to that of its digital counterpart to provide relevant data for the design of future 3D printed structures and any upcoming certification requirements for 3D printed structures.

The University of Twente is collaborating on this project with its BRIDE program, which uses the bridge's data to study how people interact with it, provide feedback on the design, and assess the project's impact on the community by examining the interactions between people, place, activities, and technology [19].

11.2 Wellington Digital Twin by Buildmedia

Another interesting case study that implements digital twins at the urban level is the Wellington Digital Twin in New Zealand developed by Buildmedia in collaboration with the Wellington City Council. The project offers a new place for interaction powered by real-time city data. This digital twin makes use of smart city technology and real-time data to provide transportation statistics for buses, trains, ferries, bicycles, and vehicles, as well as visualisations of air traffic, cycle sensor data, traffic load, and available parking spaces. Considering that human beings typically process information through visual means, this model integrates data visualisation with a simulated real-world metropolitan context to achieve a strong visual appeal. To create an urban-scale Digital Twin with this level of complexity, it was necessary to combine data from different agencies [20] (Fig. 9).

The project started with the building of the 3D landscape using heightmaps derived from geographical data made available by Land Information New Zealand. More than 18,000 structures with 3D models and textures were built using a photogrammetry model of the Wellington Central Business District provided by the Wellington City Council. In this project, it is possible to understand the importance of adopting

Fig. 9 Image of the Wellington Digital Twin developed by Buildmedia (Reproduced from Buildmedia 2021)

appropriate visualization technologies to construct Digital Twins, where the adoption of game engines allows the management of large quantities of data. The model was detailed using a variety of animation and 3D components, and the entire scene was handled within a physics engine capable of producing the atmosphere system that gives the scene a feeling of realism and enables the simulation of various environmental conditions. Then, this realistic replica of a real city was connected to a network of IoT sensors spread around the city. These data on traffic, pollution, and temperature are communicated using a custom REST API that can manage a variety of data formats, including JSON, XML, and HTTP [20].

Although the Wellington Digital Twin will continue to evolve, it is already being used to show a range of citywide initiatives for public consultations and meetings with stakeholders. Having all the information in one model simplifies and enhances interactions with local businesses and public stakeholders. The capacity to trace decision-making is also crucial; it is possible to start collecting a history of proposals and choices that will affect the city, enabling more participation and transparency with local inhabitants.

12 Potential Impact of Digital Twins in the Real Estate and AEC Industry

The impact that digital twins will have over the years in the Real Estate and Architecture, Engineering and Construction (AEC) Industry depends on several factors and it is difficult the predict since the technology will evolve and adapt over the years.

On the other side, it is possible to identify some specific areas in which Digital Twin could have a considerable impact.

One of the first aspects to consider is the impact that this technology may have on our society especially if digital twins are deployed on an urban or even national scale, which could lead to new ways of interacting with the built environment. For example, one of the goals of the National Digital Twin programme in the United Kingdom is to create a network of digital twins connected through shared data [21]. The establishment of such a building ecosystem might enable new services for citizens and could open new market opportunities related to real estate assets. Thanks to this technology in the near future buildings and structures could be augmented beyond their physical nature, becoming accessible online and providing new ways of experiencing them in the digital realm. This could also affect the value of physical buildings, introducing new digital economies linked to physical buildings.

The influence of digital twins on building sustainability and life cycle management is another essential scenario to consider. Thanks to their monitoring capabilities, digital twins can help reduce energy consumption and carbon dioxide emissions related to building assets. On the other hand, thanks to the data collected by digital twins during the life cycle of a building, they can become very useful to inform and activate recycling strategies at the end of a building's life cycle, promoting a circular economy and sustainable use of resources. Even small improvements gained via data-driven decision-making that can be enabled by digital twins might have substantial social, environmental, and economic consequences in a sector like real estate which represents a considerable percentage of the total carbon footprint of an entire country and a considerable portion of a country's gross domestic product (GDP).

Another field where we can expect digital twins to have a considerable impact is the field of software used to design and manage the construction processes. As the industry has evolved from CAD to BIM, we can expect to see new software becoming necessary to implement and manage digital twins in professional practice. Understanding the difference between BIM and digital twins is important to understand what the impact of digital twins could be in this field. The fundamental distinction between a BIM model and a digital twin is that the latter exchanges data in real time with the physical asset. A digital twin is an interactive platform that collects and visualises data, a living model that visualises information in real time and can be used to simulate user scenarios and environmental conditions based on the acquired data. While a BIM model transfers information primarily in one direction based on user input and the amount of information remains constant over time if users do not add new information to the model, a digital twin is a two-way information model that aims to generate feedback loops between the physical and virtual worlds to optimise the system performance. The accuracy of the digital twin should increase over time based on feedback loops between digital and physical assets. Due to the difference in purpose, a digital twin focuses more on dynamic data, whereas a BIM model focuses more on static data. This is not due to a technological constraint, in fact it would be possible to build a digital twin from a detailed BIM model. A digital twin during its life cycle should be able to provide more information than initially entered by the user. Digital twins should provide information on an asset's performance, as well as relevant feedback for improving the asset's design depending on the data collected. As more data is collected and comparable assets are deployed with their own digital twins, the accuracy of the representation of the digital twin improves.

Finally, it is important to consider how public and private organizations may develop or acquire the wide range of skills that digital twins necessitate in order to create a successful digital twin ecosystem. The skills required to build a digital twin range from data science to IoT, AI, BIM, and data visualization. This may require interdisciplinary partnerships, but also highlights the need for competence frameworks to assist governmental and commercial institutions in exploiting the full potential of digital twins. As a result, another potential effect of digital twins is that they can necessitate the establishment of new specialities and skill sets tailored to digital twin projects.

In conclusion, the application of digital twins is not yet widespread enough to disrupt the real estate industry, but as full digital twins of buildings become more common and standardised, companies will be able to optimise entire buildings, asset portfolios and lifecycles through a constant flow of data and information.

New business models and market opportunities will then emerge, as well as a change in the way places are conceived and created. Buildings could become more dynamic and responsive to human behaviour and the natural environment. In the future, it may be necessary to design not just the physical space of a building, but also its cyberspace.

References

1. Grieves, M., Vickers, J.: Digital twin: mitigating unpredictable, undesirable emergent behavior in complex systems. In: Kahlen, F.-J., et al. (eds.) Transdisciplinary Perspectives on Complex Systems, pp. 85–113. ©Springer International Publishing, Switzerland (2017)
2. What is a digital twin?. IBM. https://www.ibm.com/topics/what-is-a-digital-twin. Last accessed 15 Dec 2022
3. Apollo 13: the first digital twin. Simcenter. https://blogs.sw.siemens.com/simcenter/apollo-13-the-first-digital-twin/ (2020). Last accessed 15 Dec 2022
4. Shafto, M., Conroy, M., Doyle, R., Glaessgen, E., Kemp, C., LeMoigne, J., Wang, L.: Draft modeling, simulation, information technology & processing roadmap. Technology Area 11, NASA, Washington (2010)
5. The technology. In: 37th America's Cup. https://www.americascup.com/the-technology. Last accessed 12 Dec 2022
6. Farrell, P.: Flying high at the America's Cup. Siemens Eng. Innov. (7), 46–53 (2021)
7. Huang, J.: NVIDIA to build Earth-2 supercomputer to see our future. NVIDIA Blog. https://blogs.nvidia.com/blog/2021/11/12/earth-2-supercomputer (2021). Last accessed 05 Dec 2022
8. Destination earth. In: Shaping Europe's Digital Future. https://digital-strategy.ec.europa.eu/en/policies/destination-earth. Last accessed 02 Dec 2022
9. Oligschlager, P., Strang, M., Winter, T., Bergsma, L., Van der Koelen, O.: Real Estate Innovations Overview, 6th edn. KPMG Real Estate Advisory (2021)
10. Hamilton, I., Kennard, H., Rapf, O., Kockat, J., Zuhaib, S., Toth, Z., Barrett, M., Milne, C.: 2022 Global Status Report for Buildings and Construction. United Nations Environment Programme, Nairobi (2022)

11. Hackney, W., Williams, K., et al. (eds.): Digital Twins For All. The Australia | New Zealand Digital Twin Blueprint, Smart Cities Council, Wellington (2021)
12. One third of the world's remaining safe carbon budget could be determined by urban policy decisions in the next five years. C40 Cities. https://www.c40.org/news/one-third-of-the-world-s-remaining-safe-carbon-budget-could-be-determined-by-urban-policy-decisions-in-the-next-five-years/ (2015). Last accessed 15 Dec 2022
13. National Digital Twin Programme. https://www.cdbb.cam.ac.uk/what-we-did/national-digital-twin-programme. Last accessed 10 Dec 2022
14. Plummer, D., Kearney, S., Monagle, A., Collins, H., Perry, V., et al.: Skills and Competency Framework—Supporting the Development and Adoption of the Information Management Framework (IMF) and the National Digital Twin. Centre for Digital Built Britain and Digital Framework Task Group, United Kingdom (2021)
15. Lawton, G.: Microsoft paves digital twins' on-ramp for construction, real estate. VentureBeat. https://venturebeat.com/business/microsoft-paves-digital-twins-on-ramp-for-construction-real-estate (2021). Last accessed 12 Dec 2022
16. Baanders: What is an ontology?. Azure Digital Twins | Microsoft Learn. https://learn.microsoft.com/en-us/azure/digital-twins/concepts-ontologies (2022). Last accessed 10 Dec 2022
17. Hart, B.: How game engines are revolutionizing digital twins. https://industrytoday.com/how-game-engines-are-revolutionizing-digital-twins/ (2021). Last accessed 30 Apr 2022
18. MX3D Bridge. MX3D. https://mx3d.com/industries/infrastructure/mx3d-bridge/ (2021). Last accessed 10 Dec 2022
19. Digital twin of the world's first 3D printed stainless steel bridge. The Alan Turing Institute. https://www.turing.ac.uk/research/research-projects/digital-twin-worlds-first-3d-printed-stainless-steel-bridge (2017). Last accessed 30 Apr 2022
20. Wellington Digital Twin. Buildmedia. https://buildmedia.com/work/wellington-digital-twin. Last accessed 02 Dec 2022
21. Bolton, A., Enzer, M., Schooling, J., et al.: The Gemini Principles: Guiding Values for the National Digital Twin and Information Management Framework. Centre for Digital Built Britain and Digital Framework Task Group, United Kingdom (2018)

The Right Algorithm for the Right Shape

An Algorithmic Framework for Efficient Design and Conception of Building Facades

Inês Caetano, António Leitão, and Francisco Bastos

Abstract Buildings are a critical element of civilization, within which we spend over around 70% of our lifetime, but also one of the main contributors to the greenhouse effect. It is therefore important to ensure their design guarantees good indoor conditions, while minimizing the environmental footprint. Among the different building elements, the facade is one that most influences these two requisites and thus its design requires, in addition to the traditional aesthetic and functional requirements, the integration of performance criteria from early design stages. However, there are still some barriers to this integration, such as the limited flexibility of design tools, the need for multiple analysis and optimization tools, and their high computational cost. Recent computational design approaches, such as Algorithmic Design (AD), have been facilitating the combination of creative processes with the search for better performing and more sustainable design solutions. However, these approaches require programming skills, which most architects do not have. To maximize its potential for architectural design, efforts should be made to reduce the complexity of AD and approximate it to the architects' design practice. We address this by proposing an AD methodology and algorithmic framework for facade design that encompasses its different stages, from conceptual design to manufacturing, and requirements, such as aesthetics, environmental performance, comfort, and costs, among others, while supporting the variability and diversity typical of architectural design problems. By combining the framework's ready-to-use algorithms, multiple design scenarios can be considered, and various design requirements addressed, helping to achieve the goals established by both the 2030 Agenda and Industry 4.0.

I. Caetano (✉) · A. Leitão
INESC-ID/Instituto Superior Técnico, University of Lisbon, 1000-029 Lisbon, Portugal
e-mail: ines.caetano@tecnico.ulisboa.pt

A. Leitão
e-mail: antonio.menezes.leitao@tecnico.ulisboa.pt

F. Bastos
CiTUA/Instituto Superior Técnico, University of Lisbon, Lisbon, Portugal
e-mail: francisco.bastos@tecnico.ulisboa.pt

Keywords Computational and parametric design · Performance-based architecture · Algorithmic workflow · Computational design analysis · Digital fabrication

United Nations' Sustainable Development Goals 9. Build resilient infrastructure, promote inclusive and sustainable industrialization and foster innovation · 11. Make cities and human settlements inclusive, safe, resilient and sustainable · 12. Ensure sustainable consumption and production patterns

Abbreviations

AD	Algorithmic Design
ADO	Architectural Design Optimization
AEC	Architecture, Engineering, and Construction
CNC	Computerized Numerical Control
DF	Digital Fabrication
GUI	Graphical User Interface
IEQ	Indoor Environmental Quality
SDGs	Sustainable Development Goals

1 Introduction

Buildings are one of the main CO_2 emitters and spenders of energy resources [1–3] but also a critical element of civilization, within which we spend over around 70% of our lifetime [4]. Half of these emissions results from the operational costs of buildings, such as heating, cooling, lighting, and ventilation; and the other half from the construction processes, namely the production, transport, and manufacture of building elements, and material disposal [2]. To meet the United Nation's Sustainable Development Goals (SDGs) [5] building design must provide good indoor conditions while minimizing the environmental footprint.

Among the different building elements, the facade is one that most influences the buildings' Indoor Environmental Quality (IEQ) and environmental performance [4, 6–10]. Therefore, in addition to aesthetic and functional requirements, facade design requires the integration of performance criteria from early design stages, as well as manufacturing-related strategies assessing the construction viability of the developed solutions.

Unfortunately, there are still barriers to such integration. One is the limited flexibility of design tools, hampering the application of iterative design changes in the search for improved solutions. Another is the large computational resources and specialized knowledge needed by analysis and optimization tools [11, 12], resulting in time-consuming and computationally-expensive processes whose results are difficult to interpret. A third barrier emerges when it becomes necessary to use multiple tools that hardly interoperate with each other, increasing the propensity for information loss and error accumulation [12]. A last critical barrier is the time and effort needed to address manufacturing and cost-related constraints [13], especially when dealing with unconventional design solutions. All these obstacles make the integration of performance and manufacturing-related design variables incompatible with project deadlines and resources, hindering the development of more ambitious facade designs whose performance goes beyond minimum regulatory requirements.

Advancements in computational design approaches have improved the integration of performance criteria within the design practice, making it easier for architects to combine creative processes with the search for better solutions in terms of environmental performance, IEQ, production costs, among others [14–16]. Algorithmic Design (AD), a design process based on algorithms [17], is one such approach that allows for (1) greater design flexibility, (2) the automation of repetitive and error-prone tasks, (3) the integration of different types of data in a single model, (4) the coordination of different design tools, such as modelling, analysis, and fabrication tools, and (5) the automatic extraction of technical information. AD has large potential for reducing the environmental footprint and meeting the 2030 SDGs Agenda for a more sustainable built environment [5].

Nevertheless, AD is an abstract approach that requires programming experience, which most architects do not have. To maximize AD's potential for architectural design and motivate the search for more sustainable design solutions, efforts should be made to reduce the technical complexity of AD strategies and make them more accessible to a wider audience. Considering the current state of the art, it is therefore important to systematize and structure the algorithmic generation of design solutions in an architectural-oriented methodology that considers the diversity of architectural design problems and their different aesthetic, performance, and construction requirements. To successfully deal with the variability and context-specificity typical of architectural practice, the proposed methodology must have enough flexibility to adapt to multiple design briefs and workflows, while providing control over the coordination of different types of information and over the translation of digital designs into actual constructions [18–20].

This chapter addresses this goal by placing particular emphasis on the field of facade design due to the aesthetic and performance relevance of this building element, as well as due to its design complexity and impact on the projects' feasibility. Given the universality and problem-solving capabilities of mathematics, this research uses its language to address the above-mentioned goal, adopting a strategy that encompasses:

- The architects' creative process, by systematizing the design complexity of current facade design strategies, while helping with the algorithmic development of new facade design solutions.

- The coordination of different conceptual and performance requirements, by automating analysis and optimization processes from early design stages, providing reliable feedback on the solutions' performance and aesthetical quality.
- The context specificity of architectural design problems, by guiding the selection of geometry- and performance-related algorithms based on the type, scale, and complexity of the problem addressed.
- The materialization of the resulting solutions, informing about their viability in terms of cost, waste, and resources, while automating the production of technical documentation for the selected fabrication strategy.

To that end, an extensive investigation on contemporary architectural processes is first presented, which discusses the impact of the growing environmental awareness and new computational design means on both facade design strategies and the increased complexity of construction processes. A mathematics-based methodology is then proposed, together with its implementation in an AD framework containing different algorithms embracing the complexity of procedural modelling, performance simulation, design optimization, and fabrication techniques. To evaluate the potential of the proposal to support design workflows encompassing creative intents, performance requirements, and fabrication constraints, from conceptual design to manufacturing stages, a set of case studies is then presented and discussed. Finally, the chapter concludes with considerations on the case studies' results, elaborating on the proposal's suitability to meet the 2030 SDGs Agenda and approximate the Architecture, Engineering, and Construction (AEC) field to the Industry 4.0 paradigm.

2 Facing New Challenges

The need to successfully respond to the growing concern with the buildings' ecological footprint motivated the increasing integration of design performance in architecture, triggering new design approaches that go beyond aesthetic and functional levels [21, 22]. As a result, design practices that are more environmentally-aware emerged [23]. Under the names of performance-based, performance-driven, performance-oriented, or even performative design, this design paradigm has been gaining ground in the literature [24–27] as well as in architectural practice, particularly in the design of facades due to the aesthetic and environmental relevance of this building element [7].

2.1 Environmental Concerns

We are currently facing an environmental problem and the AEC sector is one of the main contributors. According to the literature, buildings are responsible for 50% of natural resources consumption, 42% of the total energy consumption, and 35% of greenhouse gas emissions [2], which are among the main contributors for climate change and global warming [28]. Half of these emissions result from the operational costs of buildings, such as heating, cooling, lighting, and ventilation, and the other half from construction processes, namely the production, transport, and manufacture of building elements and the disposal of building materials [2]. In 2012, for instance, buildings were responsible for around 75% of Europe's energy consumption, and almost 70% of it was for space heating [3].

Given the urgent need to minimize the buildings' ecological footprint [3, 4] and stop the growing trend of CO_2 emissions [3], several regulations and incentives were established worldwide [29]. The Building Research Establishment Environmental Assessment Method (BREEAM), first published in 1990, is the oldest method for assessing, rating, and certifying the buildings' sustainability regarding a wide range of environmental issues. This method was followed by several other regulations, including the Kyoto protocol signed in 1997, one of the first initiatives to limit CO_2 emissions; the European Union's Energy Performance of Buildings Directive (2002/91/EC, 2010/31/EU, and COM/2016/0765) targeting the improvement of buildings' energy performance; the U.S. Green Building Council's certification Leadership in Energy and Environmental Design (LEED); and the Japanese Comprehensive Assessment System for Built Environment Efficiency (CASBEE).

The existing legislation, however, requires architects to evaluate the performance of their designs regarding different criteria [11] to ensure the proposed metrics are met [12]. As a result, building design has become an even more demanding task, since it must simultaneously respond to the already existing aesthetic, structural, and IEQ requirements, plus the increasing number of performance metrics established worldwide.

2.2 *The Role of Performance Analysis*

In the last decades, several analysis tools were released to help architects evaluate the performance of their designs regarding different criteria. Examples of analysis tools include EnergyPlus and TRNSYS for whole-building energy simulations, Radiance for (day)lighting analysis, and DAYSIM for climate-based daylight simulations. Using these tools, architects become more aware of their designs' ecological footprint, as well as the impact of design changes on the solutions' environmental performance [4].

Unfortunately, many practitioners still do not use any kind of digital analysis tool in their design processes and those who do rarely benefit from these tools to support their creative process, using them instead to validate the performance of already well-defined solutions [12, 28, 30]. Moreover, obtaining accurate analysis results remains a challenging task due to the wide variety of factors that affect building performance

[4], including external ones (e.g., climate, geographic location, site conditions, etc.) and internal ones (e.g., occupants' behavior [31], spatial orientation [32], envelope transmittance [33], etc.).

The need for specialized knowledge, the lack of intuitive Graphical User Interfaces (GUI), the long computation times, the poor interoperability with the architects' preferred tools, and the idea that analysis tools restrict creative processes are some of the barriers to their widespread adoption [4, 11, 12]. Furthermore, the performance evaluation of a building requires the laborious and time-consuming production of an analytical model containing only the data needed for the intended analysis. Additionally, most analysis tools are single domain, forcing the use of multiple tools to evaluate different criteria, each one requiring a specific analytical model [30], resulting in a process that is prone to information loss and error accumulation [12]. Since a performance-based design process needs to analyze multiple design instances, the number of analytical models required grows considerably. Given the constant pressure for short deadlines in the AEC industry, evaluating an acceptable sample of possible solutions is, usually, an impracticable scenario [11].

To address the need for a faster and more reliable data-flow process suiting the iterative nature of architectural practice and minimize the alternation between different analysis tools, some modelling tools started to integrate their own analysis strategies [34]. However, the proposed solutions present some limitations in terms of (1) modelling flexibility and accuracy, particularly in representing and analyzing less conventional solutions and construction schemes, and (2) information support, often making no suggestions about which design direction to follow and how to translate analysis results into design changes.

2.3 Searching for the Best Performance

Architectural Design Optimization (ADO) was motivated by the need to more effectively explore the design space in the search for better-performing design solutions [11]. By minimizing the buildings' ecological footprint, ADO can contribute to reduce the negative impact of the AEC sector [5].

In addition to simplifying real-world complex problems, often dealing with multiple conflicting requirements [15], into mathematical ones, ADO requires the iterative remodeling of designs and their subsequent performance evaluation to check if the fitness goals are met [28]. As multiple conflicting goals often result in a set of possible solutions that are not optimal for all requirements [15], ADO does not ensure the global optimum is reached [11]. Nevertheless, it increases the chances of finding it or, at least, of getting close to it [29]. In any case, the probability of obtaining more sustainable solutions is still much higher than that of traditional practices where no optimization is applied [11, 35]. Moreover, these processes are important to remind architects of possible design solutions that might otherwise not occur to them [36].

To successfully reduce the environmental impact of buildings, architects need to adopt either passive or active design strategies that consider both the existing performance requirements and the design variables affecting these [28] from early design stages, where major performance improvements can be achieved [37]. Nevertheless, the strategies employed often do not affect the buildings' shape, let alone guide the exploration process in an environmentally driven way. Moreover, only a few requirements are usually considered at initial design stages, e.g., aesthetic and functional ones, the others being typically postponed to later stages, where the design idea is already well-established [28, 38]. However, at later stages, most design changes are quite complicated, or even impossible, due to the lack of flexibility of most design models [38]. As in most cases architects explore the design space by manually adapting the solutions according to a few analysis results [23], only a few and small design changes are often evaluated due to time and effort constraints [12, 14]. As a result, the efficacy of the ADO process often becomes compromised [38], potentially leading to unrealistic or low performance results [28].

3 Sketching Through Algorithms

The technological evolution of the last decades has provided architects with the means to explore unprecedented design solutions, facilitating the production of unconventional facade elements of different shapes, patterns, and materials [39]. In addition to enhancing the architects' creative process, emerging design technologies have been allowing the integration of building performance requirements in an environmentally aware perspective. Combined with the architects' innate desire to go beyond conventional geometries and the usually tight time and economic constraints of architectural projects [12], the process of designing building facades grew in complexity [40]. Fortunately, part of this complexity can be reduced through AD, which provides the flexibility needed to coordinate multiple design requirements and data in the search for more sustainable facade design solutions [3, 4, 6].

3.1 Extending Design Creativity

AD is a design approach based on algorithms that has been gaining prominence in architecture [41–46] and that greatly contributes to the movement of the AEC sector towards both the Industry 4.0 and the United Nations' 2030 Agenda goals. The increased design efficiency, flexibility, accuracy, and automation of this design approach have been motivating a new generation of architects to increasingly adopt AD strategies in their design practice [43]. To do so, however, architects have to acquire new skills, such as learning programming techniques, which often ends up being a barrier to AD's widespread adoption [47].

Two main AD paradigms currently stand out, the main difference between them being the type of algorithmic representation used, which can be textual or visual.

Among the two, the visual paradigm is more popular within the architectural community but current practices have evidenced several shortcomings that hinder the development of large AD programs [45, 48, 49], making the long-term use of visual-based AD strategies difficult [50]. The absence of abstraction mechanisms addressing scalability issues [49, 51], the poor intelligibility and difficult manipulation of the resulting AD programs [48, 52–54], the accentuated drop in performance when executing large AD programs [55–57], and the lack of version control mechanisms supporting collaborative design practices [52, 58] are among the limitations of visual AD. All these shortcomings have motivated the transition from this paradigm to the textual paradigm to benefit from the expressiveness and scalability of textual programming strategies [49]. Accordingly, some of the existing visual-based AD tools were extended with textual programming mechanisms [49, 51], such as loop iterations, recursive functions, and higher-order functions. Nevertheless, in most cases, it remains difficult to develop large-scale AD programs in these extended tools due to their inability to support the division of an AD file into multiple files.

Although learning textual programming has become an evident need, the transition from the visual to the textual paradigm is not trivial, often lacking the consolidation of important theoretical bases of textual programming strategies; a drawback that gets even worse when addressing more complex design problems. Moreover, as it still takes time to achieve the level of programming proficiency needed to deal with large-scale design problems, which typically involve multiple context-specific design requirements, architects usually spend more time solving programming problems than in creative and design exploration processes. Also, given the uniqueness and variability of architectural design problems, architects often face unanticipated changes that are not contemplated in the structure of their AD program, lacking the parameters needed to modify the model in the desired way and thus forcing its redesign and delaying the design process [59].

3.2 Algorithmic Analysis and Optimization

The growing awareness on climate change and the need to reduce the AEC sector's ecological footprint has motivated the implementation of several building regulations aiming at meeting the United Nations' SDGs [5]. This has forced architects to increasingly resort to analysis tools to check if their designs meet the established criteria and, when they don't, to rethink their designs and repeat the process, considerably increasing the complexity of their design processes.

With the advent of new computational design approaches, such as AD, these processes can be improved. Besides facilitating design changes and supporting higher levels of design complexity, AD allows automating labor-intensive and error prone tasks such as those of analysis processes, particularly, the generation of analytical models [60] and the setup of analyses whenever the design changes. Therefore, by using AD, it becomes possible to not only perform several iterative design analyses with less time and effort [15], but also automate them in optimization routines and evaluate larger design spaces in the search for better-performing solutions [61].

To address the need to use multiple design tools and evaluate the solutions regarding different metrics, several AD tools started to integrate functionalities that embrace different analysis and optimization strategies. One example is Grasshopper's plugins for (1) structural analysis and form-finding, e.g., Kangaroo, Karamba3D, Millipede, and Peregrine; (2) lighting and thermal analysis, e.g., Ladybug, Honeybee, DIVA, and ClimateStudio; and (3) design optimization, e.g., Galapagos, Goat, Octopus, Wallacei, and Opossum. Other examples include Dynamo's addons for structural, energy, daylighting, and thermodynamic analysis, and optimization.

Nevertheless, despite AD facilitating the application of design analysis and optimization routines since early design stages, their use remains difficult due to (1) the uniqueness and conflicting nature of most design requirements [29]; (2) the need to explore wide design spaces in order to achieve acceptable results [29]; (3) the technical complexity and poor intuitiveness of most analysis/optimization tools; and (4) the need to convert ADO problems into abstract, mathematical formulations [30, 62]. Despite the existing AD tools to solve some of these barriers, they are mostly based on the visual programming paradigm and thus quickly become unable to cope with the complexity typical of large-scale architectural design problems.

The need to make design analysis and optimization strategies more accessible to architects from early design stages has been increasingly addressed in the literature [4, 29, 35, 63, 64]: Schlueter and Thesseling [65], for instance, assessed the integration of a prototype tool to assist energy/exergy calculations from early stages; Petersen and Svendsen [66] presented a proposal to help designers make informed design decisions at early stages regarding energy and inside spaces' environmental performance [66]; Madrazo et al. [67] proposed a method to recover information from repositories, hold calculation results, and support early-stage design decisions; Attia et al. [68] developed an energy-oriented software tool to support the design of zero energy buildings in an Egyptian context; Lin and Gerber [69] presented a framework to guide early-stage design exploration and decision-making processes based on energy performance; Negendahl [70] proposed an alternative method to the current IFC implementation to support early stages design processes combining different design, AD, and analysis tools; Finally, Konis et al. [71] presented a framework to improve daylighting and natural ventilation performances at early design stages.

The proposed solutions, however, do not entirely solve the challenges of early-stage ADO problems due to still presenting (1) limited modelling flexibility; (2) reduced interoperability; (3) few performance criteria; and (4) a narrow scope of application.

4 The Case of Building Facades

Architectural design problems are unique because, besides involving several design requirements that can be global or context-specific, straightforward or abstract, and fixed or evolving, they must respond to unpredictable creative intents and design briefs [3, 4, 40]. The design of building facades is a particularly relevant case because of the environmental impact of this architectural element [7, 8], as well as its design complexity [6, 40] and critical role in improving the IEQ of buildings. Nevertheless, it is often the case that the design means used do not provide the flexibility needed to quickly explore and evaluate a wide range of design solutions, hindering not only the architects' creative process but also the search for more sustainable solutions.

Among its different applications, AD has proved to be particularly advantageous for facade design processes, providing the flexibility needed to coordinate multiple design requirements [21, 72] and thus achieve design solutions with reduced environmental impact and minimum energy demands. Moreover, AD motivated the gradual shift towards Industry 4.0, allowing not only the automation of manufacturing and construction processes, reducing their production times, energy consumption, and resource waste, but also the production of unconventional facade elements whose manufacture was previously not viable [73–75].

4.1 Geometric Exploration

To reduce the complexity of AD and facilitate the algorithmic development of facade design solutions, several AD tools were released. One example is ParaCloud Gem, a generative 3D design tool that provides features to (1) map 3D elements on a mesh, (2) subdivide and edit surfaces, (3) integrate fitness requirements, and (4) 3D print the resulting solutions. Another example is Dynamo's packages Quads from Rectangular Grid, Ampersand, Clockwork, LunchBox, MapToSurface, Pattern Toolkit, and LynnPkg, which include features for surface paneling, mapping elements on a surface, and pattern creation. A last example is Grashopper's multiple plugins, such as: (1) PanelingTools, which provides surface paneling functionalities and rationalization techniques for analysis and fabrication; (2) LunchBox, which integrates functionalities to explore mathematical shapes, surface paneling, and wire structures; (3) Weaverbird, which contains mesh subdivision procedures and mechanisms to help prepare meshes for fabrication; (4) Parakeet, which provides functionalities to develop algorithmic patterns resulting from tiling, geometric shapes and grid subdivisions, edge deformation, etc.; and (5) SkinDesigner, which includes mechanisms to produce facade designs made of repeating elements.

Despite facilitating algorithmic activities typical of facade design processes, such as creating point-grids on a surface, mapping elements in different ways, manipulating the elements' size, shape, rotation, etc., by applying rules or attractors, these tools still present some limitations. The first shortcoming is the fact that most of these tools are based on visual programming and thus suffer from the limitations of this AD paradigm [48, 49], particularly, scalability. Another limitation is the tools' limited ability to directly address relevant facade design concepts such as materiality and tectonic relation between elements, often only addressing generic panelization, subdivision, and population of surfaces. Finally, most of these tools are limited by the available predefined operators, which can hardly be configured by the user to respond to more specific problems [45].

4.2 Analysis and Optimization

Building facades are one of the most optimized elements in architecture because of their important role in the buildings' environmental performance [4, 11, 12, 29, 63, 76–79], their design greatly contributing to meet the United Nations' goal of making the production and consumption of cities sustainable. Figure 1 presents a set of architectural examples whose building envelope design was guided by performance.

To deal with the design complexity of this building element and its multiple and context-specific design requirements, several facade-oriented optimization methodologies have been proposed in the literature. These include Bouchlaghem's [80] computer-based model to design building facades based on their thermal performance; Wang et al. [81] multi-objective optimization model to design green buildings; Ochoa and Capeluto's [79] model to materialize design ideas based on climate

Fig. 1 Institut du Monde Arabe (©authors); Campus Kolding of the University of Southern Denmark (©Henning Larsen); Al Bahr Towers (©Andrew Shenouda); Louvre Abu Dhabi (©authors); City Hall (©authors)

and visual comfort strategies; Gagne and Andersen's [82, 83] tool to guide facade design exploration based on illuminance and glare levels; Jin and Overend's [84] optimization prototype to identify optimal facade designs regarding functional, financial, and environmental requirements; Gamas et al. [85] study on the use of evolutionary multi-objective algorithms to optimize building envelopes in terms of thermal and daylight performance; Elghandour et al. [86] method to improve facade design daylight performance; and finally, Pantazis and Gerber's [87] agent-based framework for generating, evaluating, and optimizing facade designs from early design stages.

Despite the extensive literature, most proposals are (1) context-specific [79, 81, 83], (2) have limited modelling flexibility [80, 81, 83, 84], (3) address a single requirement (mostly energy consumption) [80, 83], and (4) require knowing in advance which optimization technique best suits a specific problem (a task that needs experience and specialized knowledge) [81, 84]. These limitations therefore make the proposals' widespread application often difficult, forcing architects to master and use an extensive range of tools/strategies to address the multiplicity of requirements guiding facade design problems. Moreover, some proposals do not present a GUI displaying the resulting solutions [80, 81, 84], making their use little intuitive and insufficiently user-friendly, or do not directly communicate with the design tools architects use [79–81, 84], often leading to interoperability issues and increasing the efforts associated with the transition between tools. The existing exceptions [82, 86, 87], however, are based on visual programming, thus sharing its limitations [49].

5 Making Digital Real

The desire for unprecedented shapes and structures has always been present in architecture, becoming further accentuated with the emergence of digital design tools, which provided architects with the freedom to design any shape they wanted. However, given the limitations of traditional construction methods, the realization of such shapes is often compromised. Digital Fabrication (DF) strategies are gradually changing this reality, albeit still with some limitations [88]. By combining DF with AD strategies, architects can control the entire design-to-manufacturing process in an informed way, reducing the distance between design thinking and making [89, 90], which is critical to bring the AEC sector closer to the Industry 4.0 and ensure sustainable industrialization and production.

5.1 Digital Fabrication Strategies

DF encompasses fabrication strategies based on Computerized Numerical Control (CNC) machines that automate manufacturing processes and the production of building elements of varying geometries and materials. These methods allow architects to control the entire design-to-fabrication process in an entirely digital manner [89] and not only achieve higher levels of design complexity and accuracy, but also produce nonstandard building elements that would otherwise be unviable to produce [91].

Ideally, DF strategies would enable the conversion of traditional manufacturing processes, where only the mass production and assembly of standard elements is economically viable [89], into new ones benefiting from mass-customization strategies to produce multiple unique elements at low costs [92]. This scenario, however, remains a challenge in the AEC industry because of the uniqueness of architectural projects, which require the production of multiple context-specific elements [92], and the limitations of the available manufacturing technologies in terms of cost, machining time, scale, material waste, and special spatial conditions. Nevertheless, their gradual cost decrease is motivating their increasing use in architecture [93].

DF encompasses a wide variety of manufacturing strategies that vary in terms of (1) the process used to shape the elements, e.g., by adding, removing, cutting, or deforming materials; (2) materials supported, (3) element shapes and scales allowed; and (4) types of surface finishing, e.g., smooth, textured, printed, perforated, bumped, etc. DF strategies are generally categorized into five groups of manufacturing processes, namely *additive, subtractive, formative* [91, 93–96], *cutting* [89, 97, 98], and *robotic* [99].

The first one, *additive*, is based on the addition of material layers to produce the desired shape [91], requiring the translation of the digital model into a sequence of two-dimensional paths [89]. 3D printing is the most popular *additive* process in architecture but there are other techniques available, such as stereolithography, fused deposition modelling, laser sintering, and digital light processing [91, 96]. These methods have the advantage of directly converting digital models into physical elements without requiring additional devices and allowing the production of a wide range of shapes in a viable way [96]. Moreover, the available machines are often silent, produce reduced material waste, and do not require programming expertise [89]. Among their limitations are the difficulty to produce large-scale building elements [93], the poor surface finishing quality achieved, and the large production times [96]. Examples of 3D printed facade elements include those of the *Arachne project* in China (Fig. 2 left), the *Cabin of 3D printed Curiosities* in California (Fig. 2 middle); and the *Europe Building* in Amsterdam (Fig. 2 right).

The second group, *subtractive*, uses electro-, chemically, or mechanically reduced techniques to remove or separate particles of raw material from an existing solid [91, 98] to achieve the desired shape [93]. In architecture, CNC milling and routing processes are the most applied techniques [93]. Compared to *additive* processes, these technologies support a wider range of element scales and materials, have a higher geometric precision and production efficiency [89], but also produce a lot more material waste [91]. These methods have been applied in architecture, for instance, to (1) carve facade elements, e.g., the cork panels of a house in Aroeira, Portugal (Fig. 3 left); (2) perforate and bump sheet facade panels, e.g., the metal panels of *de Young Museum* in San Francisco (Fig. 3 middle); and (3) produce customized molds to cast facade elements, e.g., the concrete facade of *MaoHaus* in Beijing (Fig. 3 right).

Fig. 2 Additive manufacturing: Arachne 3D printed facade (©Archi-Solution Workshop); House of 3D Printed Curiosities (©Matthew Millman Photography/Emerging Objects); EU Building 3D printed facade (©Ossip van Duivenbode)

Fig. 3 Subtractive manufacturing: cork facade (©GenCork and Sofalca); De Young Museum (©David Basulto via flickr); MaoHaus facade (©XiaZhi)

The third group, *formative*, uses mechanical forces to reshape or deform materials into the intended shape [93], often resorting to heating to make the material adapt to the new geometry and then to cooling to keep the new geometry stable [89]. CNC folding, CNC bending, CNC punching, hydro morphing, and welding are some examples [91]. In architecture, these techniques have been mostly applied in the manufacturing of (1) unconventional metal panels [92], such as those of the *Experience Music Project* in Seattle (Fig. 4 left), and (2) heat-slumped or heat-bended glass facade elements [100], such as those of the *Holt Renfrew flagship store* in Vancouver

Fig. 4 Formative manufacturing: Experience Music Project (©Jon Stockton/CC BY-SA 3.0); Holt Renfrew flagship store in Vancouver (©Marc Simmons/FrontInc); Elbphilharmonie Hamburg (©Bahman Engheta/CC BY-SA 4.0)

(Fig. 4 middle) and the *Elbphilharmonie* in Hamburg (Fig. 4 right), respectively. Nevertheless, given the still expensive price of both the machines and materials used by these methods [91], their use remains limited in the field.

The next group, *cutting*, involves the extraction of two-dimensional planar elements from surfaces or solids by using strategies like contouring, triangulation, and unfolding, among others [91]. These processes follow a set of instructions provided by the digital model to produce flat components with the desired shape [93], often resorting to laser-beam, plasma-arc, and waterjet technologies. *Cutting* is a very popular strategy in the field, probably the most used one [89, 98], especially to produce complex facade panel patterns. Among its advantages are its geometric precision and both its reduced cost and production times. Among its limitations are the limited range of materials and thicknesses supported and the need to adapt the technology used accordingly [89, 98]. Examples of *cutting* applications include the ceiling of the *Trumpf Campus Gatehouse* in Stuttgart [101] and the facade panels of the *Megalithic Museum* in Mora, Portugal (Fig. 5 left examples).

The last group involves the use of *robotic* arms or drones to accurately place elements in layers by controlling their location and position. These methods make it possible to reduce or even remove the lack of accuracy typical of manual assembly processes, allowing a rigorous correspondence between the intended design and its final product [95]. Examples of *robotic* strategies include the brick facades of the *Chi She Gallery* and the *Winery Gantenbein* in Switzerland (Fig. 5 right examples).

Fig. 5 Cutting and robotic manufacturing: House 77 by dIONISO LAB (©FG|SG Fotografia de Arquitetura); Megalithic Museum by CVDB Arquitetos (©Fernando Guerra|FG + SG); Chi She Gallery (©Su Shengliang); Winery Gantenbein (©Christoph Kadel via flickr)

5.2 Balancing Creativity and Feasibility

Despite the currently available mass-production techniques to manufacture non-conventional elements at low cost, none of them is entirely suitable to deal with the geometric diversity of architectural design, which usually requires the manufacturing of hundreds or thousands of non-standard elements [92] that are often project-specific. To make the construction of free-form shapes and complex facade patterns possible, architects have been increasingly adopting *geometric optimization* techniques [102] in their design practice. These strategies allow architects to gain more insight and control over their designs [40], facilitating the latter's gradual adaptation until reaching the desired feasibility [13]. Popular examples of geometric optimization strategies for architectural design include *design rationalization* and *surface paneling*.

Design rationalization is an example of a *geometric optimization* strategy that focuses on subtly adjusting the building elements that are expensive to produce until meeting the established economic and construction requirements and without compromising the design's aesthetics [92, 103]. Based on the literature [13, 40, 104–106], design rationalization can vary in terms of temporal application in the design process, i.e., before, during, or after the design development process, and target of the rationalization process, i.e., the building elements to which it is applied, e.g., frames, facade panels, wall tiles, and shading devices. Figure 6 presents some architectural examples resulting from the application of rationalization strategies at different design stages.

Panelization, or *paneling*, is a geometric optimization strategy that focuses on dividing a large surface into smaller panels of constructable size and acceptable cost, while preserving the design intent [92, 106, 107]. This strategy involves two dependent tasks: the segmentation of the original shape into smaller pieces and the approximation of each smaller piece to a shape that can be manufactured at a reasonable cost.

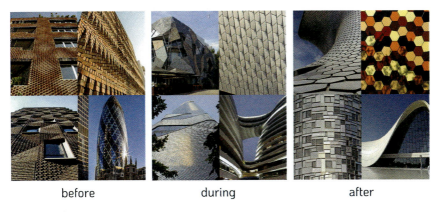

Fig. 6 Architectural rationalization (from top-left to bottom-right): DIY For Architects (©Sstudiomm); South Asian Human Rights Documentation Centre (©AnagramArchitects); Federation Square by LAB architecture studio (©authors); MAAT (©Hufton + Crow/AL_A); Museo Soumaya by FR-EE/Fernando Romero Enterprise (©naturemyhome via Needpix); Spanish Pavilion 2005 (©Edmund Sumner); 290 Mulberry Street by SHoP Architects (©Amy Barkow); 30 St Mary Axe by Foster + Partners (©Suhail Akhtar via flickr[1]); Bangkok Central Embassy (©Hufton + Crow/AL_A); Galaxy SOHO by Zaha Hadid Architects (©authors); 100 11th Avenue by Ateliers Jean Nouvel—delivery architect: Beyer Blinder Belle Architects (©Philippe Ruault); Heydar Aliyev Center by Zaha Hadid Architects (©Aleksandr Zykov via flickr)

Dividing a surface into smaller planar surfaces of different polygonal shapes, like triangular, quadrilateral, and hexagonal, is the cheapest paneling strategy. Another strategy involves the division of the original surface into smoothly bent stripes, also known as single-curved panels or developable surfaces, that can be produced by simply bending a flat piece of metal sheet; a technique that has been showing gradual improvements over time [106]. Another paneling strategy focuses on dividing the surface into double-curved perfectly fitting panels, resulting in smooth curved surfaces with a high finishing quality. Among the three, the last strategy is the most precise but also the most expensive as it often requires the production of several customized molds [107]. Figure 7 presents some examples of paneling strategies organized by type.

Despite the variety of existing technologies and architectural examples, the production of large-scale free-form facades with either unconventional or intricate geometric patterns remains a challenging, and often expensive, task [107–110]. Moreover, the existing literature and practical examples mostly focus on simple patterning techniques, i.e., triangular, quadrangular, and hexagonal panels, rarely considering other shapes or geometric patterns.

To make the digitally produced design solutions feasible, several tools were released to facilitate the manufacturing and assembly of shapes with higher levels of design complexity. These include, among others, the Grasshopper's plugins HAL, FabTools, BowerBird, OpenNest, Kuka|PRC, Xylinus, Droid, RoboDK, Robot

---

[1] https://www.flickr.com/photos/192540662@N04/.

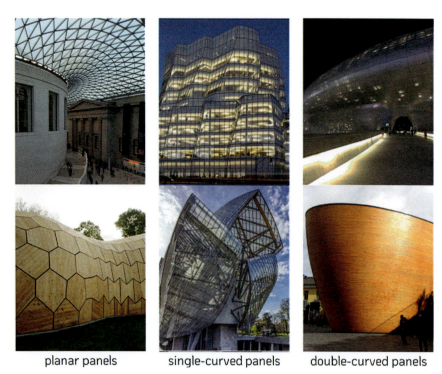

planar panels   single-curved panels   double-curved panels

Fig. 7 Paneling strategies (from top-left to bottom right): British Museum Great Court (©authors); IAC building (©Peter Miller via flickr); Dongdaemun Design Plaza (©authors); Landesgartenschau Exhibition Hall (©ICD/ITKE/IIGS University Stuttgart); Foundation Louis Vuitton (©BarrieT via flickr); Kamppi Chapel (©authors)

Components, Robots, Bark beetle, and Ivy; Dynamo's addons DynaFabrication, Fabrication API, 3BMLabs.DigiFab, and ParametricMonkey; and Blender's addon Laser Slicer. However, as these plugins only target the visual programming paradigm, the manufacturing of more complex solutions is often difficult [45, 48, 49, 51, 55–57]. Moreover, these plugins are mostly tool-specific, requiring the use of tools to assess different construction schemes. Additionally, they do not entirely automate the design-to-fabrication conversion nor the extraction of technical documentation, since they often depend on manual- or script-based interventions that are laborious, time-consuming, and error prone. Given the uniqueness of architectural design problems, these interventions can hardly be reused in different projects without major modifications, thus hindering the testing of different manufacturing possibilities to assess their aesthetic and environmental impact.

6 The Mathematics of Facades

AD has the potential to improve AEC's ecological footprint. However, it has also a higher level of complexity and abstraction that hampers its widespread use. To motivate the adoption of AD, it is critical to provide strategies systematizing and structuring the algorithmic generation of design solutions in an architectural-oriented way. We address this by proposing a mathematics-based methodology and framework to support the algorithmic development of building facades from conceptual to later design stages. The proposed solution considers the wide variety of design briefs as well as different aesthetic intents, performance requirements, and fabrication strategies. Its use promises to decrease the time and effort spent with the algorithmic implementation of new facade designs, while providing the flexibility needed to handle the design complexity and variability of facade design processes.

With this proposal we aim to promote informed design practices aligned with the need to reduce the environmental footprint of the AEC sector. By facilitating design experimentation and the coordination of different types of data and requirements, we expect architects to evaluate wider design spaces within an acceptable time and effort, increasing the chances of achieving more sustainable solutions in terms of energy consumption, environmental impact, waste production, etc. We also aim to increase the control over design-to-manufacturing processes, reducing the gap between what can be digitally explored through AD and its subsequent manufacturing. Lastly, by democratizing design exploration and manufacturing in an AD workflow entirely driven by architects, we also expect to increase the accuracy and quality of the produced solutions, as well as the perception of how different design strategies and DF technologies can lead to more sustainable design outcomes.

6.1 Structuring a Design Theory

To successfully handle the variability and context-specificity of architectural design through AD, we must address the practice's challenges through a computational perspective. Considering that computational tools operate by following a set of instructions, described through programming languages that are increasingly imitating the universally understood language of mathematics, we propose using the latter formalism to (1) structure the AD methodology, (2) define the different AD strategies, and (3) implement both (1) and (2) in an AD framework targeting facade design processes.

During this process, it is important to ensure the resulting methodology supports the *variability* of architectural practice, the *uniqueness* of design briefs, and the *diversity* of existing design requirements in a coherent and flexible way. Clearly, covering all possible design scenarios would be an impossible task. Nevertheless, we believe that by providing the solutions that more frequently occur, or whose application is more generic, we can not only reduce the initial investment required

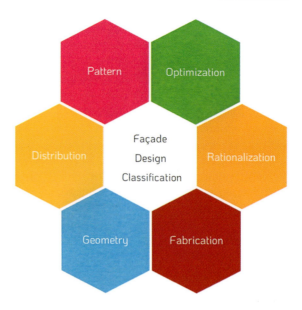

Fig. 8 Algorithmic-based classification of facade design strategies according to their role in the design process

by AD, but also embrace a large range of design scenarios and problems. Moreover, by benefiting from these strategies since early design stages, we expect architects to spend much less time and effort with programming and debugging tasks, sparing them from having to write all the algorithms from scratch each time they start a new project.

Based on a previous analysis of a wide range of contemporary building facades and different facade-oriented classifications [111–115], the proposed methodology and framework is organized in a six-fold structure containing different types of facade design strategies, whose categorization follows an algorithmic perspective (Fig. 8):

1. Geometry: to shape the building facade.
2. Distribution: to differently distribute the facade elements.
3. Pattern: to geometrically manipulate the facade elements.
4. Optimization: to adapt the facade design according to one or more fitness criteria.
5. Rationalization: to control the facade design's feasibility.
6. Fabrication: to prepare the facade design for manufacturing.

By using this categorization, the architect is guided towards the most suitable algorithms in terms of design intent, performance requirements, available resources, and construction means. Not only does this resolve many of the limitations found when using AD, particularly those related with the programming task, but it also facilitates the architects' response to the context-specificity and variability of design processes. The idea to use generic solutions to recurrent design problems draws inspiration from previous works [47, 116–121], which focused on providing sets of predefined reusable algorithms to reduce AD's initial investment. Nevertheless, our solution has the novelty of (1) focusing on the textual programming paradigm, benefiting from its scalability and expressiveness; and (2) going beyond initial design stages, integrating relevant design strategies and specialized tools beyond those of geometric exploration processes, such as analysis, optimization, and fabrication. Some of the predefined AD strategies are illustrated in Fig. 9, their mathematical structure and implementation being further elaborated in [122].

Given the diversity and uniqueness of most facade design requirements, it is not reasonable to expect that this matching process yields a complete algorithmic solution. Our proposal therefore assumes the architect as the one responsible for (1) dividing the whole design into parts, (2) establishing the dependencies between them, (3) instantiating and combining the different strategies dealing with each part, (4) implementing additional algorithms when needed, and (5) evaluating the results. Even so, we believe our proposal will increase the architects' design freedom, while improving the design process precision and ability to adapt to different design briefs. Additionally, by smoothing the design-to-fabrication transition, we expect

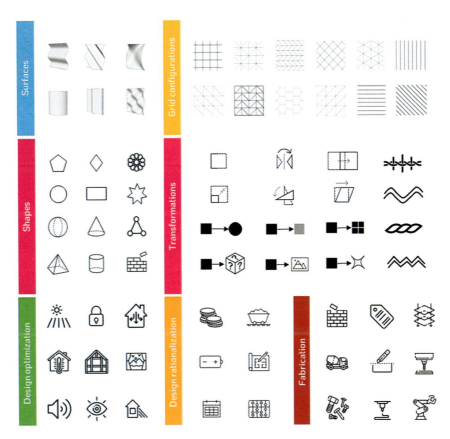

Fig. 9 AD framework conceptual representation: some of the implemented facade design strategies organized by category

our solution to improve the coordination between the geometry-, performance-, and fabrication-related information, and thus support more informed design processes towards environmentally aware solutions.

6.2 Conscientiously Driven Design Workflows

As mentioned in the beginning of this section, the goal of this research is to simplify the use of AD. To that end, a methodology and framework are proposed, focusing on the field of facade design. To evaluate the proposal's suitability for architectural practice and its ability to support more conscientious design processes, we applied the framework in the development of a set of case studies in collaboration with practice-based architectural design studios without AD skills. Figure 10 presents an overview of the resulting AD workflows by establishing a correlation between the different design stages and the algorithmic strategies used in each one. Further details on the selected case studies and their results can be found in [123–126].

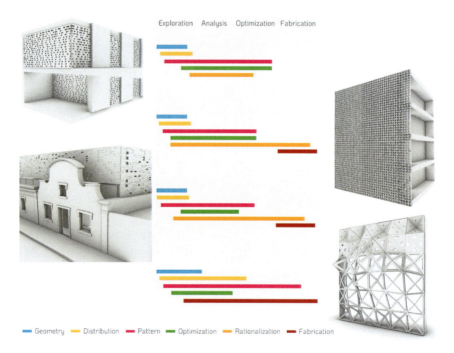

Fig. 10 Case studies' AD workflow: the design stages encompassed (top) and the algorithmic categories used in each one identified with different colors (below)

The analysis of the previous results shows the ubiquity of the different algorithmic categories in the studios' design process, continuously supporting the architects' different design tasks, while actively coordinating the multiple types of data and tools involved.

Regarding the first example, the aim was to develop a set of facade shading panels made of horizontal wood elements whose size and position addressed different daylight/shading requirements and privacy levels. In the first stage (Geometric exploration), the design studio implemented the design intent and explored different variations of it by benefiting from different geometry-related algorithms (Fig. 11, blue, yellow, and pink lines), as well as performance and rationalization ones (Fig. 11, green and orange lines, respectively). While the first algorithms allowed the architects to set geometric constraints and dependencies between them according to their design intent, i.e., creating wood bars of alternating sizes, with the smaller bars randomly varying their length and position; the second algorithms allowed them to iteratively adjust the parameters and dependencies of the geometry-related algorithms, either increasing or decreasing the smaller bars' length and in-between distances according to the existing daylight/shading and privacy requirements. Finally, the rationalization algorithms enabled the architects to gradually decrease the design's geometric freedom to ensure the solution fit the budget, restricting the range of possible sizes for the small bars.

As it is visible in Fig. 11, a similar scenario occurred in the following design stages (Design analysis and optimization), since both processes were guided by

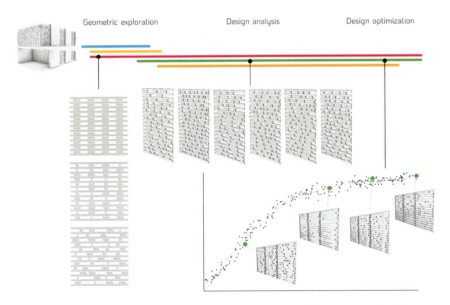

Fig. 11 Design workflow of a set of facade shading panels with some key moments of its AD process (from left to right): design intent implementation, environmentally driven geometric variation, and design optimization

performance requirements and analysis results (green line) as well as aesthetic and cost considerations (pink and orange lines). The result was a set of design solutions that comply with the architects' creative intent and the existing cost, shading, and privacy requirements, from which the architects could then select the one that most pleased them.

In the second example the aim was to develop a set of building facades entirely made of *cobogó*-inspired elements of unique sizes and shapes, whose geometric characteristics created a complex, apparently random, visual effect and simultaneously adapted to the building inside functions. Like the previous case study, the first stage (Geometric exploration) benefited from geometry-related algorithms (the blue, yellow, and pink lines in Fig. 12) as well as performance and rationalization ones (the green and orange lines in the same figure).

The same can be said about the second stage (Design analysis), which benefited from the same categories of algorithms to iteratively adjust the design's geometric characteristics according to the existing daylight, privacy, and natural ventilation needs. The result was a set of *cobogó*-inspired elements of different opacities, whose random spatial distribution met the performance needs of each adjacent area (Fig. 12, Design analysis).

Regarding the Design optimization stage, it focused on minimizing the solution's fabrication costs (Fig. 12, orange line), reducing the variety of facade elements without compromising the design intent and the existing performance requirements (Fig. 12, pink and green lines). This concern was carried over to the ensuing design stage (Fabrication) and coordinated with the available resources and manufacturing means (Fig. 12, red line).

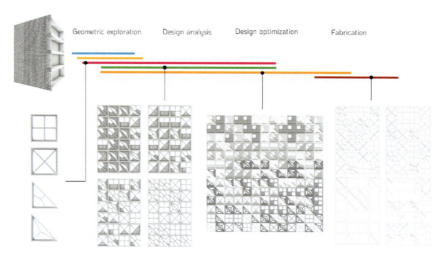

Fig. 12 Design workflow of a set of *cobogó*-inspired facade panels with some key moments of its AD process (from left to right): *cobogó* elements geometric exploration, aesthetic- and performance-based design variations of the facade panels, control of manufacturing costs, and automatic extraction of technical drawings for fabrication

Regarding the third case study, the aim was to create a visually dynamic brick facade responding to different privacy and daylight requirements. As in the previous examples, all design stages (exploration, analysis, and optimization) resulted from a coordination between aesthetic intents, performance requirements, and economic constraints. As illustrated in Fig. 13, in the first stage (Geometric exploration), the architects used different categories of algorithms to implement their design intent, namely (1) geometry-related algorithms (blue, yellow, and pink lines) to generate a facade pattern made of differently sized and randomly distributed and protruded bricks creating punctual voids, and (2) cost-related algorithms to reduce the design geometric freedom, restricting the range of possible brick sizes and protrusion positions. In the second stage (Design analysis), the architects added some performance algorithms to the previous combination to test different design variations with varying levels of permeability and ratios between brick sizes and protrusion positions. In the third stage (Design optimization), the architects used the available rationalization algorithms to control the range of configurations allowed, reducing the solution's manufacturing costs and waste. In the last stage (Fabrication), the architects took advantage of the existing manufacturing-related algorithms (Fig. 13, red line) to detail the solution as well as to extract information about the quantities and position of the existing brick typologies.

Regarding the last example, the architects adopted the workflow of Fig. 14, where the different geometry-, performance-, and fabrication-related algorithms all contributed to the design development of a set of unconventional facade panels, from conceptual exploration to manufacturing preparation. In this case study, the aim was to produce a facade design prototype made of different metal panels, whose varying shapes created a visually complex and irregular surface stereotomy and whose different levels of permeability responded to the existing performance requirements.

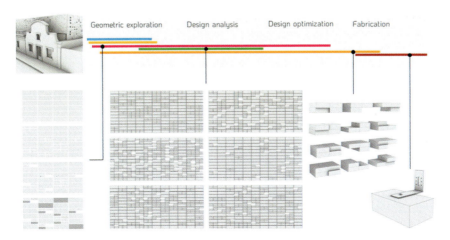

Fig. 13 Design workflow of a brick facade pattern with some key moments of its AD process (from left to right): pattern geometric evolution, aesthetic- and performance-based design exploration, design rationalization, and design detailing

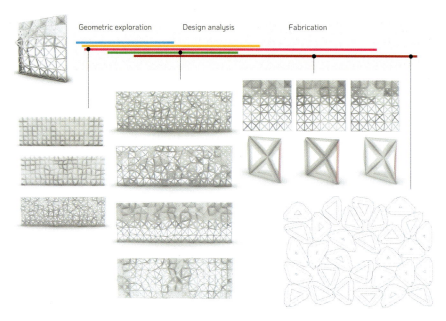

Fig. 14 Design workflow of a facade design prototype with some key moments of its AD process (from left to right): design intent implementation, aesthetic- and performance-based design exploration, design detailing, and manufacturing documentation

As illustrated in Fig. 14, the workflow started with the implementation of the design intent, which benefited from geometry-related functionalities, and proceeded with the panels' geometric exploration, coordinating the previous algorithms with performance-related ones. The result was a set of facade panels with the desired geometric irregularity and volumetry and simultaneously presenting different levels of permeability that met the existing shading and privacy needs. The process continued with the integration of additional manufacturing-related strategies, allowing the architects to (1) segment the facade design into smaller parts to facilitate the subsequent assembly and transportation processes; (2) create connection details between the panels, and between the panels and the facade structure; and (3) automatically produce technical drawings for manufacturing (in this case, laser cutting), while benefitting from labeling and nesting strategies to reduce both material waste and assembly complexity.

As it is visible in Fig. 14, the architects also benefited from the available fabrication strategies during the geometric exploration of this case study. By coordinating the design intent with manufacturing constraints, the architects could extend their creative process, not only increasing the range of construction schemes considered, but also gaining a better insight on the impact of each manufacturing scenario on the solution's aesthetic quality.

7 Discussion and Final Considerations

Buildings are one of the main contributors to the greenhouse effect, which explains the growing concern about reducing their environmental footprint. Given the great influence facades have on the environmental performance of buildings, their design has been increasingly integrating performance requirements in addition to the traditional aesthetic and functional constraints. Recent computational design strategies, such as Algorithmic Design (AD), have lowered the barriers to the adoption of performance-based design strategies, but their use remains shy in the field mostly because of their technical complexity and level of abstraction.

Nevertheless, to meet the 2030 Agenda's and Industry 4.0 goals of creating more sustainable built environments, architects must adopt AD strategies. To help such adoption, we propose reducing the complexity of AD approaches by structuring a methodology and framework containing different algorithmic strategies that can be easily combined in the development of new design solutions. Given the environmental relevance of building facades, the proposal focuses on this building element, encompassing not only its different design stages and requirements but also the variability and diversity of architectural design problems. The aim is to support AD workflows that coordinate aesthetic, performance, and construction requirements in a flexible and responsive way since early design stages, promoting the search for more sustainable solutions.

7.1 Design Workflow

To evaluate the proposal, we applied the framework in a set of case studies developed collaboratively by architects with and without AD skills. Multiple design scenarios were considered, and various design requirements addressed. Based on the results, we conclude that the use of the AD framework improved:

1. The design freedom—given the ease with which architects could select and combine different algorithmic strategies and apply iterative design changes, multiple design variations were tested in all case studies, and a wide range of design scenarios was considered.
2. The coordination between different requirements since early design stages—given the solutions' algorithmic nature and parametricity, the architects could easily apply performance-and manufacturing-related principles to drive geometric exploration processes and the other way round, i.e., using geometry-related principles to guide design optimization and fabrication.
3. The architects' decision-making processes—the design freedom combined with the flexible integration of different types of data provided architects with more insight on the quality of the solutions as well as on the impact of design changes.

4. The design space explored—because of the greater design freedom and coordination between different requirements, architects could devote more time to creative exploration, increasing the range of solutions considered.
5. The quality of the solutions achieved—as the architects' decisions were more informed, they could more easily guide the design development process towards better solutions.
6. The control over design-to-manufacture—the ability to automatically extract technical documentation for manufacturing, combined with the flexibility to test multiple design variations, allowed architects to consider multiple construction scenarios, while assessing the impact each one had on the solution's aesthetic and sustainability.

Despite not considering all possible design scenarios, our proposal was successful in responding to the most common design problems, while adapting to more specific circumstances when needed. In these cases, the developed extensions were then incorporated into the AD framework, becoming available for future use. The possibility to incrementally extend the framework with the results of its practical application is intentional, allowing us to not only increase the range of predefined strategies available, but also refine and adapt the existing ones to real case requirements and constraints.

7.2 Meeting Sustainable Development Goals

Despite the simplicity of the presented case studies, they demonstrate the potential of the proposed methodology and framework to support AD workflows and promote more informed design processes towards more sustainable facade design solutions. As the integration and coordination of different performance requirements is facilitated, architects are left with more time to explore the design space and consider other solutions beyond those initially imagined. This in turn allows architects to gain more control over the design development process, increasing the chances of achieving more sustainable solutions that meet the 2030 SDGs Agenda of making the built environment sustainable in terms of production and consumption [5].

This ability is visible in the first case study, when the architects combined performance analysis results with their aesthetic preferences, ensuring the obtained solutions complied with both the design intent and the performance requirements. Besides guiding the design space navigation towards the architects' preferences, the proposed solution facilitated the analysis of the trade-offs between aesthetic and performance requirements.

The case studies also proved the ability of the proposal to facilitate the concretization of less conventional design solutions, while increasing their production efficiency and sustainability by minimizing both energy consumption and waste. This was demonstrated by the second and third case studies, where the architects gradually reduced the cost and material waste resulting from the solutions' manufacturing without compromising the design intent and performance requirements.

To sum up, we conclude that the proposed AD methodology and framework allowed architects to approximate their creative thinking with the design making, a critical step towards the Industry 4.0 and its goal for sustainable industrialization and production.

Acknowledgements This work was supported by national funds through *Fundação para a Ciência e a Tecnologia* (FCT) (references UIDB/50021/2020, PTDC/ART-DAQ/31061/2017) and the PhD grant under contract of FCT (SFRH/BD/128628/2017).

References

1. Ruiz-Geli, E.: It is all about particles. In: Kolarevic, B., Parlac, V. (eds.) Building Dynamics: Exploring Architecture of Change. Routledge (2015)
2. Dillen, W., Lombaert, G., Mertens, R., Van Beurden, H., Jaspaert, D., Schevenels, M.: Optimization in a realistic structural engineering context: redesign of the Market Hall in Ghent. Eng. Struct. **228** (2020)
3. Boeck, L., Verbeke, S., Audenaert, A., Mesmaeker, L.: Improving the energy performance of residential buildings: a literature review. Renew. Sustain. Energy Rev. **52**, 960–975 (2015)
4. Huang, Y., Niu, J.: Optimal building envelope design based on simulated performance: history, current status and new potentials. Energy Build. **117**, 387–398 (2015)
5. United Nations: Sustainable Development Goals (2015)
6. Boswell, C.K.: Exterior Building Enclosures: Design Process and Composition for Innovative Facades. Wiley (2013)
7. Schittich, C.: Building Skins. Birkhäuser (2006)
8. ElGhazi, Y.S.: Building skins in the age of information technology. Faculty of Engineering, Cairo University (2009)
9. Picco, M., Lollini, R., Marengo, M.: Towards energy performance evaluation in early stage building design: a simplification methodology for commercial building models. Energy Build. **76**, 497–505 (2014)
10. Knaack, U., Bilow, M.: Façades: Principles of Construction. Birkhäuser Verlag (2007)
11. Machairas, V., Tsangrassoulis, A., Axarli, K.: Algorithms for optimization of building design: a review. Renew. Sustain. Energy Rev. **31**, 101–112 (2014)
12. Touloupaki, E., Theodosiou, T.: Performance simulation integrated in parametric 3D modeling as a method for early stage design optimization—a review. Energies (2017)
13. Austern, G., Capeluto, I.G., Grobman, Y.J.: Rationalization methods in computer aided fabrication: a critical review. Autom. Constr. **90**, 281–293 (2018)
14. Oxman, R.: Performance-based design: current practices and research issues. IJAC **06**, 1–17 (2008)
15. Henriksson, V., Hult, M.: Rationalizing freeform architecture. Chalmers University of Technology (2015)
16. Figliola, A., Battisti, A.: Feedback on the design processes for the materialization of informed architectures. In: Post-industrial Robotics: Exploring Informed Architecture, pp. 155–173. Springer Singapore (2021)
17. Caetano, I., Santos, L., Leitão, A.: Computational design in architecture: defining parametric, generative, and algorithmic design. Front. Arch. Res. **9**, 287–300 (2020)

18. Garber, R.: Information modelling today. In: Garber, R. (ed.) BIM Design: Realising the Creative Potential of Building Information Modelling, pp. 14–27. Wiley (2014)
19. D'Agostino, D., D'Agostino, P., Minelli, F., Minichiello, F.: Proposal of a new automated workflow for the computational performance-driven design optimization of building energy need and construction cost. Energy Build. **239** (2021)
20. Muehlbauer, M.: Typogenetic design—aesthetic decision support for architectural shape generation. RMIT University (2018)
21. Kolarevic, B.: Towards the performative in architecture. In: Kolarevic, B., Malkawi, A.M. (eds.) Performative Architecture. Beyond Instrumentality, pp. 203–214. Spon Press (2005)
22. Fasoulaki, E.: Integrated design: a generative multi-performative design approach. MIT University (2008)
23. Anton, I., Tănase, D.: Informed geometries. Parametric modelling and energy analysis in early stages of design. Energy Procedia **85**, 9–16 (2016)
24. Zuk, W., Clark, R.H.: Kinetic Architecture. Van Nostrand Reinhold (1970)
25. Frazer, J.: An Evolutionary Architecture. Architectural Association Publications (1995)
26. Kolarevic, B., Malkawi, A.: Performative Architecture: Beyond Instrumentality. Spon Press (2005)
27. Hensel, M.: Performance-Oriented Architecture: Rethinking Architectural Design and the Built Environment. Wiley (2013)
28. Ciardiello, A., Rosso, F., Dell'Olmo, J., Ciancio, V., Ferrero, M., Salata, F.: Multi-objective approach to the optimization of shape and envelope in building energy design. Appl. Energy (2020)
29. Evins, R.: A review of computational optimisation methods applied to sustainable building design. Renew. Sustain. Energy Rev. **22**, 230–245 (2013)
30. Belém, C.G.: Optimization of time-consuming objective functions: derivative-free approaches and their application in architecture. IST, University of Lisbon (2019)
31. D'Oca, S., Hong, T., Langevin, J.: The human dimensions of energy use in buildings: a review. Renew. Sustain. Energy Rev. **81**, 731–742 (2018)
32. Heiselberg, P., Brohus, H., Hesselholt, A., Rasmussen, H., Seinre, E., Thomas, S.: Application of sensitivity analysis in design of sustainable buildings. Renewable Energy **34**, 2030–2036 (2009)
33. Pisello, A.L., Castaldo, V.L., Rosso, F., Piselli, C., Ferrero, M., Cotana, F.: Traditional and innovative materials for energy efficiency in buildings. Key Eng. Mater. **678**, 14–34 (2016)
34. Schodek, D., Bechthold, M., Griggs, J.K., Kao, K., Steinberg, M.: Digital Design and Manufacturing: CAD/CAM Applications in Architecture and Design. Wiley (2005)
35. Nguyen, A.-T., Reiter, S., Rigo, P.: A review on simulation-based optimization methods applied to building performance analysis. Appl. Energy **113**, 1043–1058 (2014)
36. Kalay, Y.: Architecture's New Media: Principles, Theories, and Methods of Computer-Aided Design. MIT Press (2004)
37. Han, T., Huang, Q., Zhang, A., Zhang, Q.: Simulation-based decision support tools in the early design stages of a green building—a review. Sustainability (2018)
38. Shi, X.: Performance-based and performance-driven architectural design and optimization. Front. Arch. Civ. Eng. China **4**, 512–518 (2010)
39. Picon, A.: Ornament: The Politics of Architecture and Subjectivity. Wiley (2013)
40. Dritsas, S.: Design-built: rationalization strategies and applications. IJAC **10**, 575–594 (2012)
41. Garber, R.: BIM Design: Realising the Creative Potential of Building Information Modelling. Wiley (2014)
42. Alfaris, A., Merello, R.: The generative multi-performance design system. In: Proceedings of the 28th ACADIA Conference, pp. 448–457 (2008)
43. Terzidis, K.: Algorithmic design: a paradigm shift in architecture? In: Proceedings of the 22nd eCAADe Conference, pp. 201–207 (2004)
44. Bukhari, F.A.: A hierarchical evolutionary algorithmic design (HEAD) system for generating and evolving building design models. QUT (2011)

45. Zboinska, M.A.: Hybrid CAD/E platform supporting exploratory architectural design. CAD **59**, 64–84 (2015)
46. Oxman, R.: Thinking difference: theories and models of parametric design thinking. Des. Stud. 1–36 (2017)
47. Woodbury, R.: Elements of Parametric Design. Routledge, New York (2010)
48. Janssen, P.: Visual dataflow modelling: some thoughts on complexity. In: Proceedings of the 32nd eCAADe Conference, pp. 305–314 (2014)
49. Leitão, A., Santos, L., Lopes, J.: Programming languages for generative design: a comparative study. IJAC **10**, 139–162 (2012)
50. Noone, M., Mooney, A.: Visual and textual programming languages: a systematic review of the literature. J. Comput. Educ. **5**, 149–174 (2018)
51. Janssen, P., Li, R., Mohanty, A.: Möbius: a parametric modeller for the web. In: Proceedings of the 21st CAADRIA Conference, pp. 157–166 (2016)
52. Cristie, V., Joyce, S.C.: 'GHShot': a collaborative and distributed visual version control for Grasshopper parametric programming. In: Proceedings of the 37th eCAADe and 23rd SIGraDi Joint Conference, pp. 35–44 (2020)
53. Harding, J.E., Shepherd, P.: Meta-parametric design. Des. Stud. **52**, 73–95 (2017)
54. Davis, D.: Modelled on software engineering: flexible parametric models in the practice of architecture. RMIT University (2013)
55. Wortmann, T., Tunçer, B.: Differentiating parametric design: digital workflows in contemporary architecture and construction. Des. Stud. **53**, 173–197 (2017)
56. Nezamaldin, D.: Parametric design with visual programming in dynamo with Revit: The conversion from CAD models to BIM and the design of analytical applications. KTH Skolan för arkitektur och samhällsbyggnad (2019)
57. Leitão, A., Lopes, J., Santos, L.: Illustrated programming. In: Proceedings of the 34th ACADIA Conference, pp. 291–300 (2014)
58. Feist, S., Ferreira, B., Leitão, A.: Collaborative algorithmic-based building information modelling. In: Proceedings of the 22nd CAADRIA Conference, pp. 613–622 (2017)
59. Davis, D., Burry, J., Burry, M.: Understanding visual scripts: improving collaboration through modular programming. IJAC **09**, 361–376 (2011)
60. Aguiar, R., Cardoso, C., Leitão, A.: Algorithmic design and analysis fusing disciplines. In: Proceedings of the 37th ACADIA Conference, pp. 28–37 (2017)
61. Mueller, C.T.: Computational exploration of the structural design space. MIT University (2014)
62. Wortmann, T., Nannicini, G.: Introduction to architectural design optimization. In: City Networks. Springer Optimization and Its Applications. Springer, Cham (2017)
63. Stevanović, S.: Optimization of passive solar design strategies: a review. Renew. Sustain. Energy Rev. **25**, 177–196 (2013)
64. Yang, X.S., Koziel, S., Leifsson, L.: Computational optimization, modelling and simulation: recent trends and challenges. Procedia Comput. Sci. **18**, 855–860 (2013)
65. Schlueter, A., Thesseling, F.: Building information model based energy/exergy performance assessment in early design stages. Autom. Constr. **18**, 153–163 (2009)
66. Petersen, S., Svendsen, S.: Method and simulation program informed decisions in the early stages of building design. Energy Build. **42**, 1113–1119 (2010)
67. Madrazo, L., Massetti, M., Font, G., Alomar, I.: Integrating energy simulation in the early stage of building design. In: Proceedings of the 3rd BauSIM Conference, pp. 175–182 (2010)
68. Attia, S., Gratia, E., De Herde, A., Hensen, J.L.M.: Simulation-based decision support tool for early stages of zero-energy building design. Energy Build. **49**, 2–15 (2012)
69. Lin, S.E., Gerber, D.J.: Designing-in performance: evolutionary energy performance feedback for early stage design. In: Proceedings of the 13th BuildingSimulation Conference, pp. 386–393 (2013)
70. Negendahl, K.: Building performance simulation in the early design stage: an introduction to integrated dynamic models. Autom. Constr. **54**, 39–53 (2015)

71. Konis, K., Gamas, A., Kensek, K.: Passive performance and building form: an optimization framework for early-stage design support. Sol. Energy **125**, 161–179 (2016)
72. Menges, A.: Fusing the computational and the physical. AD Mag. **85** (2015)
73. Iwamoto, L.: Digital Fabrications—Architectural and Material Techniques. Princeton Architectural Press (2009)
74. Dent, A., Sherr, L.: Material Innovation: Architecture. Thames & Hudson (2014)
75. Gramazio, F., Kohler, M. (eds.): Made by Robots: Challenging Architecture at a Larger Scale. AD Mag. **84** (2014)
76. Loonen, R.C.G.M., Favoino, F., Hensen, J.L.M., Overend, M.: Review of current status, requirements and opportunities for building performance simulation of adaptive facades. J. Build. Perform. Simul. **10**, 205–223 (2016)
77. Kolarevic, B., Parlac, V.: Adaptative, responsive building skins. In: Kolarevic, B., Parlac, V. (eds.) Buildings Dynamics: Exploring an Architecture of Change. Routledge (2015)
78. López, M., Rubio, R., Martín, S., Croxford, B.: How plants inspire facades. From plants to architecture: Biomimetic principles for the development of adaptive architectural envelopes. Renew. Sustain. Energy Rev. **67**, 692–703 (2017)
79. Ochoa, C.E., Capeluto, I.G.: Advice tool for early design stages of intelligent facades based on energy and visual comfort approach. Energy Build. **41**, 480–488 (2009)
80. Bouchlaghem, N.: Optimizing the design of building envelopes for thermal performance. Autom. Constr. **10**, 101–112 (2000)
81. Wang, W., Zmeureanu, R., Rivard, H.: Applying multi-objective genetic algorithms in green building design optimization. Build. Environ. **40**, 1512–1525 (2005)
82. Gagne, J., Andersen, M.: A generative facade design method based on daylighting performance goals. J. Build. Perform. Simul. **5**, 141–154 (2012)
83. Gagne, J.M.L., Andersen, M.: Multi-objective optimization for daylighting design using a genetic algorithm. In: Proceedings of the 4th SimBuild Conference (2010)
84. Jin, Q., Overend, M.: A prototype whole-life value optimization tool for façade design. J. Build. Perform. Simul. **7**, 217–232 (2014)
85. Gamas, A., Konis, K., Kensek, K.: A parametric fenestration design approach for optimizing thermal and daylighting performance in complex urban settings. In: Proceedings of the 43rd ASES Conference, pp. 87–94 (2014)
86. Elghandour, A., Saleh, A., Aboeineen, O., Elmokadem, A.: Using parametric design to optimize building's façade skin to improve indoor daylighting performance. In: Proceedings of the 3rd BSO Conference, pp. 353–361 (2016)
87. Pantazis, E., Gerber, D.: A framework for generating and evaluating façade designs using a multi-agent system approach. IJAC **16**, 248–270 (2018)
88. Austern, G., Elber, G., Capeluto, I.G., Grobman, Y.J.: Adapting architectural form to digital fabrication constraints. In: AAG 2018, pp. 10–33. Klein Publishing GmbH (Ltd.) (2018)
89. Dunn, N.: Digital Fabrication in Architecture. Laurence King Publishing (2012)
90. Overall, S., Rysavy, J.P., Miller, C., Sharples, W., Sharples, C., Kumar, S., Vittadini, A., Saby, V.: Direct-to-drawing: automation in extruded terracotta fabrication. In: Fabricate 2020. UCL Press (2020)
91. Castañeda, E., Lauret, B., Lirola, J.M., Ovando, G.: Free-form architectural envelopes: digital processes opportunities of industrial production at a reasonable price. J. Facade Des. Eng. **3**, 1–13 (2015)
92. Lee, G., Kim, S.: Case study of mass customization of double-curved metal façade panels using a new hybrid sheet metal processing technique. J. Constr. Eng. Manag. **138**, 1322–1330 (2012)
93. Aksamija, A.: Integrating Innovation in Architecture: Design, Methods and Technology for Progressive Practice and Research. Wiley (2016)
94. Soar, R., Andreen, D.: The role of additive manufacturing and physiomimetic computational design for digital construction. AD Mag. **82**, 126–135 (2012)
95. Paio, A., Eloy, S., Rato, V.M., Resende, R., de Oliveira, M.J.: Prototyping vitruvius, new challenges: digital education, research and practice. Nexus J. **14**, 409–429 (2012)

96. Jančič, L.: Implications of the use of additive manufacturing in architectural design. Univerza v Ljubljani (2016)
97. Kolarevic, B.: The (risky) craft of digital making. In: Manufacturing Material Effects: Rethinking Design and Making in Architecture. Routledge (2008)
98. Afify, H.M.N., Elghaffar, Z.A.S.: Advanced digital manufacturing techniques (CAM) in architecture. In: Proceedings of the 3rd ASCAAD Conference, pp. 67–80 (2007)
99. Bayram, A.K.Ş.: Digital fabrication shift in architecture. In: Architectural Sciences and Technology, pp. 173–193 (2021)
100. Simmons, M.: Material collaborations. In: Manufacturing Material Effects: Rethinking Design and Making in Architecture (2008)
101. Barkow, F.: Cut to fit. In: Manufacturing Material Effects: Rethinking Design and Making in Architecture (2008)
102. Mesnil, R., Douthe, C., Baverel, O., Léger, B., Caron, J.F.: Isogonal moulding surfaces: a family of shapes for high node congruence in free-form structures. Autom. Constr. **59**, 38–47 (2015)
103. Pottmann, H.: Architectural geometry as design knowledge. AD Mag. **80**, 72–77 (2010)
104. Hesselgren, L., Charitou, R., Dritsas, S.: The Bishopsgate Tower case study. IJAC **5**, 61–81 (2007)
105. Whitehead, H.: Laws of form. In: Kolarevic, B. (ed.) Architecture in the Digital Age: Design and Manufacturing, pp. 116–148. Spon Press (2003)
106. Pottmann, H., Eigensatz, M., Vaxman, A., Wallner, J.: Architectural geometry. Comput. Graph. 145–164 (2015)
107. Eigensatz, M., Kilian, M., Schiftner, A., Mitra, N., Pottmann, H., Pauly, M.: Paneling architectural freeform surfaces. ACM Trans. Graph. **29** (2010)
108. Eigensatz, M., Deuss, M., Schiftner, A., Kilian, M., Mitra, N., Pottmann, H., Pauly, M.: Case studies in cost-optimized paneling of architectural freeform surfaces. In: AAG 2010, pp. 47–72. Springer (2010)
109. Andrade, D., Harada, M., Shimada, K.: Framework for automatic generation of facades on free-form surfaces. Front. Arch. Res. **6**, 273–289 (2017)
110. Flöry, S., Pottmann, H.: Ruled surfaces for rationalization and design in architecture. In: Proceedings of the 30th ACADIA, pp. 103–109 (2010)
111. Moussavi, F., Kubo, M. (eds.): The Function of Ornament. Actar (2006)
112. Pell, B.: The Articulate Surface: Ornament and Technology in Contemporary Architecture. Birkhäuser GmbH (2010)
113. Fox, M., Kemp, M.: Interactive Architecture. Princeton Architectural Press (2009)
114. Velasco, R., Brakke, A.P., Chavarro, D.: Dynamic façades and computation: towards an inclusive categorization of high performance kinetic façade systems. In: Proceedings of the 16th CAADFutures Conference, pp. 172–191 (2015)
115. Waseef, A., El-Mowafy, B.N.: Towards a new classification for responsive kinetic facades. In: Proceedings of the MIC 2017 Conference (2017)
116. Alexander, C., Ishikawa, S., Silverstienm, M.: A Pattern Language: Towns, Buildings, Construction. Oxford University Press (1977)
117. Woodbury, R., Aish, R., Kilian, A.: Some patterns for parametric modeling. In: Proceedings of the 27th ACADIA Conference, pp. 222–229 (2007)
118. Qian, Z.C.: Design patterns: augmenting design practice in parametric CAD systems. Simon Fraser University (2009)
119. Chien, S., Su, H., Huang, Y.: PARADE: a pattern-based knowledge repository for parametric designs. In: Proceedings of the 20th CAADRIA Conference (2015)
120. Lin, C.-J.: The STG-framework: a pattern-based algorithmic framework for developing generative models of parametric architectural design at the conceptual design stage. Comput.-Aided Des. Appl. **15**, 653–660 (2018)
121. Su, H., Chien, S.: Revealing patterns: using parametric design patterns in building façade design workflow. In: Proceedings of the 21st CAADRIA Conference, pp. 167–176 (2016)

122. Caetano, I., Leitão, A.: Mathematically developing building facades: an algorithmic framework. In: Eloy, S., Leite Viana, D., Morais, F., Vieira Vaz, J. (eds.) Formal Methods in Architecture: Advances in Science, Technology & Innovation. IEREK Interdisciplinary Series for Sustainable Development (2021)
123. Caetano, I., Leitão, A.: Integration of an algorithmic BIM approach in a traditional architecture studio. J. Comput. Des. Eng. **6**, 327–336 (2019)
124. Caetano, I., Ilunga, G., Belém, C., Aguiar, R., Feist, S., Bastos, F., Leitão, A.: Case studies on the integration of algorithmic design processes in traditional design workflows. In: Proceedings of the 23rd CAADRIA Conference, pp. 129–138 (2018)
125. Caetano, I., Leitão, A., Bastos, F.: From architectural requirements to physical creations. J. Façade Des. Eng. **8**, 59–80 (2020)
126. Caetano, I., Leitão, A., Bastos, F.: Converting algorithms into tangible solutions: a workflow for materializing algorithmic facade designs. In: Correia, A., Azenha, M., Cruz, P., Novais, P., Pereira, P. (eds.) Trends on Construction in the Digital Era. ISIC 2022. Lecture Notes in Civil Engineering, vol 306. Springer, Cham (2023)

Volatile Data: Strategies to Leverage Datasets into Design Applications

Edoardo Tibuzzi and **Georgios Adamopoulos**

Abstract As the AEC industry is approaching a stage of maturity in the digital transformation journey, AKT II's p.art team has been pioneering it since its inception over 25 years ago. Data as an underlying driver of design, informing decisions earlier on and addressing issues from the macro scale of social impact to the micro scale of structural, environmental, and sustainable optimization has been the principal focus of this practice driven research team. Below 3 main examples are chosen to describe how tapping into intangible knowledge hidden in internal or external datasets, helped exploiting it into targets, processes and design solutions. The intention is to critique the current availability of datasets, how to understand and avoid data bias, and finally the hurdles to overcome into getting from raw data to implemented design drivers. Those pioneering exercises are exploring the novel opportunities provided by the hybridization of processes and cross disciplinary datasets, to enhance the built environment and to learn from the more granular availability of relevant data. In an effort to provide support to the architectural industry, the examples covered below are showcasing how technology can be leveraged to expedite the achievement of some of the Sustainable development goals set by the U.N., specifically in "Part 1", we will demonstrate how accessing an existing dataset and using state of the art software visualization techniques is supporting the team in highlighting issues and potential mitigations of goals 11 (sustainable cities and communities), 13 (climate action), 14 (life below water) and 15 (life on land). "Part 2" is showcasing the opportunity on one side to make existing datasets available to the public through a mobile app, and on the other end, to use the same app to gather specific user data. In "Part 3" we will demonstrate how novel design techniques helped us design a waterless garden in the desertic climate of Sharjah, proving that using an inter-disciplinary approach, mixing architectural design, building physics knowledge, computational fluid dynamics simulation and parametric modelling, helped the team predicting the best geometric output for the garden landscaping that provided a recreation of

E. Tibuzzi (✉) · G. Adamopoulos
P.art, AKTII, London EC1Y 8AF, UK
e-mail: edoardo.tibuzzi@akt-uk.com

G. Adamopoulos
e-mail: georgios.adamopoulos@akt-uk.com

a natural environment to facilitate indigenous plants growth, effectively targeting U.N. goals 2 (zero hunger), 3 (good health and wellbeing), 7 (affordable and clean energy), 11 (sustainable cities and communities), 12 (responsible consumption and production), 13 (climate action) and 15 (life on land).

Keywords Data driven design · Bioclimatic design · Computational fluid dynamics · GPU accelerated data visualization · Data farming · AR · VR

United Nations' Sustainable Development Goals 9. Build resilient infrastructure, promote inclusive and sustainable industrialization and foster innovation · 11. Make cities and human settlements inclusive, safe, resilient and sustainable · 13. Take urgent action to combat climate change and its impacts

1 Part 1: Geoscope 2 at the 2021 Venice Biennale

Starting from a macro scale in 2021 we have been involved in the "Geoscope 2".

The architects Daniel Lopez-Perez and Jesse Reiser (Principal of RUR architecture) have together re-thought R. Buckminster Fuller's iconic Geoscope concept for the 2021 Venice architecture biennale "How we live together".

The installation featured a dynamic video animation which has been designed and produced by AKT II's computational team.

The video was conceived to newly visualise the proliferation of atmospheric carbon, globally, that's occurred so far during our 21st century. The video is based on satellite data that's freely available online. Yet for many visitors, this has been a first experience of truly seeing the formerly intangible CO_2 catastrophe. The climate crisis is a shared problem, and to solve it we need a shared understanding (Fig. 1).

1.1 Visualizing 20 years of Daily Global CO_2 Data in Real-Time 3D

To produce this film, we designed and programmed in the Unity game engine a real-time particle simulation of global CO_2 emissions for the period 2000–2018 using publicly available data.

The narrative device we employed was that of a globe surrounded by violently shifting gases traversing borders, continents, and hemispheres. The main point, delivered implicitly, was those acts of any single country, eventually become the problem of everyone, as nothing ever stays local in the stratosphere.

During the film, millions of luminous particles flow rapidly around a three-dimension model of the Earth, following historical wind velocities. The particles'

Fig. 1 Carbon concentration dataset visualised on the terrestrial globe

colour and transparency are mapped to historical CO_2 concentrations sampled at each particle's position.

The colorization range remains fixed throughout the animation period. The result is, sadly, striking A globe that appears mostly dark and blue during the first years of the millennium, with a few dissipating plumes of CO_2 produced by the developed nations of the north hemisphere, turns into a bright red hellscape towards the end of our decade, trapped underneath a turbulent veil of CO_2 levels that remain constantly high everywhere (Fig. 2).

1.2 Visualization Philosophy

We deliberately attempted the shift from a static, data-science-oriented presentation style, to a dynamic, artistic interpretation using techniques and tools borrowed from the visual effects and gaming industries.

In most, if not all, presentations of climate-related datasets, a common visualization strategy is employed: The data are presented on a globe or some form of rectangular projection, using a colormap to colorize and accentuate the underlying values.

The narrative in that case, is a presentational one: the truth is laid to the observer as is, with a colour legend being a crucial, indispensable feature. The qualitative evaluation of how "good" or "bad" the situation presented is, is left to the observer's ability to associate the presented values with their physical, real-world meaning. In other words, the people able to decipher the meaning of the data, are the ones that are already familiar with the science behind.

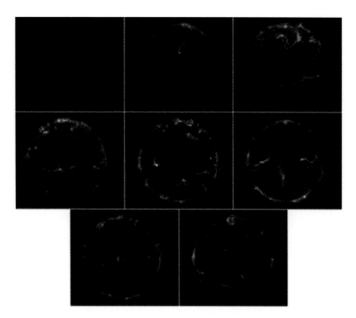

Fig. 2 Frames from the game-engine, showing the carbon concentrations at various years. From top to bottom, left to right: 2000, 2001, 2002, 2005, 2008, 2010, 2014, 2017

In our case, we moved towards a dramatized interpretation of the data, that prioritizes the visual impact and the viewer's emotional response, while staying faithful to the underlying values.

The understanding of the data in this case is a perceptual one. Simply put, one does not need to have prior knowledge of the data or even climate change itself, to perceive how radically the situation has changed in the last 20 years, and how a relatively local problem has engulfed the whole planet.

A legend with numbers and colours in this case is unnecessary, because the perception of the direness of the situation is direct, present in the emotional response of the viewer to the visuals. One does not need to know "how much", because they simply see it.

1.3 Motivation

Our motivation for this work was the deep belief that important data, like global CO_2 emissions, must be presented in ways that maximize their impact in people unfamiliar with the scientific field that produced them. One should not need to develop a solid grasp of climate science to form an educated opinion about climate, or to be convinced about the criticality of our current situation. Projections of future CO_2 levels in the atmosphere and the associated climate forcing, as well as our ability to control CO_2

levels, depend substantially on our scientific understanding of the natural carbon cycle [3]. As an example, in 2003, growing vegetation in North America removed approximately 500 million tons of carbon per year ($\pm$ 50%) from the atmosphere and stored it as plant material and soil organic matter [4].

In other words, the natural and inevitable knowledge difference between experts and no-experts, must not stand obstacle in the process of forming a common ground about important, urgent, global issues, such as climate crisis, global pandemics, or world warfare. Knowing how the structure and function of arctic terrestrial ecosystems are responding to recent and persistent climate change is paramount to understanding the future state of the Earth system and how humans will need to adapt [5].

We strongly believe that visual storytelling can form bridges across knowledge fields and transform scientific data into real-world impact. This project was an opportunity to put this belief to the test.

1.4 Datasets

Although there's a wealth of real-world carbon observatories, such as NASA's OCO-2 (Orbiting Carbon Observatory-2), the datasets they typical produce are sparse and point-like, unable to cover the entirety of the globe. For the purposes of our simulation, we required dense, texture-like datasets with 100% coverage of the Earth. We therefore turned to the U.S. NOAA's (National Oceanic and Atmospheric Administration) Global Monitoring Laboratory, and their Carbon Tracker tool, a CO_2 measurement and modelling system developed to keep track of sources and sinks of carbon dioxide around the world.

Carbon Tracker uses atmospheric CO_2 observations from a host of collaborators and simulated atmospheric transport to estimate these surface fluxes of CO_2 [1].

The "co2_total" dataset we chose, sums all available CO_2 components (background, land biosphere excluding fires, fossil fuel, wildfire, and air-sea exchange fields) in a global 3 $\times$ 2-degree grid. The dataset contained 7017 daily files in netCDF format, covering the period from 2000 to 2018, and totalling at 554.8 Giga Bytes.

Each file contained 3-hourly data for the day of the year it represented. The data were volumetrically structured, shaped in a grid 120 voxels in longitude, 90 voxels in latitude and 25 voxels in elevation, following the atmospheric levels of the TM5 (Transport Model 5) model, ranging from 0 to 80 km from the surface of the Earth [2].

Out of the 25 atmospheric levels available, we chose to use and animate only level 1 (elevations up to 34.5 m) due to its immediate relevance to the ground level air-quality where most of human activity takes place. The temporal resolution of 3 h was far too detailed for our time-lapse animations, so the eight daily 3-h intervals were combined and averaged in singular daily files. Finally, from the data fields included in each file, the relevant ones in our case were the CO_2 mole fractions, and the U

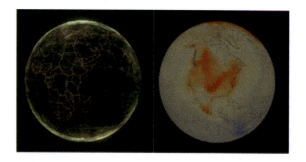

Fig. 3 Wind velocity vector field (left) and CO_2 mole fractions (right) mapped to spherical coordinates

and V wind velocities for the real-time simulation of aerosol particle flows. All other fields were dropped from the dataset.

The above operations were performed using the open-source tool nco hosted in Ubuntu Linux.

1.5 Game Engine

The particle simulation was programmed in the Unity game-engine. To make the original dataset compatible with the game-engine's native entities, an encoding process had to take place, to turn the cleaned-up netCDF files into GPU-friendly Texture objects (Fig. 3).

1.6 File Parsing

The primary parsing of the files inside the Unity environment was done using Microsoft's Scientific Dataset Lite (SDSLite) C# library. The netCDF data was loaded in memory as standard C# arrays of double precision numbers, representing CO_2 mole fractions and wind vectors.

Data Encoding

One obvious and memory-efficient way to sequence the daily data slices, would be to encode the files as video frames, and then play-back the video into a Render Texture object. The limitation of this solution is that interpolation between frames is not possible during play-back, and thus undesirable "sleepiness" would be apparent as each frame would sharply snap to the next.

To smoothly transition through the daily data, we decided to encode the whole dataset into a Texture3D object, to take advantage of the game-engine's native ability to bilinearly interpolate in three dimensions. This way even slow animation speeds still produced pleasing transitions.

Unfortunately, a Texture3D resource has an upper bound in its resolution on any axis. Specifically, it cannot exceed 2048 × 2048 × 2048 voxels. Our data contained 7017 slices on the temporal (Z) axis, clearly impossible to fit in a single Texture3D. To address this, we split the data in multiple Texture3D objects. For logical clarity and simplicity, we decided to split the dataset by year, producing 18 volumes, with a resolution of 120 × 90 × 365 voxels each. At any time, a maximum of 2 volumes were loaded in video memory, in the transition of one year to another.

Particle Simulation

The actual particle simulation takes place entirely on the GPU, utilizing Unity's Visual Effect Graph.

Each particle moves around the globe using spherical coordinates, which are remapped into normalized texture space and used to sample the velocity and CO_2 field at the current temporal slice. In every frame, the sampled value of the 2D velocity field is remapped to spherical coordinates and used to push each particle to its future position. The sampled value of the CO_2 field is used to colorize the particle and adjust its alpha (transparency) value. The particles are colorized using a modified Kelvin range gradient, and rendered as single-pixel-wide lines, using additive blending.

2 Part 2: Beyond the Map, a Live Environmental and Social Data Visualizer for London

Another example on how we have demonstrated the potential of accessing, understanding visualizing and using data is the prototype augmented-reality service Beyond the Map (2019) In this speculation for the LFA, we have built an app that makes several, formerly intangible social, economic, and environmental qualities newly visible and accessible for the public. AKT II led the computational R & D—which leveraged existing data and bespoke analysis, to encompasses metrics such as air pollution, wind comfort, pedestrian safety, and property value—to make the information visible and user-rateable, through augmented reality, using a standard smartphone camera. The development focused on engaging with real world 'key activators', tapping into the GPS functionality of the mobile devices, to begin the experience at designated locations. Specific data sets then were illustrated through augmented reality displayed on the device screen, allowing space to reflect and experience intangible threshold that surrounded the user. The overall experience aimed to start a conversation on the role of data gathering and visualisation in today's society (Figs. 4 and 5).

Fig. 4 A diagrammatic visualization of beyond the map app

Fig. 5 Real time pollution concentration visualized through the AR capability of the app

3 Part 3: Becoming Xerophile, an Experiment to Design with Data

Lastly, we reflected on the application and role of technology to create green spaces in an extreme environment. In support to Cooking Sections and as part of 2019 Sharjah Architecture triennial we developed a new model for a non-irrigated urban garden using Bioclimatic design tools and live sensors to design and monitor the performance of the gardens. As a prototype, this project extended past the initial Triennial and has been monitored to gauge success. We only used technology, in this case, the bioclimatic tools we normally use for master planning and envelope design, have been fundamental to define the geometry of the gardens and enhance the microclimate of the earth mound structures (Fig. 6).

From the Greek terms 'xeros' (meaning dry) and 'philos' (meaning loving) 'Becoming Xerophile' sets out to challenge the idea of the desert as a bare landscape and instead develop a new model for a non-irrigated urban garden. Nine earth mounds have been carved on the site of an old, disused school. These structures with various microclimates enable the desert plants to have the optimal environment they require to flourish. The 'water without watering' model is based on ancient techniques of cultivation. The earth mounds make use of the soil and rubble from the local school's renovation and contain (between them) over 40 different species of desert plants.

With this region facing environmental challenges in which high temperatures and infrequent precipitation contribute to water scarcity, pilot projects that embed adaptive research into design such as Becoming Xerophile are urgent and extremely relevant.

Fig. 6 Becoming Xerophile garden in Sharjah two years after construction

To design a series of optimal spaces in the harsh conditions of the city and promote the proliferation of vegetation we have used Computational fluid dynamics and a bespoke digital tool developed in-house to assess the optimal diameter and depth of the different "sand bowls" by running microclimate analysis on a series of different shapes and iterations. A full microclimate study was performed looking at the impact of wind and solar radiation over the proposed scheme by varying the defined geometries of the sand bowls and select the best performers.

The study investigated wind flow patterns in the garden as influenced by its layout and proposed plant shelter geometries. The analysis relied on probabilistic local weather data to calculate and visualize wind flow paths and velocities for 16 main incoming directions. The local weather was based on the Sharjah International Airport data measured roughly 15 km West of the site.

Urban location of the garden by a major road along the same direction can thus rendered it exposed to relatively strong wind flows incoming from the sea (North-West), potentially causing discomfort for garden users. The aim of the analysis was to assess impact of these flows on local conditions and the tempering effects of its proposed geometry.

A closer look over the garden area reveals that even in exposed conditions wind speed remains low in areas where the plants are based in deeper pits. In the largest pit with seating provision the conditions are shown to remain practically unchanged throughout (Fig. 7).

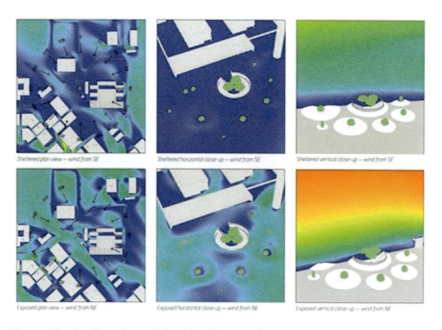

Fig. 7 CFD wind simulation analysis of the different design proposals for the garden geometry

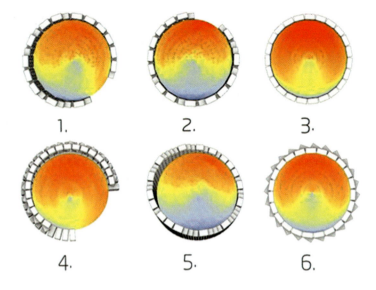

Fig. 8 Radiation analysis of different brick typologies

Subsequently to the wind comfort a radiation analysis was performed on 6 brickwork bowl typologies.

For reference the bowls were numbered and colour coded as shown in Fig. 8.

Examined parameters included bowl orientation, cone depth and bowl diameter.

Incident radiation was tested for 360° range of bowl orientations. The results indicated that radiation within the bowl is minimized when the shading area that is oriented towards South-East is maximized. In other words, increasing wall area in SE direction has the greatest effect in creating shade within the bowl.

Changing bowl depth changed the side angle and thus the angle at which radiation hits the surface. Five cone depths were studied, downwards from the ground level in 0.5-m increments.

The analysis found that incident radiation per unit area reduces when bowl depth is increased.

Increasing bowl diameter increases incident radiation per unit area. This is due to the bowl slope angle approaching the ground plane, increasing the amount of radiation falling on unit area in the process.

The analysis was revised for the updated bowl designs. Annual, summer solstice and winter solstice analysis periods were considered.

During the summer solstice the baseline incident radiation per m^2 is roughly double that of the winter solstice and is the expected annual maximum radiation day. High sun angle in summer period will make shading with vertical walls relatively less effective than horizontal shading.

Growing foliage in the garden will block an increasing fraction of incident radiation from reaching the bowl surface. That fraction is dependent on crown geometry and density. For existing plants, crown density this is normally assessed using density cards. For future estimates it is best approximated on basis of existing examples growing in similar environmental conditions. Increased crown size and density reduce incident radiation on the ground below the plant.

Analysis results for each bowl in the garden are shown in Fig. 8, whilst a Comparison between annual radiation with and without foliage can be observed in the Fig. 9 (Fig. 10).

After the above study, the garden bowls were constructed and planted. The garden was also equipped with a suite of sensors that measure the small microclimates generated by the earth mound structures. The sensors measure rainfall, solar radiation, wind speed and direction, air temperature and relative humidity, soil moisture, and leaf wetness. Through their materiality, shading, depth, and positioning, the structures optimise both air humidity and moisture drawn from the water table. The condition of the plants inside and outside the earth mounds are monitored at fifteen-minute intervals, and data has been gathered for the next three years (Figs. 11 and 12).

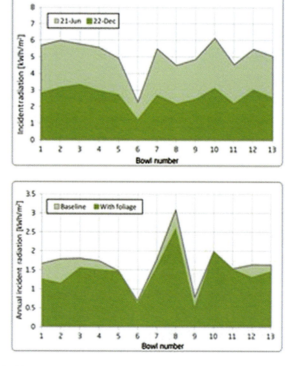

Fig. 9 Annual incident radiation results for each bowl

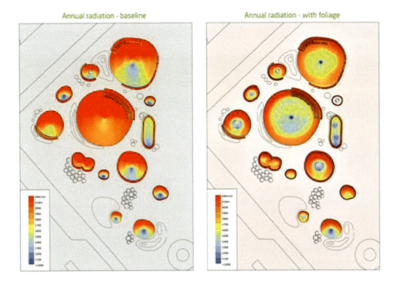

Fig. 10 Comparison between annual incident radiation with and without foliage

Fig. 11 Total station sensor installed at the center of the bowl

The intention of this chapter has been to focus and explore, with the help of delivered research project, how blending data farming and interpretation, with advanced computational skills under the design lead approach can help Architects, engineers, and Environmental designers to overcome the challenges our industry is currently facing.

Fig. 12 Shot of the large central bowl during final phases of installation

References

1. Janssen, J.: A bioclimatic comfort design toolkit for high-rise buildings. CTBUH J. **2017**(II), 20–26 (2017)
2. Navarro, A.L., Cadena, J.D.B., Favoino, F., Donato, M., Poli, T., Perino, M., Overend, M.: Occupant-centred control strategies for adaptive facades: preliminary study of the impact of shortwave solar radiation on thermal comfort. In: Conference: Building Simulation 2019: 16th Conference of IBPSA (2019)
3. Peters, W., Jacobson, A.R., Sweeney, C., Andrews, A.E., Conway, T.J., Masarie, K., Miller, J.B., Bruhwiler, L.M.P., Pétron, G., Hirsch, A.I., Worthy, D.E.J., van der Werf, G.R., Randerson, J.T., Wennberg, P.O., Krol, M.C., Tans, P.P.: An atmospheric perspective on North American carbon dioxide exchange: CarbonTracker. Proc. Natl. Acad. Sci. **104**(48), (2007)
4. Pacala, S., Birdsey, R.A., Conant, R.T., Davis, K., Hales, B., Jenkins, J.C., Johnston, M., Marland, G., Paustian, K., Wofsy, S.C.: In: King, A.W., Dilling, L., Zimmerman, G.P., Fairman, D.M., Houghton, R.A., Marland, G.A., Rose, A.Z., Wilbanks, T.J. (eds.) The First State of the Carbon Cycle Report (SOCCR): The North American Carbon Budget and Implications for the Global Carbon Cycle, pp. 69–91. US Climate Change Science Program, Washington, DC (2007)
5. Hinzman, L.D., Bettez, N.D., Bolton, W.R., Chapin, F.S., Dyurgerov, M.B., Fastie, C.L., Griffith, B., Hollister, R.D., Hope, A., Huntington, H.P., et al.: Clim. Change **72**, 251–298 (2005)
6. NOAA Homepage. https://gml.noaa.gov/ccgg/carbontracker/

Simulating Energy Renovation Towards Climate Neutrality. Digital Workflows and Tools for Life Cycle Assessment of Collective Housing in Portugal and Sweden

Rafael Campamà Pizarro, Adrian Krężlik, and Ricardo Bernardo

Abstract This chapter compares two digital workflows and tool selection for best practice renovation simulation in Portugal and Sweden. Both workflows are part of ongoing research projects that seek to provide a robust workflow adapted to different user needs and scale implementation. While in Portugal, the focus is on the building, the assessment is scaled up to the neighbourhood in Sweden. Geographic information systems datasets are used to streamline the modelling of buildings. The resulting models are used for the simulation and optimisation of renovation scenarios. These scenarios are evaluated as an equilibrium fit between user well-being (thermal and visual comfort), planet (entire carbon life cycle) and cost-effectiveness (energy and cost efficiency). The workflows presented are an example of computational architecture at work. Both workflows successfully interconnect different databases and disciplines to help the design teams and be a useful working tool for the different stakeholders in a renovation project. A complex context in which being flexible and transparent is necessary to make better and more informed decisions.

Keywords Renovation · Automation · Urban modelling · LCA · Digital workflows · Decarbonization · Performance-based decision-making

United Nations' Sustainable Development Goals 3. Ensure healthy lives and promote well-being for all at all ages · 11. Make cities and human settlements inclusive, safe, resilient and sustainable · 13. Take urgent action to combat climate change and its impacts

---

R. Campamà Pizarro (✉) · R. Bernardo
Energy and Building Design Division, Faculty of Engineering, Lund University, Lund, Sweden
e-mail: rafael.campama_pizarro@ebd.lth.se

A. Krężlik
Faculty of Architecture, University of Porto, Porto, Portugal

1 Introduction

1.1 Background

The urgent need to reduce carbon emissions and mitigate climate change has put the spotlight on buildings and the construction sector, which is responsible for nearly 40% of global greenhouse gas emissions. The European Union (EU) has committed to becoming the first climate-neutral continent by 2050, with Sweden aiming to achieve this goal by 2045. Despite this, around 75% of buildings in the EU are not energy efficient [1], and deep energy renovations remain rare, with an annual rate of only 0.2% in the EU28 [2, 3]. However, energy-efficient renovations and on-site renewable energy production are critical to achieving decarbonisation goals, and scaling up these efforts to entire neighborhoods can have an even more significant impact [1, 4, 5]. Fortunately, digital tools and databases have become accessible to practitioners in the last decade, providing new opportunities to estimate the impact of buildings in their lifecycle.

1.2 The Renovation Wave

The Renovation Wave [6] initiative by the European Commission aims to achieve a deep renovation of existing buildings, with a focus on energy poverty, public buildings, and decarbonization of heating and cooling [7]. The Renovation Wave initiative and the EU's goal to become climate-neutral provide an impetus for the AEC sector to fully embrace the 4.0 paradigm. With the digitization of traditional processes, the AEC sector has an opportunity to implement decarbonization targets and transition from a primary focus on creating new architecture to redesigning the existing buildings through comprehensive renovation workflows. This requires a shift in focus from formal research to a holistic approach that involves the entire supply chain, from first-level training courses to the creation of innovative companies.

Each European region is facing different challenges with renovation. They are related to building tradition, skills, age of the building stock, legislation, climate, and available resources, among the most critical factors [8–10]. From an Industry 4.0 point of view, new renovation strategies should integrate new technologies and make them accessible to the different actors involved [11].

For this transition to Industry 4.0 to be successful, the new workflows should allow different users, whether they have technical knowledge or not, to make use of them. In addition, these processes must be flexible to cover the different stages, from digital design work, through digital manufacturing and construction processes, to smart buildings and their operation, as summarised in Fig. 1.

At the moment, there is no generic digital workflow for all European locations. As there is no homogeneous architecture or construction market. At national level it

Fig. 1 Workflows 4.0 from design to operation

differs in various aspects, such as interpretation of the EU directives to the local policies, maturity of understanding the challenges of sustainability and decarbonisation workflows, digitalisation level, size of companies, and education of professionals. Therefore, any workflow must be flexible and transparent, but at the same time case-specific to be effective. A balance between top-down and bottom-up approaches is necessary to accelerate the implementation of more extensive renovation projects. Authors do not exclude the fact that some of the elements of the proposed workflow may not be accurate for some of the markets in the short term.

1.3 Measuring Carbon in a Building's Lifecycle

Measuring carbon in a building's lifecycle involves using Life Cycle Assessment (LCA), an analysis technique that allows for assessing environmental impacts associated with all the stages of a product's life. LCA assesses the Global Warming Potential (GWP), which associates the amount of carbon monoxide and other greenhouse gases released into the atmosphere with undertaken activities. It is measured using unified metrics for all the gases, CO_2eq. Digitalizing the LCA workflow requires all the stakeholders to register and publish data in an interoperable and transparent way. There is no universal LCA method; however, ISO 14040 standards have been developed as a base for LCA.

LCA 14040 is a framework that evaluates carbon used by the industry, starting from sourcing and manufacturing through design and construction, and operation until demolition. Figure 2 shows LCA 14040 divides a building life cycle into four parts: Stage A (Production and Construction), Stage B (Operation of the Building), Stage C (Deconstruction and Waste Processing), and Stage D (Building Afterlife).

Various aspects of the LCA have been researched in Portugal [12], including material studies, comparison between refurbishment and demolition [13], roof retrofitting, embodied and operational carbon of a steel frame building [14], embodied carbon of cladding [15], and the cost and embodied carbon of renovation of traditional architecture [16]. After careful revision of articles on the carbon impact of the LCA stages in Portugal [17, 18], it appears that the most critical are A1–A3 (related to material production) and B6 (related to building operation), which encompass more than 95% of GHG emission of the LCA. Therefore, a balance between these two stages was selected to use for the presented workflow.

In Sweden, studies conducting LCAs for different building typologies have shown that the construction stage (A1–A5) accounts for over 50% of the total carbon impact, followed by the operational energy stage (B6) with around 40%. The trend

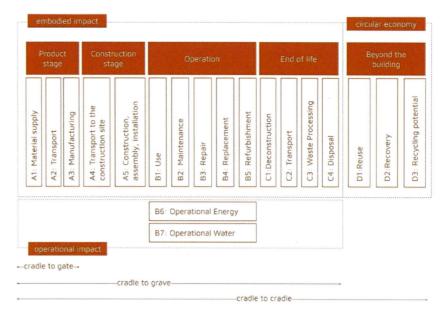

Fig. 2 Life cycle assessment stages in ISO 14040

of construction, especially in the Nordic region, has an increased focus on reducing the carbon footprint of buildings by reducing heat loss to lower operational energy. It is important to find the balance between the two, by using biogenic and low-carbon materials, sourcing locally where possible to minimize transport impacts, and optimizing the construction process to renovate buildings and decrease the existing high energy demand of the current building stock.

1.4 Carbon Neutral Definitions

Buildings and their construction contribute to approximately one-fifth of Sweden's greenhouse gas emissions, making it crucial for the sector to adopt methods that support low emissions and carbon neutrality [19, 20]. While the Swedish government has introduced a climate declaration requirement for all new buildings constructed after January 1, 2022, and plans to include the end-of-life stage in climate declarations from 2027, there is a lack of definitions and guidelines on how to achieve carbon neutrality for specific buildings, especially during renovation projects.

To address this, various carbon–neutral definitions have been developed and implemented around Sweden, Europe and globally. The Swedish Green Building Council (SGBC) [21] created NollCO$_2$ [22, 23] as a method for certifying new buildings targeting net-zero climate impact. According to NollCO$_2$, a building is considered "climate-neutral" when its construction, operation, and end-of-life aspects are

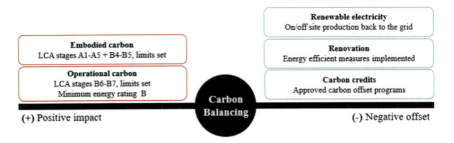

Fig. 3 NollCO$_2$ carbon balance considerations. *Source* Adapted from [19]

weighed against any climate actions or offsets that balance the total climate impact to zero. NollCO$_2$'s carbon balance considerations are shown in Fig. 3.

However, it is important to note that carbon neutrality is not well-defined, and the lack of clarity is even more critical in building renovation projects. It is unclear which stages should be considered and which compensation measures should be used [24]. Therefore, flexible processes that can adapt to different definitions and success factors of various stakeholders are necessary.

Existing tools and databases could empower the entire renovation process, from design to decision-making, and assist in achieving carbon neutrality [25–27]. Moreover, the use of renewable energy sources and the adoption of energy-efficient practices during construction and operation can significantly reduce emissions.

In summary, while the adoption of carbon–neutral certification methods is a step in the right direction, it is crucial to have clear guidelines and definitions for achieving carbon neutrality during building renovation projects. The use of flexible processes, existing tools, and databases can support decision-making and assist in achieving carbon neutrality goals.

2 Creating a Robust Workflow for Renovation

The proposed workflow for Portugal and Sweden needs to reflect the countries specifics, starting from the climate conditions, building culture and tradition to ownership structure.

2.1 Portugal

To accelerate the transformation, and to enhance thermal comfort levels and reduce energy consumption of an existing building stock a new digital workflow is needed. The workflow proposed for deep renovation and retrofitting in Portugal is divided into five phases.

Virtual Model (VM) Building. The model is based on the existing documentation (e.g. original project, evacuation roots drawings), on-site survey and photogrammetry. The 3D model is constructed to reflect the requirements of (a) energy and daylight modelling, (b) calculation of the quantity of materials used in the building.

Building Energy Model (BEM). Completes the VM with climate data, information on materials and construction systems, program, passive systems, occupation and operational schedules, users and their behavior and HVAC systems and their settings and more. All data is compiled into a BEM and later an energy performance simulation is run. The outcome of this step is the total energy consumption in the period of 30 years.

Embodied Carbon. The VM is used to quantify the materials used in the building, it is measured according to the functional units[1] declared in the Environmental Product Declaration or other available data sets.

Building Daylight Model. The VM and BEM data are completed with sensor distribution grid, daylight settings and simulation thresholds. The outcome of the simulation is spatial Daylight Autonomy.

Operational Carbon is a product of total energy consumption in the operation phase and values of the energy mix projection for the next 30 years, including the official decarbonisation plan. A simple calculation with prediction is made using native Grasshopper components.

A crucial element of the workflow is model calibration and comparison with existing benchmarks. Such a practice reduces allows to reduce errors and incorrect assumption. Only after calibration does a full-scale simulation brings result close to the actual building performance.

Once all the simulation is completed values for three metrics are obtained:

- Energy Balance (kWh/m^2/year), which allows comparing the impact of energy flows (heating, cooling, lighting, infiltration) throughout the year on the overall energy consumption, Thermal Comfort demonstrates levels of standard and adaptive comfort throughout the year and in different rooms,
- Life Cycle Carbon (kgCO$_2$ eq) combines the values of embodied carbon and operation carbon but excludes end of life carbon (due to lack of data)
- Daylight Autonomy (%) evaluates how much of the simulated spaces receive enough daylight according to pre-established thresholds.

A graphical abstract of the workflow can be seen in Fig. 4. All the phases are strictly interconnected. To perform the Energy Modelling, the Virtual Model is necessary. The Daylight Model is built on top of the Energy Model. Material data used for Energy Model and Daylight Model are used to calculate the Embodied Carbon For better understating of multiple options, a Pareto front representation is draw. There

---

[1] Functional Unit (f.u.) is a unit that allows comparing different materials with the same purpose, for example a 1 m^2 of insulation of the thermal transmittance of 1.0 W/m^2 K regardless the thickness of the material, or a column resistance of 2,222 kN regardless it thickness or material.

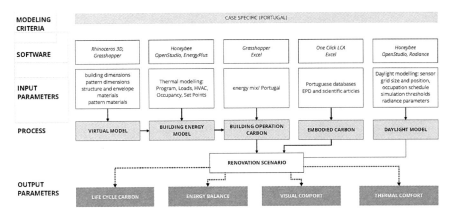

Fig. 4 Graphical abstract for the workflow of Portugal

is no best solution but a range of non-dominated best renovation scenarios. Each of them performs the best in some or other categories.

The renovation methodology workflow responds to homeownership structure in Portugal. Most of the houses are owned by the tenants, which means that renovation will happen on a building scale.

2.2 Sweden

Similarly to many other countries in Europe, Sweden experienced a period of intensive building construction in the post-war years. Even today, 40% of the country's total multi-family housing stock dates from this period [28].

Consequently, there is room for significant improvement in energy performance. In addition, after almost half a century, most of these buildings require considerable renovation. Therefore, there is an excellent opportunity to refurbish buildings of this period, recognised for having the highest energy-saving potential in the Swedish building stock [29]. This is the reasoning why the authors decided to test this group of buildings as a priority in the Swedish workflow. Indeed, these buildings have relatively simple geometrical shapes, which facilitates their modelling. Additionally, are grouped in great numbers in whole neighborhoods and districts, which helps scale the evaluation to urban areas. The process is graphically summarised according to its main stages in Fig. 5.

The workflow proposed for deep renovation and retrofitting in Sweden is divided into five phases:

Selection of the building or neighbourhood. A preliminary characterisation is carried out to analyse several relevant aspects: urban metrics and geographic and climatic aspects.

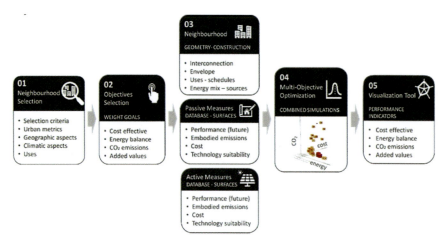

Fig. 5 Main stages of the Swedish workflow

Selection of objectives. To tailor the process of modelling, simulation, and optimisation; it is necessary to define the objectives and which indicators are most suitable and informative for the end-user. Since the focus is on the decarbonisation potential of existing neighbourhoods, the objectives are defined as an equilibrium fit between cost, energy balance, CO_2 emissions, and other added values such as thermal comfort. To evaluate the usefulness of the workflow and what goals should be considered, a group of property owners, designers and researchers with experience in renovation projects in Sweden joined the research as reference group. The objectives varied significantly when moving from one building to an entire neighborhood, where improvement of the common and outdoor spaces, along with other co-benefits became most relevant. Overall, the main objectives focused on reducing environmental impact, reducing energy consumption, and improving indoor comfort and energy resilience. Economic profitability was not considered as important as cost-effectiveness, i.e. not seeking a better economic return through the renovation, but rather making it as economically viable as possible and maximising use of the available funds.

Modelling and characterization. The parameters that will define the building model are assigned. These variables are structured into geometry-dependent and database-dependent variables. These variables are extracted from various sources and optimized before being transferred to a BPS engine. The modelling process is automated using freely available GIS sources, where the user enters the building address or selects the building on a map. The associated footprint is downloaded from OpenStreetMap, and metadata from the footprint is used to obtain the corresponding Energy Performance Certificate (EPC), from which the number of floors and heated floor area can be obtained [30]. Other necessary data for geometry generation, such as average height values and thermal properties, are obtained from existing databases.

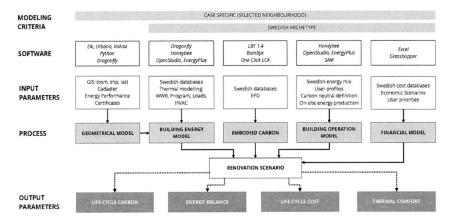

Fig. 6 Graphical abstract for the workflow of Sweden

This section provides a brief overview of the building energy modelling process, indicating that it is not an in-depth description. The authors may have chosen to provide a concise summary because they have already published an article that specifically details the energy modelling process [30].

As detailed in Fig. 6, the procedure is case specific for the geometrical model but provides approximate inputs when better data is unavailable. Therefore, a BEM can be automatically generated from the footprint and no other inputs from the user. Indeed, the monitored annual energy need for heating from the EPC is used to benchmark the resulting energy need from the simulation. The number of stories obtained from to the Energy Performance Certificate corresponding to each footprint.

Simulation and Evaluation

Resulting BEM or Urban Energy Models (UEM) are simulated. A baseline as predefined "anyway renovation" is assigned. Depending on the number of variables and steps defined as inputs, the different iterations for renovation scenarios are calculated. Various simulation tools are coupled depending on the number of buildings to simulate. Since the entire workflow is natively working in Grasshopper, other plugins can be added to the "workflow ecosystem". Indeed, different optimisation plugins, with different built-in algorithms, were studied to actively influence the number of simulations needed to obtain a specific (single) objective or combination or range of different (multi) objectives. The studied optimisation plugins are shown in more detail in the Tools sections.

Visualisation Tool

Due to the parametric simulation approach, large amounts of data are handled as outputs. To handle the input and output parameters, an interactive parallel coordinate diagram is employed.

Parallel coordinates graphs are a visualization tool that allows the user to plot multiple variables on parallel axes and visualize their relationships [31]. This technique is particularly useful for analysing large datasets and identifying patterns and relationships between variables. Several studies have demonstrated the effectiveness of parallel coordinates as a visualization tool for building performance analysis [32–34]. The advantages of parallel coordinates for building renovation scenarios include the ability to explore a large dataset in a single visualization, to interact with the data in a flexible and intuitive way, and to easily identify outliers and anomalies in the data. Based on the demonstrated strengths and capabilities of parallel coordinates graphs in the literature [35–38] we are confident that employing this visualization tool in our study will contribute to a robust and comprehensive analysis of building renovation scenarios.

Figure 7 shows a parallel coordinate diagram with the total number of iterations obtained for a number of different input and output variables (Fig. 7).

This type of diagram groups both inputs and outputs and scales the available range of data per variable in the vertical parallel axis. Each axis represents a variable, and each line crossing the axis represents an iteration (Fig. 8).

In Fig. 9, it can be seen how users can filter the results according to their interests. Maximising energy savings, minimising environmental impact or defining an available cost range are objective examples that can be selected intuitively. Once the objectives are selected, the different iterations leading to those results are highlighted.

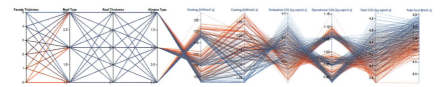

Fig. 7 Parallel coordinates diagram containing all the resulting iterations

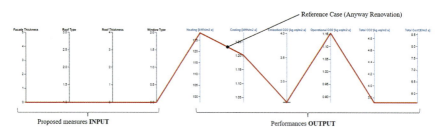

Fig. 8 Definition of the baseline condition in the parallel coordinates diagram (in this case anyway renovation) for benchmarking

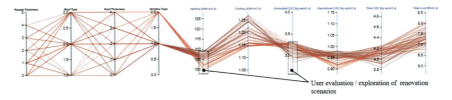

Fig. 9 Parallel coordinates diagram with two range-filters applied by the user to the building performance outputs. All the renovation scenarios complying with those outputs are highlighted

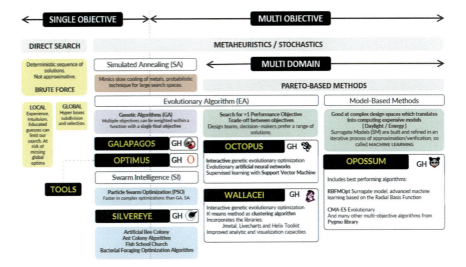

Fig. 10 Optimisation tools and plug-ins evaluated and tested in the Grasshopper 3D environment

3 Tools

3.1 Geometry Model/Virtual Model (GM/VM)

For both workflows, the models are generated from CAD software (Rhinoceros and Grasshopper). Additionally, Grasshopper, a Visual Programming tool, will serve as a platform to connect all the steps of the proposed workflow. Grasshopper is integrated in Rhinoceros 3D modeler, allowing for native interoperability. The choice of Grasshopper/Rhinoceros over Dynamo/Revit or other modelers was motivated by a greater availability of open-source plugins in Rhinoceros 3D (Food4Rhino) [38]. However, it is possible to integrate Grasshopper with Revit or other formats, although the study was limited to the Rhinoceros 3D environment. To create a Building Energy Model, Ladybug Tools (LBT) [39, 40] are employed. LBT is a set of parametric tools and a user interface that allow energy modelers and architects to define parameters such as occupancy schedules, HVAC setpoints, and programs.

Later the generated numerical model is exported to a.idf file that will be analysed by EnergyPlus. For the block and neighbourhood scales, Dragonfly's [40] LBT plugin is used in parallel to Honeybee's LBT. This translates into adding the UrbanOPT layer over Energy Plus as the simulation engine. URBANopt is a Software Development Kit (SDK)—a collection of open-source modules focused on underlying analytics for a variety of multi-building design and analysis use cases using OpenStudio and EnergyPlus [41]. EnergyPlus is a whole-building energy simulation engine to model both energy consumption—for heating, cooling, ventilation, lighting, and process and plug loads, it was developed with the support of the US Department of Energy Building Technologies Office.

To create a Building Daylight Model again LBT tools are employed, they serve as a tool to set-up parameters like occupancy schedules and material radiance. The model is connected to Radiance and Dayism. Radiance is a raytracing engine integrating a suite of tools used to perform lighting simulation, principally daylight analysis and glare. Daysim, daylighting analysis software, serves for simulation of annual daylight performance.

One-Click LCA and Bombyx are two software used for the calculation of the embodied carbon in a building. One-Click LCA uses material data from multiple partners and also has a generic database for materials. It is of the highest importance to use a national or regional database, ideally using GWP data provided by the material producer, otherwise, the calculation is likely to be inaccurate.

For the Portugal workflow, other calculations such as operational carbon and financial model are calculated with Excel, data are exported and imported to and from Grasshopper.

For the Swedish workflow, cost calculations are integrated in the Grasshopper environment and coupled to the energy and the embodied carbon models. The financial scenarios and LCC calculations are also integrated in the Grasshopper environment. The goal for this full integration in Grasshopper was to enable complete interoperability with other tools, from GIS, to optimisation and data handling plugins available in that same Grasshopper ecosystem. In Fig. 11, the optimisation tools tested in both workflows are summarised and organised according to their approach regarding objectives (single or multiple) and the used algorithms and libraries [33, 42–53].

Finally, an overall caption of the whole Swedish workflow, as seen in the Grasshopper canvas, is shown in Fig. 12. The script integrates a traditional building energy modelling process with input flows from GIS and open access databases in order to process different building(s) renovation performances. Other Grasshopper plugins can be coupled to include additional capabilities, such as (but not limited to) optimisation and automation.

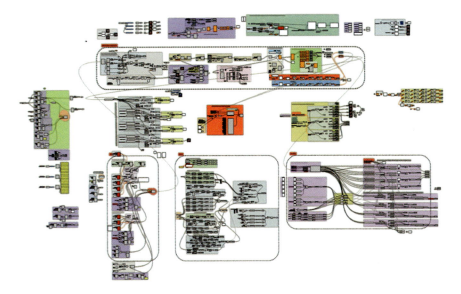

Fig. 11 Grasshopper definition script of the Swedish workflow, as seen in the Grasshopper visual programming canvas

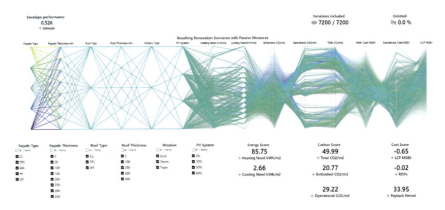

Fig. 12 Graphical user interface proposed for renovation scenarios evaluation

3.2 Graphical User Interface (GUI) for Evaluation

In this section, we introduce a Graphical User Interface (GUI) designed to facilitate the visualization and analysis of the various renovation iterations simulated in our study. The GUI is shown in Fig. 12 and employs parallel coordinates to effectively display the multi-dimensional input and output data, enabling users to efficiently compare and evaluate the performance of the renovation scenarios in terms of energy, cost, and environmental aspects.

To demonstrate the utility of the GUI, we present a step-by-step decision-making process based on an overall best performance profile, which was tested with users to ensure its effectiveness in guiding informed choices for sustainable renovation strategies.

Step 1: Optimal facade thickness is determined by balancing energy, carbon, and costs scores. Users evaluate different facade thicknesses in the GUI, identifying the most appropriate solution that meets the performance criteria (Fig. 13).

Step 2: Similarly, the optimal roof thickness is determined by evaluating the trade-offs between energy, carbon, and costs scores. Users analyze various roof thicknesses using the GUI to select the most suitable option (Fig. 14).

Step 3: Users assess whether to keep, upgrade, or change the windows based on their performance in terms of energy, carbon, and costs scores. The GUI allows users to

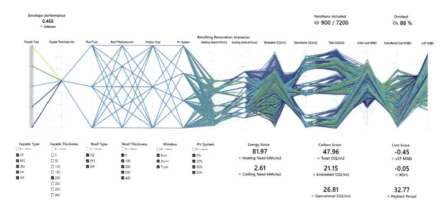

Fig. 13 Optimal facade thickness is 200 mm

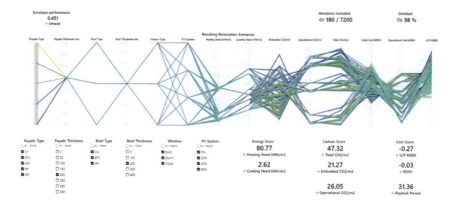

Fig. 14 Optimal roof thickness is 200 mm

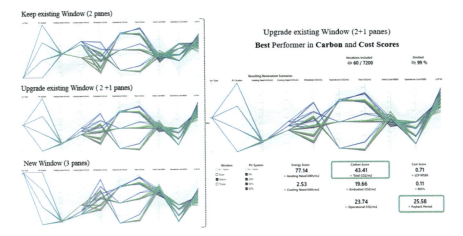

Fig. 15 Optimal window measure is upgrade existing window with an additional pane on the exterior side

visualize and compare the impact of each window option on the overall renovation strategy (Fig. 15).

Step 4: The inclusion of photovoltaic (PV) solar panels is evaluated, considering the available roof area and the potential contribution to the building's energy performance. The GUI helps users to determine the optimal PV panel coverage by visualizing the trade-offs between energy generation, carbon emissions, and costs (Fig. 16).

Step 5: Lastly, users decide on the insulation materials for the facade and roof improvements. This variable is considered last, as it has the least influence on the overall performance of the renovation strategy (Fig. 17).

The GUI allows users to assign different colors to each insulation material, which aids in the visualization of their impact. When the outputs display distinct color zones, it signifies a higher influence of the material on the variability of outputs. Conversely, a blurred mix of colors indicates a lower influence of the material on the outputs. A Figure illustrating this "coloring" difference is presented, further emphasizing the importance of material selection in the decision-making process (Fig. 18).

Through this step-by-step example, we showcase the applicability and effectiveness of the proposed methodology in guiding informed decisions for sustainable renovation strategies, using the GUI to visualize and analyze the trade-offs between energy, carbon, and costs scores in a comprehensive and user-friendly manner.

Fig. 16 Optimal roof utilization for PV panels is 80 of the available area

Fig. 17 Insulation material choice for the façade and roof

4 Discussion and Conclusion

The proposed workflows have demonstrated their effectiveness for modeling and assessing existing buildings and neighborhoods in Portugal and Sweden. In Fig. 11 we can see a graphical overview of the two workflows, where the tools used for modelling are highlighted. The Portuguese workflow excels at the individual house

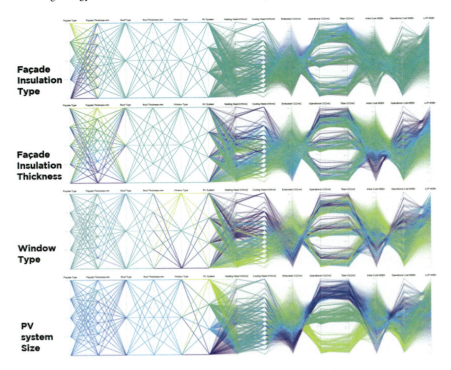

Fig. 18 Impact of different measures on the final renovation relative performances

and building scale, whereas the Swedish workflow is tailored for the neighborhood scale. These complementary approaches offer a comprehensive and adaptable framework for analysing renovation scenarios and identifying context-specific strategies for sustainable urban development (Fig. 19).

The integration of both workflows within the Grasshopper environment enables a seamless transition between the Honeybee and Dragonfly plugins from Ladybug Tools. This interoperability facilitates the exchange of energy models, allowing the two processes to benefit from each other's strengths, thereby enhancing the overall modeling capability. Despite challenges in data availability and quality, the workflows successfully provide meaningful results by incorporating average values from archetypes when required.

In conclusion, the proposed workflows offer promising solutions for modeling and assessing renovation scenarios in Portugal and Sweden, while considering the influence of Industry 4.0 on architecture and sustainable urban development. The flexibility, adaptability, and potential for future improvements of these workflows hold great promise for advancing the field in this context. Recognizing that the workflows are still in their experimental phase, further calibration of the energy and

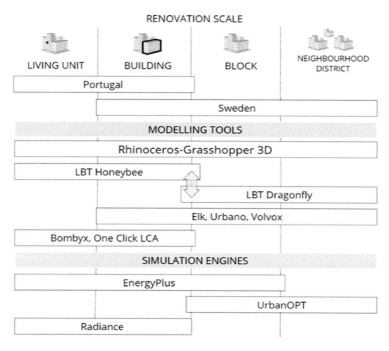

Fig. 19 Comparison of the renovation scales, modelling tools and simulation engines of each workflow

carbon models is necessary to improve their accuracy and reliability. Future work should focus on refining the workflows by:

- Incorporating additional data sources to enhance the quality and completeness of the input data, thereby reducing the reliance on average values from archetypes.
- Expanding the application of the workflows to various building types and urban contexts to validate their performance and versatility.
- Collaborating with local stakeholders, such as urban planners, architects, and policymakers, to ensure that the workflows address real-world challenges and contribute to sustainable urban development.
- Investigating the integration of other digital tools, such as BIM and other design software, to further enhance the workflows' capabilities and enable a more comprehensive assessment of renovation scenarios.
- Exploring the potential of machine learning and artificial intelligence techniques to optimize the workflows, streamline the modeling process, and generate data-driven insights.

By continuing to refine these workflows and exploring new avenues for collaboration and innovation within the framework of Industry 4.0, researchers and practitioners alike can develop more effective strategies to address the pressing challenges of urban sustainability. The integration of these workflows with Industry 4.0 technologies will ultimately contribute to the transformation of the architectural and construction sectors, fostering a more sustainable and efficient built environment.

Funding This research was jointly supported by the funding received by the authors.

Adrian Krężlik was funded by the MPP2030-FCT PhD Grant, financed by the Portuguese Foundation for Science and Technology (FCT), under the MIT Portugal Program.

Rafael Campamà Pizarro and Ricardo Bernardo were funded by The Swedish Energy Agency, as part of the project "Renovation towards Climate Neutral Neighbourhoods in Sweden" (project number: 51696-1).

References

1. European Commission: Comprehensive study of building energy renovation activities and uptake of NZEB in the EU. Brussel (2019). https://doi.org/10.2833/14675
2. Meijer, F., Itard, L., Sunikka-Blank, M.: Comparing European residential building stocks: performance, renovation and policy opportunities. Build. Res. Inf. **37**(5–6), 533–551 (2009). https://doi.org/10.1080/09613210903189376
3. Ástmarsson, B., Jensen, P.A., Maslesa, E.: Sustainable renovation of residential buildings and the landlord/tenant dilemma. Energy Policy **63**, 355–362 (2013). https://doi.org/10.1016/J.ENPOL.2013.08.046
4. Ferreira, M., Almeida, M.: Benefits from energy related building renovation beyond costs, energy and emissions. Energy Procedia **78**, 2397–2402 (2015). https://doi.org/10.1016/J.EGYPRO.2015.11.199
5. Annex 56: cost effective energy and carbon emissions optimization in building renovation project. http://www.iea-annex56.org/. Accessed 14 Sept 2020
6. European Commission: A European green deal | European Commission. https://ec.europa.eu/info/strategy/priorities-2019-2024/european-green-deal_en. Accessed 09 Sept 2020
7. European Commission: Comprehensive study of building energy renovation activities and the uptake of nearly zero-energy buildings in the EU—Publications Office of the EU. https://op.europa.eu/en/publication-detail/-/publication/97d6a4ca-5847-11ea-8b81-01aa75ed71a1/language-en/format-PDF/source-119528141 (2019). Accessed 07 Nov 2022
8. Cattano, C., Valdes-Vasquez, R., Plumblee, J.M., Klotz, L.: Potential solutions to common barriers experienced during the delivery of building renovations for improved energy performance: literature review and case study. J. Archit. Eng. **19**(3), 164–167 (2013). https://doi.org/10.1061/(ASCE)AE.1943-5568.0000126
9. Artola, I.: Directorate General for Internal Policies Policy Department A: Economic and Scientific Policy Boosting Building Renovation: What Potential and Value for Europe? (2016)
10. European Parliament: Europaparlamentets och rådets direktiv 2010/31/EU av den 19 maj 2010 om byggnaders energiprestanda (2010)
11. Pacheco-Torgal, F., Faria, J., Jalali, S.: Embodied energy versus operational energy. Showing the shortcomings of the energy performance building directive (EPBD). Mater. Sci. Forum **730–732**, 587–591 (2013). https://doi.org/10.4028/WWW.SCIENTIFIC.NET/MSF.730-732.587
12. Brás, A., Gomes, V.: LCA implementation in the selection of thermal enhanced mortars for energetic rehabilitation of school buildings. Energy Build. **92**, 1–9 (2015). https://doi.org/10.1016/J.ENBUILD.2015.01.007
13. Gaspar, P.L., Santos, A.L.: Embodied energy on refurbishment vs. demolition: a southern Europe case study. Energy Build. **87**, 386–394 (2015). https://doi.org/10.1016/J.ENBUILD.2014.11.040

14. Rodrigues, C., Freire, F.: Integrated life-cycle assessment and thermal dynamic simulation of alternative scenarios for the roof retrofit of a house. Build. Environ. **81**, 204–215 (2014). https://doi.org/10.1016/J.BUILDENV.2014.07.001
15. Librelotto, L.I., Kekez, M., Bártolo, H.M.G.: The environmental impact of an ETICs layer: a case of study with life cycle assessment (LCA) from environmental product declaration (EPD) in Portugal. MIX Sustentável **6**(2), 139–148 (2020). https://doi.org/10.29183/2447-3073.MIX 2020.V6.N2.139-148
16. Rodrigues, F., Matos, R., Alves, A., Ribeirinho, P., Rodrigues, H.: Building life cycle applied to refurbishment of a traditional building from Oporto, Portugal. J. Build. Eng. **17**, 84–95 (2018). https://doi.org/10.1016/J.JOBE.2018.01.010
17. Venkatarama Reddy, B.V., Jagadish, K.S.: Embodied energy of common and alternative building materials and technologies. Energy Build. **35**(2), 129–137 (2003). https://doi.org/10.1016/S0378-7788(01)00141-4
18. Mourão, J., Gomes, R., Matias, L., Niza, S.: Combining embodied and operational energy in buildings refurbishment assessment. Energy Build. **197**, 34–46 (2019). https://doi.org/10.1016/J.ENBUILD.2019.05.033
19. Daya, B., Nolan, H.: Energy renovation packages for the decarbonisation of Swedish multi-family buildings. http://lup.lub.lu.se/student-papers/record/9086172 (2022). Accessed 17 Apr 2023
20. Gremmelspacher, J.M., Campamà Pizarro, R., van Jaarsveld, M., Davidsson, H., Johansson, D.: Historical building renovation and PV optimisation towards NetZEB in Sweden. Sol. Energy **223**, 248–260 (2021). https://doi.org/10.1016/J.SOLENER.2021.02.067
21. Manualer och verktyg för Miljöbyggnad. Sweden Green Building Council. https://www.sgbc.se/certifiering/miljobyggnad/anvandarstod-for-miljobyggnad/manualer-och-verktyg-for-certifiering-i-miljobyggnad/. Accessed 17 Apr 2023
22. Stoll, P.: NollC02 Ny Byggnad. https://www.sgbc.se/app/uploads/2020/04/NollCO2-Remiss manual-20200417.pdf (2020). Accessed 14 Sept 2020
23. Manualer och ramverk för NollCO2. Sweden Green Building Council. https://www.sgbc.se/certifiering/nollco2/anvandarstod-for-nollco2/manualer-och-ramverk-for-nollco2/. Accessed 17 Apr 2023
24. Aive, N.T., Razna, R.: Climate-neutral buildings—impact of existing definitions on building design. http://lup.lub.lu.se/student-papers/record/9097599 (2022). Accessed 17 Apr 2023
25. Nair, G., Fransson, Å., Olofsson, T.: Perspectives of building professionals on the use of LCA tools in Swedish climate declaration. E3S Web Conf. **246**, 13004 (2021). https://doi.org/10.1051/E3SCONF/202124613004
26. Buyle, M., Braet, J., Audenaert, A.: Life cycle assessment in the construction sector: a review. Renew. Sustain. Energy Rev. **26**, 379–388 (2013)
27. Basbagill, J., Flager, F., Lepech, M., Fischer, M.: Application of life-cycle assessment to early stage building design for reduced embodied environmental impacts. Build. Environ. **60**, 81–92 (2013)
28. Hall, T., Vidén, S.: The million homes programme: a review of the great Swedish planning project. Plan. Perspect. **20**(3), 301–328 (2005). https://doi.org/10.1080/02665430500130233
29. Ministry of the Environment: Sweden's long-term strategy for reducing greenhouse gas emissions (2020)
30. Campamà Pizarro, R., Bernardo, R., Wall, M.: Streamlining building energy modelling using open access databases—a methodology towards decarbonisation of residential buildings in Sweden. Sustainability **15**(5), 3887 (2023). https://doi.org/10.3390/SU15053887
31. Inselberg, A.: Parallel Coordinates: Visual Multidimensional Geometry and Its Applications, pp. 1–554 (2009). https://doi.org/10.1007/978-0-387-68628-8/COVER
32. Fua, Y.-H., Ward, M.O., Rundensteiner, E.A.: Hierarchical parallel coordinates for exploration of large datasets. http://www.spss.com/software/diamond. Accessed 17 Apr 2023
33. Schüler, N., Cajot, S., Peter, M., Page, J., Maréchal, F.: The optimum is not the goal: capturing the decision space for the planning of new neighborhoods. Front. Built Environ. **3** (2018). https://doi.org/10.3389/FBUIL.2017.00076

34. Cajot, S., Schüler, N., Peter, M., Koch, A., Maréchal, F.: Interactive optimization with parallel coordinates: exploring multidimensional spaces for decision support. Front. ICT **5**, 32 (2018). https://doi.org/10.3389/FICT.2018.00032/BIBTEX
35. Johansson, J., Forsell, C.: Evaluation of parallel coordinates: overview, categorization and guidelines for future research. IEEE Trans. Vis. Comput. Graph. **22**(1), 579–588 (2016). https://doi.org/10.1109/TVCG.2015.2466992
36. Claessen, J.H.T., Van Wijk, J.J.: Flexible linked axes for multivariate data visualization. IEEE Trans. Vis. Comput. Graph. **17**(12), 2310–2316 (2011). https://doi.org/10.1109/TVCG.2011.201
37. Guo, P., Xiao, H., Wang, Z., Yuan, X.: Interactive local clustering operations for high dimensional data in parallel coordinates. In: IEEE Pacific Visualization Symposium 2010, PacificVis 2010—Proceedings, pp. 97–104 (2010). https://doi.org/10.1109/PACIFICVIS.2010.5429608
38. grasshopper · GitHub Topics · GitHub. https://github.com/topics/grasshopper. Accessed 28 Nov 2022
39. Ladybug Tools | Honeybee. https://www.ladybug.tools/honeybee.html. Accessed 01 Mar 2023
40. Ladybug Tools | Dragonfly. https://www.ladybug.tools/dragonfly.html. Accessed 01 Mar 2023
41. Charan, T., et al.: Integration of open-source urbanopt and dragonfly energy modeling capabilities into practitioner workflows for district-scale planning and design. Energies (Basel) **14**(18), (2021). https://doi.org/10.3390/EN14185931
42. Gutmann, H.M.: A radial basis function method for global optimization. J. Global Optim. **19**(3), 201–227 (2001). https://doi.org/10.1023/A:1011255519438
43. Longo, S., Montana, F., Riva Sanseverino, E.: A review on optimization and cost-optimal methodologies in low-energy buildings design and environmental considerations. Sustain. Cities Soc. **45**, 87–104 (2019). https://doi.org/10.1016/j.scs.2018.11.027
44. Ferrara, M., Fabrizio, E., Virgone, J., Filippi, M.: A simulation-based optimization method for cost-optimal analysis of nearly zero energy buildings. Energy Build. **84**, 442–457 (2014). https://doi.org/10.1016/j.enbuild.2014.08.031
45. Wortmann, T.: Model-based optimization for architectural design: optimizing daylight and glare in Grasshopper. Technology|Architecture + Design **1**(2), 176–185 (2017). https://doi.org/10.1080/24751448.2017.1354615
46. Waibel, C., Evins, R., Wortmann, T.: Are genetic algorithms really the best choice for building energy optimization?. In: Proceedings of the Symposium on Simulation for Architecture & Urban Design. https://www.academia.edu/33362629/Are_Genetic_Algorithms_Really_the_Best_Choice_for_Building_Energy_Optimization. Accessed 14 Sept 2020
47. Wortmann, T., Costa, A., Nannicini, G., Schroepfer, T.: Advantages of surrogate models for architectural design optimization. Artificial intelligence for engineering design. Anal. Manufact. **29**, 471–481 (2015). https://doi.org/10.1017/S0890060415000451
48. Wortmann, T., Costa, A., Nannicini, G., Schroepfer, T.: Advantages of surrogate models for architectural design optimization. In: Artificial Intelligence for Engineering Design, Analysis and Manufacturing: AIEDAM, pp. 471–481. Cambridge University Press (2015). https://doi.org/10.1017/S0890060415000451
49. Cichocka, J., Browne, W.N., Ramirez, E.R.: Optimization in the architectural practice an international survey. In: Conference: 22nd International Conference on Computer-Aided Architectural Design Research in Asia (CAADRIA 2017), Hong Kong, 2017. https://www.researchgate.net/publication/316089266_OPTIMIZATION_IN_THE_ARCHITECTURAL_PRACTICE_An_International_Survey. Accessed 14 Sept 2020
50. (10) (PDF) Multi-objective optimization for daylight retrofit. https://www.researchgate.net/publication/341776761_Multi-Objective_Optimization_for_Daylight_Retrofit#fullTextFileContent. Accessed 14 Sept 2020
51. Costa, A., Nannicini, G.: RBFOpt: an open-source library for black-box optimization with costly function evaluations. Math. Program. Comput. **10**, 597–629 (2018). https://doi.org/10.1007/s12532-018-0144-7
52. Jalali, Z., Noorzai, E., Heidari, S.: Design and optimization of form and facade of an office building using the genetic algorithm. Sci. Technol. Built Environ. **26**(2), 128–140 (2020). https://doi.org/10.1080/23744731.2019.1624095

53. Lin, S.H., Gerber, D.J.: Evolutionary energy performance feedback for design: multidisciplinary design optimization and performance boundaries for design decision support. Energy Build. **84**, 426–441 (2014). https://doi.org/10.1016/j.enbuild.2014.08.034

Configurator: A Platform for Multifamily Residential Design and Customisation

Henry David Louth, Cesar Fragachan, Vishu Bhooshan, and Shajay Bhooshan

Abstract Game technologies in Architecture, Engineering and Construction (AEC) industries are currently utilised for a variety of analytical, and single author applications such as test fitting, simulated city fabric, and evaluation of feasible solution sets. Advances in materials and fabrication technologies, design for manufacture and assembly (DfMA), and industrialised construction of building components, continue to shift the housing paradigm from standardisation toward mass customisation. These recent developments are trending toward user focus, negotiated planning, and choice prioritisation in design, production planning, and manufacture. The authors' research is motivated by game technologies' suitability to negotiate problems facing integration of design customisation, user choice, and negotiated governance in supply chain integration and procurment pipeline. The paper presents the author's research into decentralised multi-author decision making, co-authorship, contribution of digital experts, and incentivisation models. Game engine technology is outlined to deliver user-focused, participation-driven, mass-customised housing outcomes. A real-time online platform use case configurator and the corresponding digital tool-chain integration is presented and discussed. The multiplayer gameplay of such results in construction feasible customised housing developments. The footprint, unit mix, and spatial organisation of which, conventionally authored by an architect or developer, herein is authored by the participants aggregate decisions.

Keywords Gamification · Modularity · Mass customisation · Housing · Industrialised construction

---

H. D. Louth (✉) · C. Fragachan · V. Bhooshan · S. Bhooshan
Computation and Design Group (CODE), Zaha Hadid Architects (ZHA), London, UK
e-mail: henry.louth@zaha-hadid.com

C. Fragachan
e-mail: cesar.fragachan@zaha-hadid.com

V. Bhooshan
e-mail: vishu.bhooshan@zaha-hadid.com

S. Bhooshan
e-mail: shajay.bhooshan@zaha-hadid.com

© The Author(s), under exclusive license to Springer Nature Switzerland AG 2024
M. Barberio et al. (eds.), *Architecture and Design for Industry 4.0*, Lecture Notes in Mechanical Engineering, https://doi.org/10.1007/978-3-031-36922-3_40

1 Introduction

The technology platform presented, developed by the authors, is the latest build in the ongoing applied research into game technologies suitability toward architectural design and procurment planning. The research is motivated by game technologies capacity to negotiate consumer and stakeholder priorities, to democratise an emerging digital property marketplace, and to reduce housing procurement timelines. The technology platform and toolchain address problems facing implementation in downstream supply chains and procurement including:

1. modular housing design strategy for mass customisation
2. suitability of prefabricated elements to supply chain (SC) integration
3. continuity issues of an end-to-end digital fabrication toolchain
4. flexible procurement models to shifting consumer demand

The paper details an application of game technology in architectural design, a configurator for multifamily residential design as well as the systems design strategy and components implicit to its creation. Gamification principles, a multiplayer sequential turn-based gameplay, and participant incentivisation result in emergent massing properties in a residential community design development. In a configurator, the needs of the customer, in this instance the occupiers themselves, are met through a customisation process, often autonomously through a virtual interface. The paper discusses the capacity of such a configurator platform to disrupt conventional procurement processes and bring stakeholders together earlier in the design process than in a conventional delivery process. The additional roles and capabilities required by design professionals are outlined including user experience and interface design, backend asset optimisation, database integration, and cloud services deployment.

The game environment demonstrates a real-time digital platform that negotiates variations and decentralised non-expert buyer actions through consensus design planning. The paper argues stakeholder decision-making and game mechanics governed by expert participants' knowledge result in construction feasible, engineering-aware outcomes. The utilisation of such and the corresponding modular *kit-of-parts* benefit early design planning and procurement of the configurator results.

1.1 Contributions and Organisation

This paper describes the use of game technologies in architectural design, the benefits of user-centric design, and a systems design approach utilising modular content for multi-family residential through a use case configurator (Fig. 1). The main novel contributions of the paper are as follows:

1. The configuration of a community of residences by the residents themselves through a technology platform.

Fig. 1 Configurator: game screens

2. The rulesets for the game mechanics are the digitised result of the participation of the engineering disciplines' validation of feasible solution sets in the system design.
3. The curvilinear components in architectural design are procedurally created through a discretised volumetric data structure.
4. The mass customisation of multifamily residential units inclusive of positioning, cladding elements, space plan, furnishing, and contractual use rights (Fig. 2).
5. The use of timber cladding elements as a kit of parts is amenable to robotic fabrication supply chain integration (Fig. 3).

Fig. 2 Visualisation of a configuration

Fig. 3 Visualisation of a configuration

The Sect. 2 provides a brief survey of precedent in stake holder participation in the housing domain, outlining the role of industry, manufacture, and supply chain to address customisation, and discusses the motivation and engagement of participants using principles in behavioural science to enhance immersive gameplay. In Sect. 3 we present a brief survey of building industry configurators, their usages, and target audiences and the benefits of such. In Sect. 4 we discuss relevant prior work in game technologies and platform technology development at Zaha Hadid Architects (ZHA) including legacy builds, dashboard development, and content creation. Section 5 describes the configurator use case through the initial Pilot Build and corresponding session results. We discuss the analysis and present the features, functionality, and content modifications planned and validated via engineering participants for the subsequent build. The Sales Release is summarised, and its corresponding session results are presented. In Sect. 6 we present the workflow and custom toolchain for the configurator, the game mechanics, rulesets, and content created for the kit of parts. The technical components of the system architecture are subsequently discussed and detailed. Section 7 describes foreseeable challenges and future work trajectories and conclude in Sect. 8.

2 The Promise Mass Housing as a User Centric Product

Discrete modular organisation in residential design planning and systems design thinking in housing has a longstanding history. Gropius explored the combinatoric opportunity of 'spatial bodies' as a set of distinctive shape typologies in creating

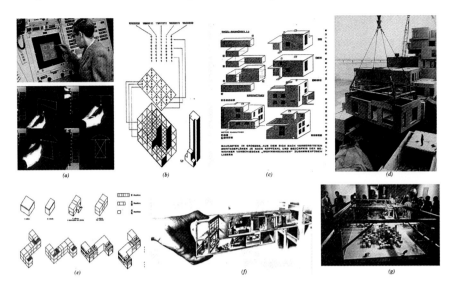

Fig. 4 Configurator seminal projects: **a** Ivan Sutherland Sketchpad [39], **b** Lionel March Binary System of Encoding Volume [71], **c** Walter Gropius and Adolph Meyer Spatial Bodies [1], **d** Habitat 67 [3], **e** Corbusier Quartier Fruges [2], **f** Corbusier Wine Rack Concept for The Unité d'habitation [2], **g** Negroponte Man–Machine [40]

unified spaces [1]. Le Corbusier implemented cellular units and standardised proportioning ratios for the working classes in Quartier Fruges and subsequently conceptualised a dual schema for 'slot-in' flexible interior layouts into a rigid structural armature in Unité d'habitation [2]. The aggregate composition of self-similar repeating units was later explored by the Metabolists and influenced the early prefabricated unit cluster arrangements and the success in Habitat 67 to form community, and access to light, air, and garden space [3] (Fig. 4).

2.1 Platform Approach: Modularity and Mass Customisation

The platform approach to construction references a strategy inherited from automotive industry [4] for system design where variations in design are achieved by adaptation and alteration of components onto a base type (BT) creating instances [5, 6]. Modularity of physical components and assemblies is adapted from fixed positions [7, 8] via a so-called *kit of parts*. Together the system, kit options, and rules augmentation define the platform.

The paradigm shift from mass production to mass customisation (MC) is rooted in the desire to focus products on the customers they serve [9]. Gilmore outlines a customisation framework comprised of strategies in assessing how best to redesign a product or process to engage customers [10]. Mass customisation of spatial qualities and cladding kits varied to demographics is evident in English worker class row housing [11]. Later examples of post war homes catalogues and plan books [12] demonstrate variation to suit a variety of stylistic and spatial demands, as well as tailor to the region and vernacular [13].

Product relevancy is a concern as consumer trends and priorities fluctuate from design to delivery stages [14–16]. The alignment of the product offered and its manufacture in the production phase—a so called agile production and lean manufacture [17, 18]—necessitates continuity of end-to-end digital toolchain in supply chain and procurement [19].

Advances in materials and fabrication technologies [20–22], design for manufacture and assembly (DfMA) [5, 23] and industrialised construction of building components [24], continue to shift the housing paradigm from standardisation using dimensional elements toward mass customisation [9, 12]. These recent developments are trending toward user focus, priority, and choice via agile manufacture [14–16].

Harnessing innovation of fabrication technologies in industrialised construction has demonstrated mixed results taking advantage of mass production capabilities in the production of premanufactured kits as housing products [25]. For instance the conversion of post-war aerospace metal-works factories toward the housing sector by Lustron yielded roughly 2500 homes [20], while Fuller's dymaxion vision net only one such structure and Prouve's Maison Tropicale only three such structures [26]. In Prouve's case the structure cost more than the market housing intended to supplant, while Fuller's suffered from inflexibility of the system to modification, and a general lack of customisation to siting and regional climatic considerations [27]. Edison's housing scheme exhibited assembly and decanting complexities of roughly two thousand unique mould components. Similarly, interiors infill kits and layouts amenable to buildings systems integration such as Matura are equally complex [28].

2.2 Gamification: The Role of Behavioural Science to Engage and Motivate

Gamification is the step of converting a design from a function-focused design (FFD) into a human-focused design (HFD) by employing concepts of behavioural science to engage human emotions through human psychology to motivate and incentivise users [29]. It is more notable for corporate applications, e-learning, and for non-game applications to engage users [30]. Research of benefits of gamification for the player include developing higher-order thinking skills [31] and problem-solving skills [32] in social cooperative settings [33]. As such, there is a changing generational attitude toward work and learning [34].

Early 'Intrinsic Instruction' frameworks identified 'heuristics for designing enjoyable user interfaces' [35]. Contemporary gamification frameworks identify *core drives* [36] while playfulness frameworks identify a broader spectrum of characteristics that exist within humans [37] and motivate us to engage in activities. Further,

Overview of Actors

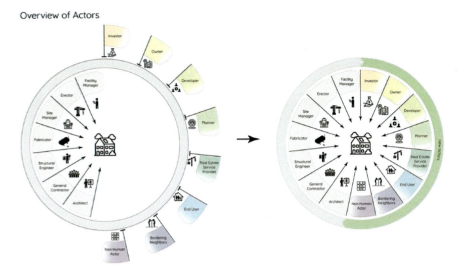

Fig. 5 Stakeholders brought together in a real-time platform

the user interface (UI) components and their integration strategy contribute to viewer perception of immersion and categories of engagement in the active world [34, 38].

User engagement and motivating strategies in the design discipline include Sutherland's Graphical User Interface (GUI) 'sketchpad' is attributed to the birth of computer graphics, drawing as a novel communication medium [39]. Likewise, Negroponte's explorations in the role of man–machine in collaborative negotiated environments [40] is consistent with the role of non-player character (NPC) role to advance, and steer gameplay and in general co-authorship of a negotiated game state (Fig. 4).

The concepts of co-authorship, mutual consensus, and civic responsibility are present in so-called 'DIY urbanism' [41]. Multiplayer dialogue is exhibited through a "panel of experts" participating on behalf of a particular motivator (Fig. 5). Each occupier, regulator, engineer, planner, etc. can be seen in analogue community board games in housing, land utilisation, and town building [42–44] and further digitisation of 'actors' in city building games.

3 State of Configurators in AEC

Configurators are comprised of components (the building blocks), processes (the rules, and taxonomies), a knowledge base (datasets for economic evaluation) and people (the users). These reduce a configurator design process to a series of transformation vectors in distinct domains. Customer attributes (CA), functional requirements (FR), design parameters of the physical solution (DP) and the variables governing the production process (PV) are in separate domains in a so called axiomatic design [45, 46].

Configurator typologies in architecture range depending upon target user audience and address different design stages, the extent of which is extensively discussed in [47]. Life cycle costs can be attributed to early design decision making [48] and therefore, design controls in conceptual stages are financially motivated. The levels of control or governing constraints asserted onto built elements is defined by actors in the system's man-built environment, the community formed by the authority, operator, developer, and occupier [49].

Most notable of the stand-alone application configurators to arrange Floor plans [45, 50, 51] and Buildings Components [52] for specific supply chains include Precast concrete [53], Curtain wall systems [54], Precast concrete panels [55], Timber walls and Floors [24, 45, 56].

Configurators have seen a rise in web-enabled apps as exploration for property search and acquisition, residential test fitting, site planning and land utilisation as early-stage planning toolkits [57–62]. Each offers a form of procedural, and generative creation of feasible solution sets, qualitative analysis, result ranking, and filtering as near real-time in early design in lieu of precision engineering [63], while few offer a collaborative co-authoring platform for project stakeholders as controlling actors [64].

There are a few commercially deployed web-based configurators of which HiStruct Steel structures [65], AGACAD Wood Precast concrete elements [66], Creatomus Private homes and apartments [67] and Projectfrog [68], are most notable.

3.1 Technical Considerations

Configurator technical considerations include the building sector, features and functionality requested. In the case of building components, these include—the geospatial coordinate system, the serialisation of volume into discrete volumetric system grid, so-called voxels allow component hosting [69–72], tiling discrete geometry [73, 89], and procedural environment creation [74, 88] and compression and encoding consideration of such [71] (Fig. 4).

3.2 Benefits

The benefits of configurators are extensively documented in [47]. These include, increases variation [6, 55, 75], ensures feasibility of solutions in the supply chain [45, 50, 53, 55, 56], minimises manual input [50, 54], smoothens on-boarding [45, 50], reduces time and cost for design production [6, 55, 56, 75], enhances coordination efficiency, [24, 53, 70] and preserves knowledge for reuse on the next product [75].

In the trend toward data-driven real estate and property technologies (PropTech), product configurators are well suited to contend as digital marketplaces for real-estate bidding, negotiation, and property valuation. As digital cloud-based applications, they are web 3.0, and democratised decentralised blockchain based collaboration ready [76].

4 Relevant Prior Works

Platform design initiatives started in 2018 and subsequent platform development was formalised through ongoing development at Zaha Hadid Architects (ZHA), Computation and Design Group (CODE) (Fig. 6), in collaboration with The Manufacturing Technology Centre (MTC), and continued teaching affiliations and enquiry in academic settings at Architectural Association, Design Research Lab (AADRL) and UCL Bartlett Research Cluster 10 (RC10) as well as workshops with Digital FUTURES (DF).

4.1 Legacy Configurator Builds

Oikos created for Architectural Association Design Research Lab AADRL, Nahmad/Bhooshan Studio is a cooperative multiplayer residential configurator exploring user-directed design and authored space plans, non-player character (NPC) interaction, and downstream supply chain considerations for furniture elements in urban infill scenarios. The game mechanics feature t-Distributed Stochastic Neighbour Embedding (t-SNE) [77] to visualise data in 144 dimensions. The user's decisions were analysed using a clustering analysis, a form of unsupervised machine learning, to

	DRL	Mobile	V2	Ikigai	Beyabu BETA	Beyabu RELEASE
Game Mode	Multiplayer – Turn By Turn Agent-based	- Sandbox - Challenges	- Procedural Arch Geom Explorator	- Configurator	- Explorator - Configurator	- Configurator
Player Actions	-Select Cluster -Select Voxels -> Expand Unit -Create Rooms - Create Furniture - Rent or Buy	- Site mass - Module Placement - Space - furniture customization - Style customization	- Unit Selection (Explorator) - Tectonic Preference - Space customization	- Cluster Selection - Unit Select and Placement - Air Rights - Space layout selection - Style Selection	- LevelSelection - Unit type and Placement - Add Ons - Air Rights - layout Selection	- Cluster Selection - LevelSelection - Unit type and Placement - Add Ons - Air Rights - layout Selection - Flip
IO	- Procedural - Static Prefabs .OBJ	- OBJ Static	- Procedural	- OBJ	- JSON + .OBJ	- JSON + .OBJ
Camera	- VR - First Person	- Rotation	- Orbit - Pan - Zoom	- Orbit - Pan - Zoom	- Orbit - Pan - Zoom	- Orbit - Pan - Zoom
Pixel streamed - Web	- No	- No	- No	- Web App	- Yes (AWS)	- Yes (AWS)
UI Framework	- Unity UI Components	- UE4 Widget BP	- UE4 Widget BP	- React JS	- Javascript, Html, CSS	- Javascript, Html, CSS

Fig. 6 Configurator legacy builds comparison

categorise data for potential end-buyers. A downstream agent-based simulation evaluated user parameters and proportions to gauge population sentiment. Content was created by discretising rooms into volumetric sub-modules (Fig. 7) for bespoke furniture configurations (Fig. 8). The proportion and scale of the modules corresponded such that downstream procedural creation of furniture objects, for instance, two modules configure a chair, four modules configure a table, and eight modules a single bed.

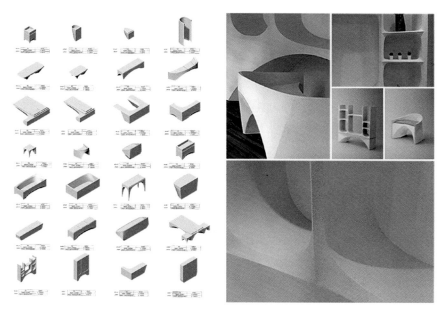

Fig. 7 Conversion of volumetric modules to customised furnishing. *Image Courtesy* Oikos, AADRL

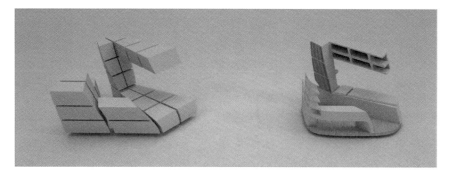

Fig. 8 Robotic hot wire cut furniture set. *Image Courtesy* Oikos, AADRL

Robotic hot-wire cutting (RHWC) technologies were leveraged in the production of 1:1 mock-up for a chair, and a bookshelf (Fig. 7). The design of architectural geometry [78] as fabrication aware and complete digital design to the production pipeline. A 6-axis robot carved polystyrene foam blocks to achieve ruled cuts following curvilinear toolpaths in a fraction of the time of CNC milling or conventional carpentry for similar shapes [79].

The Ministry of Building Innovation and Education (MOBIE) Design Challenge 2020 is a single-player residential configurator exploring human–computer interaction (HCI), and gamification principles. MOBIE features challenge modes as training sessions to familiarise with the GUI, content types, and to teach mechanics. For instance, users are asked to orient rooms vectors to face views encircling a courtyard, and in another to construct a contiguous stairway between a given set of rooms and obstacles in a section with only a fixed number of rotations or fixed number of tiles in the latter to accomplish this objective.

Subsequent ZHA platforms focused effort in integration to the zSpace framework in C++, [80] the procedural creation of geometry, the creation of building components kits in walls, floor slabs, façade panels, columns, and dashboards for a variety of user groups, as a design-assist toolkit.

4.2 User Interface and Dashboard Development

Interface design at ZHA is tuned to specific package deliveries and profiled to specific user audiences. In general, the dashboards are tuned to non-expert users, with a filtered set of design templates available for selection, with limited onscreen functionality amenable to tablet and mobile (Fig. 9). In master planning, we are exploring Microsoft PowerBi integration for parcel exploration for buyers, facade panel exploration for fabricators, interior fit-out modifications for occupiers, building components swap for architects, and site planning development. Progress bar advancement, real-time analytics, display dashboards profiled to users, and incentive mapping are notable. The gamification principles implemented in the Configurator are detailed in a sequence of storyboards that serve to set-out the backend infrastructure.

4.3 Components: Content Creation

Recent rapid advances in the field of form-finding have led to the popularity of architectural geometry [78, 81] as fabrication-aware design geometry incorporating essential aspects of function, fabrication into the shape modelling [82, 83]. The use of discrete mesh geometry representation is ubiquitous in computer graphics and animation industries, and hitherto utilised for intuitive design manipulation, and lightweight computation in a Mesh Modelling Environment (MME) [84].

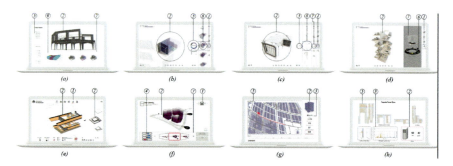

Fig. 9 ZHA dashboard design for: **a** designer, **b** occupier: house plan, **c** fabricator: house fabrication method and style, **d**, **e** occupier: house furniture selection, **f** masterplan developer, **g** masterplan investor, **h** tower facade fabricator. Dashboard elements: 1. progression stages indicator, 2. current selection, 3. analysis/performance info panel, 4. warnings/prompt info panel, 5. selection menu: home, 6. selection menu: component, 7. selection menu: variation, 8. selections filtered by availability

The shape modelling of building components is informed by innovations in fabrication becoming stylistic drivers a so called 'Tectonism' [85]. Industrialised construction building components have been developed in each RHWC concrete and singly curved columns for robotic abrasion cutting in timber (Figs. 10 and 11).

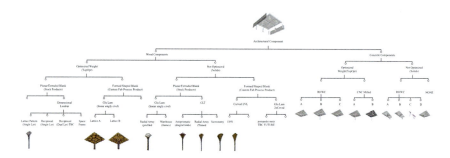

Fig. 10 Content creation: industrialised construction building components tree

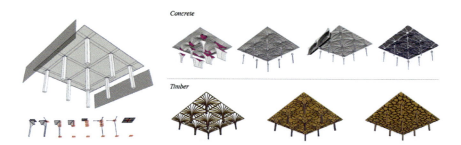

Fig. 11 Content creation: style sets of building components

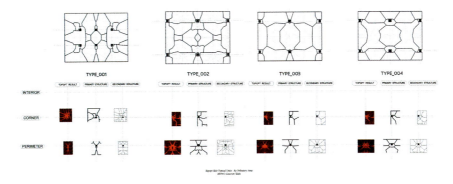

Fig. 12 Content creation: topology optimised floor slab tile components

Likewise, lightweight high-performing floor components exhibit material reduction via a so-called topology optimisation (TO) [86] extrapolated according to the grid position as interior, perimeter, or corner condition (Fig. 12).

5 Configurator

The Beyabu configurator (Configurator) is an extension of platform design at ZHA toward residential applications in remote island economies for subtropical coastal sloping terrain, leveraging mass timber in kit of parts components. It invites prospective home buyers and investors to design a customised residential development in a buyer-directed and planned community arrangement for a charter city development made possible through a Zone for Employment and Economic Development (ZEDE) governance model in Roatan, Honduras.

The game actions designed for the user experience (UX) (Fig. 13) represent an easy-to-use gameplay where buyers can configure the unit typology, aggregate placement, cladding add-ons such as roof types, development usage rights, interior space layout and fittings (Fig. 14), while cladding elements such as windows, and opaque walls, and balconies are automatically placed to maintain consistency of the overall development.

ZHA conducted two sessions of client builds using the platform separated by a 6-month period. After the first round (Pilot Build) we gathered data and user requests including datasets from the build itself and additional feature requests from a verbal questionnaire format. The verbal and numerical feedback was analysed, and insights were extracted to update the unit typologies for engineering feasibility, the game mechanics, rules, kit of parts components on offer, and functionality in gameplay. A second session (Sales Release Build) of configuration was conducted with the updated environment model, engineering feasibility, game features and functionality for thirteen buyers.

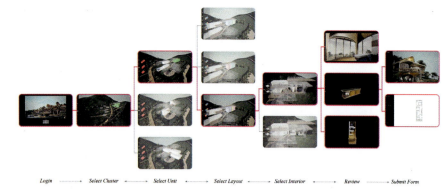

Fig. 13 Configurator game stages and progression

Fig. 14 Configuration toggles for a modular interior

5.1 Configurator: Pilot Build and Customer Needs Aquisition

The Pilot Build sessions of the platform were remotely conducted in a real-time web-based application and simultaneously the designer, client, and prospective buyer linked via video conferencing. A client representative communicated motivations and next steps not in-game, while a ZHA representative presenter explained the Configurator interface, general capabilities, and provided passive in game guidance during the configuration session. For instance, at each screen progression the presenter provided supplemental info content types on offer not yet in-game. Additionally, the presenter collated game feedback at end of the build session for content, add-ons for future selection development assessed to be likely in closing a sale. Likewise, the presenter recorded the gameplay features and functionality that would empower future autonomous use of the Configurator without video conference assistance.

An environment model with an empty system grid of the available terrain on offer—*a gameboard*—was presented to each buyer alike. After running thirteen configuration sessions, during which several configurations were tested, the final configuration options of each player were submitted, and a report of the configuration itself was generated for the buyer to retain. A total of 72 configurations were conducted. The system is distinctive from later builds in that it presented:

1. one unified system grid to position within
2. the buyers all made first placements in the development
3. three available space planning layouts per unit typology
4. air and development rights were explicitly selectable and procurable.
5. balcony projection distance was offered.

5.1.1 Session Results and Feature Planning

The selections per stage of the Configurator were automatically logged per player, per selection, to a spreadsheet. The session selections include system grid selected ID, transform matrix (position, rotation, and scale), space plan layout typology, development rights selection, balcony extension typology, and roof typology selected. The results were extracted and replicated as a 3D composite of selections and trends (Fig. 15) evaluated by the design team. User preferences and trends are as follows:

1. top-level grid spaces were preferred in comparison to lower spaces, for the expanded view vistas and for preserving the void above without having to purchase the rights to retain such.
2. end of block 'aisle' grid spaces in lieu of middle grids as it guaranteed a multi-window aspect unit, limiting the potential for demising walls to a neighbour while also preserving the immediate vista (at least to the property line itself) without having to purchase the rights to retain such.
3. space layouts that maximised the quantity of beds/number of people hosted to sleep.
4. social areas, which for Latin America included both the living and kitchen, contain panoramic views, which are predominantly the forward-facing voxels.

Buyers tended toward larger balcony upgrades including deck and cover area with the rationale the proportional cost relative to the unit total was trivial. While only a

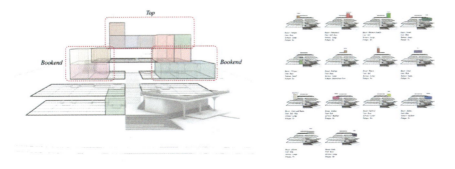

Fig. 15 Position selections in game sessions overlay

minority of buyers chose to purchase view preservation rights in neighbouring left and right voxels to maintain a neighbouring buffer and increase window aspects of their unit. Buyers tended to select development air rights to actualise private roof terrace upgrades and custom vaulted rooftop known as 'palapas' regionally (Figs. 16 and 17).

As early adopters of the planned development, buyers could steer content offerings for the subsequent Configurator build. Buyers made verbal requests during the sessions which were tallied offline and sorted as feature requests, functionality requests, or content creation requests (Fig. 18). Most requests sought to increase the occupiable gross floor area (GFA) without paying for additional voxels focused on the interior layout and space planning stage of the Configurator.

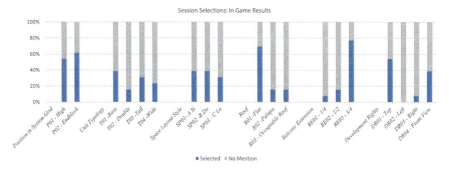

Fig. 16 Buyer selections data

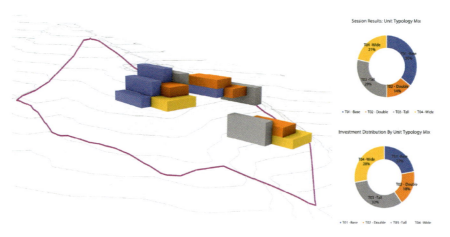

Fig. 17 Buyer unit typologies selections

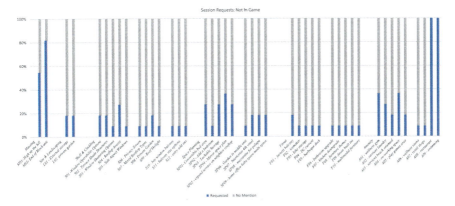

Fig. 18 Buyer requests data

The notable session trends are below:

1. Mezzanine—Height permitting, when a unit is upgraded for a palapa roof, exploit the unleveraged voxels in secion above resulting in an infill mezzanine floor option.
2. Increase occupiable interior area—To increase interior space, optimise habitable space already committed in the voxel count for the typology by swapping an exterior voxel for an interior voxel.
3. Increase sleepcount—As the primary motivation for living is long stay, there was a desire to increase sofas, increase bed count in smaller units, and increasing bedroom counts in larger units.
4. Layout context awareness—Depending on their position on the stack, the direction of side windows or the orientation of the living room relative to the view needed mirroring capability.

5.2 Configurator: Engineering Feasible Solution Sets and Rule Set Creation

The Pilot Release demonstrated low-density residential design and planning principles could be extended interactively to non-experts via a technology platform to empower customisation in real-time to participants around the world.

Construction feasible configurations planning for Sales Release necessitated resolving challenges posed by Pilot Release. Engineering integration in a fixed subset of RBU configurations development and applicability as systems design principles to broader permutations. The relative applicability and market alignment of feature and functionality requests by a small initial number of respondents.

Content such as rooftop access, bathroom size upgrades, and side balconies for multi-aspect unit configurations posed challenges to the spatial proportioning system as sub-voxel ratios. While side facing unit plans, balcony projections, and tall units posed structural challenges to the system design conceptually in its panelised assembly, assembly build-up, and primary spanning direction allowed in the 5 × 5 grid.

The compositing of the Pilot Build results exposed several preferential tendencies. Since few users selected horizontal view rights, and predominantly users selected end-of-block grid positions. The Pilot Release results challenged the idea of explicit sale of development rights and challenged the default relative equality of grid positions in the game board. Taken in combination, the results suggested implicitly capturing the result of horizontal view rights by way of increasing the quantity of end-of-block cells in the game board would result in greater monetisation potential, and reconciliation of positioning demand experienced in Pilot Release.

Considerations for regulatory feasibility rules in the game board creation were planned between releases. For instance, the maximum height positioning on existing cells considers the building type regulation for mass timber construction in residential occupancy. Similarly, the consideration for fire regulation is demonstrated in the minimum distances between community grids set out. Each affect the system grid creation prior to game play and active cells available for during game play.

The period between Pilot Release and Sales Release resulted in several functional improvements to the game mechanics, but more importantly the content and feasibility of content offerings on offer. The kit of parts, superstructure, space planning, and land planning principles were validated across the engineering spectrum via Structural, Mechanical, Plumbing, Electrical, and Fire consultations (Figs. 21 and 24).

The use of identical gameboards in the Pilot sessions necessitated a reconciliation operation to understand the implications of overlapping units as community spatial preference and understand the implications of regulatory feasibility for planning the Sales Release. Reconciliation consists of extraction of elevation height and end-of-block view aspect to sort the selections (Fig. 19). These preferences are superimposed with the regulatory land planning principles for mass timber residential construction. Together with a Sales Release amenable grid, the position principles are redeployed manually to provide a suitable basis community amenable to Sales Release development in parallel to engineering development on a basis community aggregate.

This approach exposed the need for more end of block unit positions, which implicitly retained side view rights offered in Pilot explicitly. It also exposed a general tendency toward reduced density for the overall development driven by buyer preference as void 'aisles' between clusters reduced the developable footprint possible compared to a conventional elongated frontage coastal condominium development (Fig. 20).

A component strategy was established between build releases to encode construction feasible engineering principles in the system design ruleset planned for Sales Release. For instance, incoming electrical service walls, fixture positioning in voxel relative to timber cassette knockout panel corners, and number of air handling

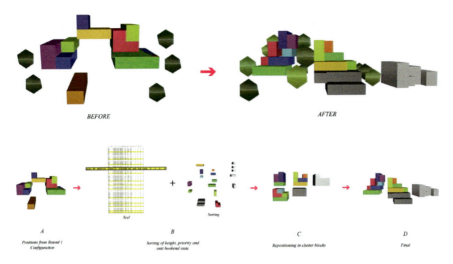

Fig. 19 Reconciliation of buyer positions

Fig. 20 Game mechanics unit rulesets visualised. **a** place wide units toward bottom, **b** avoid multi-story co-planarity, stagger units in plan, **c** maintain unit-to-grid parallel orientation, **d** avoid unit cantilevering overextension, **e** stagger units in elevation, **f** avoid void spaces on plinth, lower center of gravity, **g** place vertical units in contact with plinth, **h** limit excessive vertical access, increase horizontal aspect ratio, **i** maximum cluster height: 3 levels

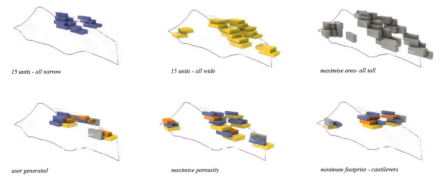

Fig. 21 Feasible permutations of community positioning directed toward specific objectives

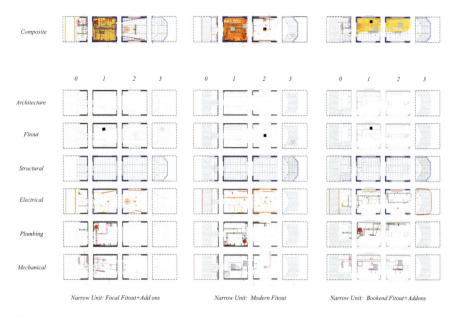

Fig. 22 Engineering principles validated in systems design

cassettes per voxel were developed in conjunction with engineering disciplines as examples of Configurator principles (Fig. 21). The Configurator results in subsequent platform build Configurator sessions are engineering aware resulting in construction feasible solution sets due to the system design platform approach to the engineering discipline's design. Likewise, the general rigid body physics, stacking, and unit associations allowed at the community aggregate level are structurally aware in game play. System design principles for stacking rulesets explicitly encoded for future game board terrain typologies exhibiting varied sloping properties (Fig. 22).

The more notable of structural principles includes clusters must be wider than tall, tall units must reside on the concrete plinth for steel interfacing, and the units are parallel in placement not perpendicular or skewed, though may translate away or toward the hillside. These assume a mass timber panellised structural design for remote coastal sloping deployments.

Likewise, the space planning kit of parts rulesets are developed for modular voxel-based interior layouts in 3 style sets (Fig. 23). Notable Sales Release content add-ons and upgrade selections include bathroom upgrades, such as exterior sheltered bathing and freestanding fixture-oriented bathrooms, added mezzanine floor kits, truncated 1/2 voxel floors inside balcony view units, and rooftop access via private stairwells (Fig. 24).

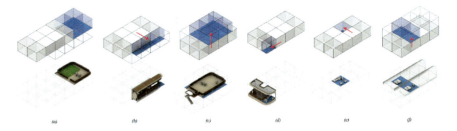

Fig. 23 Customised interior fitout raft as style set

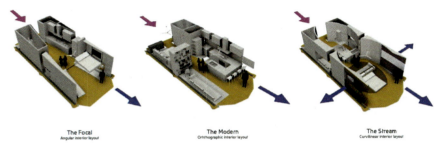

Fig. 24 Construction feasible unit upgrades and add-ons **a** extended terrace over bottom unit, **b** side balcony, **c** roof terrace, **d** entryway voxel customisation, bathroom expansion, **e** internalised outdoor living, **f** mezzanine

5.3 Configurator: Sales Release Build

During the Beyabu Sales Release the client broadened the audience pool to include early adopters, and investment buyers. The Configurator was packaged as a stand-alone application identical to Pilot. The sessions were conducted without the presence of the design team. A client-side representative was trained in general technical considerations, such as the need to restart a session, or reboot the application remotely. After running forty-two sessions, during which several configurations were tested, the final configuration options of each player were submitted, and a report of the configuration itself generated for the buyer to retain. A total of twelve configurations were conducted. The system grid generation was updated to reflect a new property boundary with more sloping terrain.

The Sales Release took into consideration the turn-based placement by previous buyers in the gameboard presented in the current game session. On game initialisation all previous player actions were procedurally reloaded in the scene. As such the sequence of the sessions influenced the decisions of the following sessions. The first buyer had priority to choose positions from a blank gameboard, much like the Pilot Release (Fig. 25), but subsequent sessions the aggregate of units became more and more tailored to the direction of the community of buyers. Once an order was processed the system locked all the voxels reserved in session. Sessions could see

Beta Release Sales Release

Fig. 25 Gameboard initial screen comparison

both the on-screen units by their previous placed neighbours as well as who owned it, whether it was an investment property or would be occupied by owner and the configuration options they had selected.

The system is distinctive in that it presented:

1. Multiple system grid sets around a hillside to position within
2. buyers were presented all prior session RBU's at initialisation
3. air and development rights were implicitly managed via the system grid creation number of grids wide
4. balcony projection distance was discontinued
5. Engineered upgrades not implemented in the Configurator platform at time of release were offered through a supplemental buyer catalogue
6. Selecting a palapa roof upgrade unlocked an additional fourth space plan option to optimise GFA via an infill mezzanine floor.
7. Space plans could be mirrored depending on their location and sunlight preferences.

5.3.1 Session Time

On the Sales Release hosted on AWS users on average spent more than 5 min per session which is less than the United States average of 6–10 min on e-commerce websites [87] (Fig. 26).

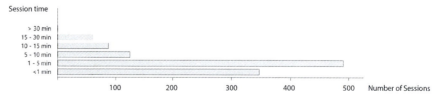

Fig. 26 Session time

6 Workflow and System Design

The Configurator was produced in Unreal Engine 4 (UE4). A bespoke toolchain was developed and implemented across a small team of six as detailed in (Fig. 27).

Content creation, data preparation, storyboarding and GUI design, and software front-end and back-end development were delineations of internal task splitting.

6.1 Mechanics

The Configurator session guides a buyer through several configuration session states, from positioning, to RBU cladding options, to contractual options, to interior layout options. The gameplay progression states are detailed in (Fig. 28). The session is advanced with each selection. The configuration state is visible and toggled selections comparable in first person perspective as walkthrough or globally in an aerial camera.

Several gamification core drives enhance gameplay as outlined in (Fig. 29). Info panels displaying the community selections, and what positions have been frequented provide a sense of inclusion and urgency (Figs. 30 and 31). Cost reporting of the selections is calculated in real time. The daily shadow casting preview gives a visual

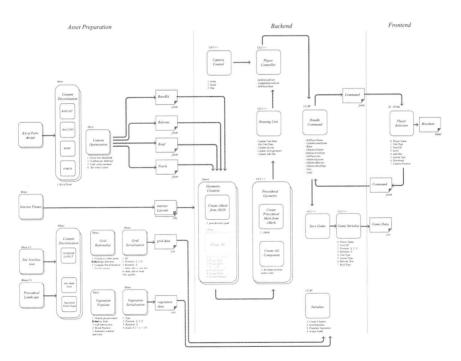

Fig. 27 Bespoke tool-chain and workflow

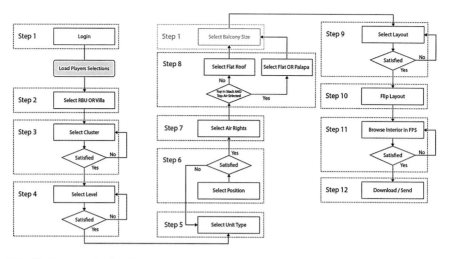

Fig. 28 Game mechanics diagram

indication of the overall community massing upon the selected position, as well as the illumination falling upon the windows, balconies, and roof. The Configurator concludes in a downloaded report of the unit configuration.

While the system grid is created outside the platform and statically loaded as a 'gameboard' each session, the active positions within the session are managed dynamically (Fig. 32). There are several positional dependencies which result in an active grid cell highlighting on rollover. The system analyses the available voxels

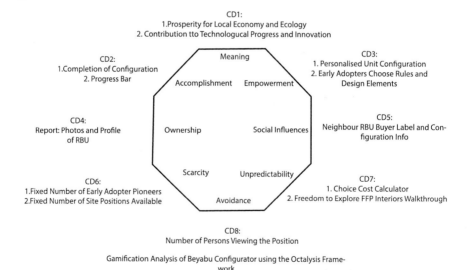

Fig. 29 Configurator gamification analysis using octalysis framework

Configurator: A Platform for Multifamily Residential Design ... 793

Fig. 30 Overlay of kit of parts selections of neighbouring units

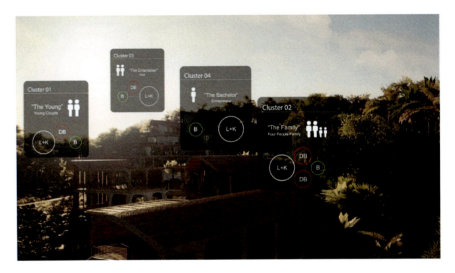

Fig. 31 Overlay of space organisation arrangement of neighbouring units

required for the specified unit typology voxel count, computing collisions, and if the unit overextends the available cluster grids in the gameboard (Fig. 33). If the constraints are satisfied and the unit highlights as a valid system placement.

Similarly, the system enforces rules for the kit of parts dependencies. For instance, if the player selects "top air rights" in session and no unit exists above the selection, the system 'unlocks' a palapa roof type content option in the roof type selection stage. While there are 54 permutations of an RBU in the Pilot Release, there are

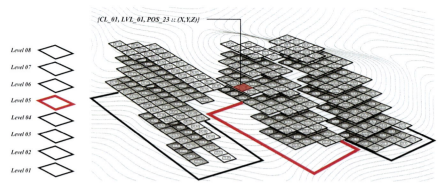

Fig. 32 System grid data-structure

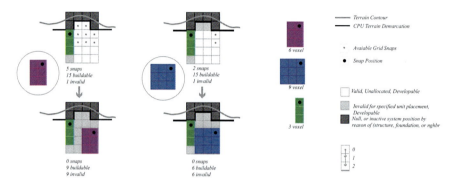

Fig. 33 Unit placement dependencies and dynamic grid state change

roughly 300 active grid cells available in the gameboard when initialised, resulting in over sixteen thousand feasible community aggregations (Fig. 34).

The shadow analysis function gives buyer feedback on exposure of the unit envelope, glazing. It is further extended as a colour coded depth map to nearby unit obstructions such as property vegetation and neighbouring units to give relative visual feedback on the selected position relative to the overall community mass and line of sight to various site features (Fig. 35).

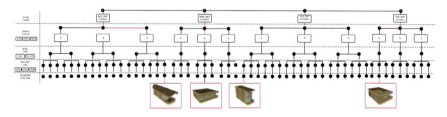

Fig. 34 Possible variations

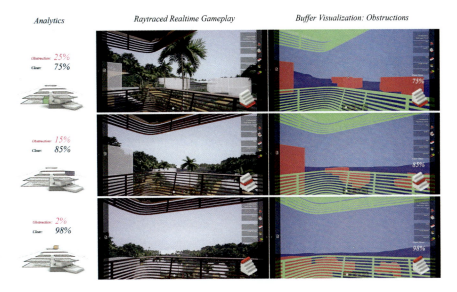

Fig. 35 Unit obstruction and depth mapping in-game

6.2 Assets and Content Creation

The environment design terrain mesh, system grids gameboard creation relative to sloping terrain, access-ways, and context building and water were procedurally generated in Rhino and exported to UE4 as separate OBJ's. While the vegetation, entourage, lighting, and materials were natively created in UE4. The RBU geometries consisting of cladding panels, structural assemblies, interior partitioning, built-ins and furnishing elements were modelled procedurally in Autodesk Maya and broken into a kit of parts in Rhino to be hosted to a voxel face edge or vertex using a half-edge data structure (Fig. 27).

The environment terrain and system grid are procedurally created in Rhino (Figs. 36 and 37). A custom utility was created to discretise the terrain into volumetric voxels corresponding to the unit floor-to-floor height and with a buffer distance to the intersection to the sloping terrain. The cells are controlled in grid groupings of between 6 and 12 cells wide radially arranged about the hill faces toward the water. The lowest datum allowed is pre-set, but all subsequent datums 6 voxels away from the hill intersection (corresponding to a unit maximally 10 m forward of the hill) are returned as viable system grid positions for the gameboard (Fig. 38).

Manual polygon modelling and parametric data preparation were used to organise and transfer assets to the Configurator. The kit of parts consisting of cladding, structure, interior layout, concrete plinths, roofs is detailed in Fig. 39. The combinatorics of such results in variations in unit shell and spatial arrangements (Fig. 38).

They are manually modelled and broken into per-face components and assigned coloured attributes per vertex. Interior partitions and furniture are combined in UE4

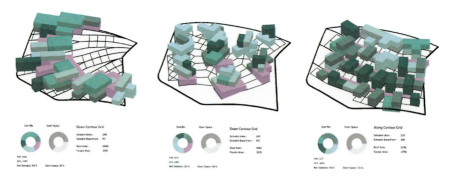

Fig. 36 Variations in a system grid generation and resulting aggregates

Fig. 37 Massing qualitative analysis of unit positions. *Image Courtesy* Hilson Moran

Fig. 38 Unit variation from a digital kit of parts

Configurator: A Platform for Multifamily Residential Design ...

Fig. 39 Configurator kit of parts

per spatial layout typology and assigned the corresponding material per face. Assets are manually exported from Rhino about the origin and composited in UE4 for collision checks.

In addition to the 3D kit of parts, the corresponding 2D technical drawings for 13 unique unit variations leveraging every kit of parts component per produced to a RIBA stage 3 for architecture, interiors, structural, mechanical, plumbing, electrical, and fire protection with particular attention to the system design principles for deployment across multiple sites in the lifespan of the technology platform.

Lastly, the GUI and buyer downloadable report (Fig. 40) consisted of renderings, diagrams, illustrations, and iconography developed for the Configurator and from various stages in platform technology development.

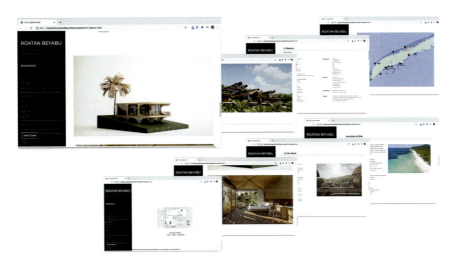

Fig. 40 Configuration downloadable report

6.3 System Architecture

The system architecture of the technology components of the configurator are as follows:

1. Data preparation
2. Asset Transfer API
3. Backend Server
4. Database
5. Client API
6. Client Frontend

6.3.1 Data Preparation

The grid gameboard is imported into UE4 and its corresponding analytics data is created per cell. Each grid geo-location holds information not only about its physical transformation in 3D space, position, rotation and scale, but also further contextual data analytic such as distance to the sea, distance to the road, distance to the clubhouse, number of steps required to climb, height above sea level, view quality and radiation exposure. This data was serialised into CSV format and plugged into the platform corresponding to the data structure needed for game initialisation (Fig. 41).

Utilising a half-edge mesh data structure hosted at a voxel face, edge, or vertex. Attributes such as materiality were given through vertex colours which could be transferable as a JSON format across multiple software. Every asset is manipulated through matrix transformations during gameplay, so all the assets had to share the same origin relative to the host voxel.

Space plan layouts and interior layouts are also organised per unit type in UE4 and with a curated style of finishes. They also need to be organised with a common origin matrix matching the parent voxel hosting them.

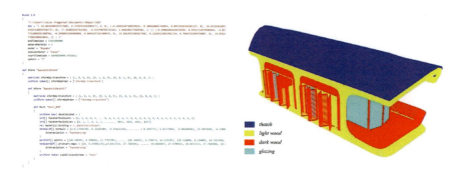

Fig. 41 Asset coordination and data preparation

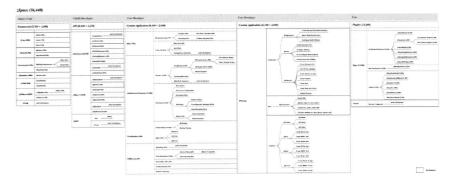

Fig. 42 zSpace C++ framework

6.3.2 Asset Transfer API

To keep a non-destructive workflow, we have developed a data-transfer protocol to manage the platform assets. The components are exported as JSON objects holding vertex and face attributes based on materiality. The interior layouts are exported as separate high-resolution polygon meshes in OBJ format with its corresponding finishing (material) file.MTL. The grid gameboard and the vegetation information from the landscape design are exported as a CSV file that holds per-voxel attributes per row and positional data, respectively.

6.3.3 Backend Server

The platform back-end is the core of the system developed in-house. *VaRest*, a UE4 plugin that handles REST[1] server communications at runtime, and Pixel Streaming (PS),[2] a UE4 internal plugin to enable a packaged UE4 application to run on a server in the cloud. The *zSpace* libraries handle the procedural generation of geometry at run-time while processing the JSON asset files (Fig. 42) [80]. The UE application runs on a server, either a local host or Amazon Web Services (AWS) host, and it handles the Client's request and updates the game state per action.

---

[1] A RESTful API is an architectural style for an application program interface (API) that uses HTTP requests to access and use data.

[2] Pixel Streaming allows a packaged Unreal Engine application to run on a server in the cloud, and stream its rendered frames and audio to browsers and mobile devices over WebRTC, a free, open project that enables web browsers with Real-Time Communications (RTC) capabilities via simple JavaScript APIs.

6.3.4 Database

Once a player reserves and RBU and the session completes, the system packages a brochure available to download as an HTML file with variables and collects images corresponding to each of them from a cloud-based database. The Client front-end collects every action or buyer decision and keeps a record of the variables as an order of selection for each customised RBU. These include Player Name, Selected Cluster ID, Selected Level Number, Selected Unit Type, Geo-location of placed Unit (Voxel ID, X, Y, Z), Selected Roof type, Selected interior Layout, and other Voxel-specific data such as distance to the sea, distance to the road, and distance to the Clubhouse.

6.3.5 Client API

The Client's request is handled by PS and communicates as a two-way stream between the Client front end and the server. Since they are separate applications written in different languages, they communicate through a JSON response that is parsed and processed at either end.

The Configurator was deployed on an Amazon Web Services (AWS) Elastic Compute Cloud (EC2) G4 instance shared through a secured link.[3] We extended UE native API for PS to avoid client-side hardware requirements and technical support. Leveraging Eagle3DStreaming[4] cloud orchestration, controlled instance creation on demand for buyers. Using native UE4 GUI and Widget components during PS caused latency issues. We developed the front end using a traditional web development framework using HTML, JavaScript and CSS styling, to enhance the user experience.

6.3.6 Client Front-End

The Configurator is built on top of Unreal 4.26 game engine leveraging real-time ray tracing technology. UE4 handles all the UI, and client-side code written in HTML, and Javascript (JS) and styled in CSS, manages the assets, handles the Client's request, procedurally instantiates the in-game actors and controls the rules of the "game." It uses C++, Blueprint (BP) classes, and has a dependency on zSpace.

---

[3] Amazon EC2 G4 instances are cost-effective and versatile GPU instances for deploying machine learning models such as image classification, object detection, and speech recognition, and for graphics-intensive applications such as remote graphics workstations, game streaming, and graphics rendering.

[4] Eagle3DStreaming is a third-party cloud hosting orchestration provider for Unreal Pixel Streamed applications.

7 Outlook and Future Work

The paper has demonstrated that AEC industries can engage stakeholders, to design, co-author, effectively crowd source and democratise the design process. The following areas of work could be improved:

1. **Incentivisation**—Value maps could be implemented on system grid positions and adjusted in session in response to various environmental, and session criteria to enhance feedback available onscreen to steer and inform unit positioning. This feature would be scalable to address multiple data streams and assimilated using visual cues much like theatre or flight seating maps. Currently only active and inactive positions receive onscreen alerts.
2. **Supply Chain Integration**—The digital end-to-end pipeline can be extended to the level of detail of fabrication parts, supply chain planning, and manufacturing data of timber structure and cladding components to improve design awareness, reduce the time of design-to-production pipeline leading to occupation (Fig. 43). The digital pipeline currently terminates at the reported configuration.
3. **Procedural Geometry**—The geometry of the units, interiors, and materials could be procedurally generated in the software stack. Currently, the system has been developed as an API to receive such methods in the correct data structure, but the content is loaded as static meshes.
4. **Asset Management**—The content database could be dynamically managed, assemblies coordinated during creation, and directories automatically synchronised. Over one hundred content organising directories responsible for the configuration kit components is difficult to maintain and coordinate manually and are not a scalable protocol for more complex game scenarios, or more robust functionalities and selection sets. 3D Scene Descriptions The compiled scene could leverage Universal Scene Description (USD) or GL Transmission Format (GLTF) to capture more information per asset and the relationship between parts. Currently, the software stack utilised UE4 to compile, coordinate, and data enrich objects.

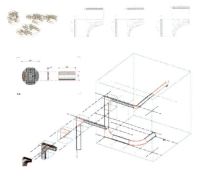

Fig. 43 Digital supply chain integration via robotic fabrication of timber components. *Image Courtesy* Circular Factory

5. **E-Commerce Features**—The Configurator content could present a local marketplace of digital assets licensed to different regional or local suppliers and artisans who could commercialise their products in the platform. Currently only ZHA-designed content is in the Configurator.
6. **Domain Extension**—The platform framework could be extended to non-residential use domain applications. For instance, urban planning, commercial, retail, and interiors present similar consensus planning and customisation negotiated design problems the platform could be adapted to address.

8 Conclusions and Summary

The paper articulates the relevance of game technologies utilisation in the AEC pipeline to achieve a user-centric multifamily residential design. In particular, the contributions as described, demonstrate decentralised buyers can engage in emergent consensus behaviour in property planning and land development. Likewise, the participation of design experts in a self-guided non-expert game environment resulting in construction feasible community is detailed. The resulting session outputs demonstrate mass customisation of components, add-ons, and options each equally exhibits a construction feasible configuration. The timeline required for future configuration outputs is dramatically reduced given the system design principles and engineering aware game mechanics and component creation. Furthermore, bringing together key stakeholders, whether encoded as non-player character rulesets or active gameplay participants, in a shared digital platform can deliver high value, locally relevant, resource effective, supply chain integrated design solutions for residential living.

Acknowledgements This research was carried out collaboratively between the authors and with the support of the respective organisations: Zaha Hadid Architects, Epic Games, AKT II, Hilson Moran Partnership, Circular Factory, AADRL Nahmad/Bhooshan Studio, The Manufacturing Technology Center, and Prospera. We would like to thank those in the Computation and Design Group contributing to documentation and design acumen, Jianfei Chu (Truman), Edward Meyers, Cheolyoung Park (Nick), Efthymia Douroudi, Taizhong Chen, with Chun-Yen Chen (James), and formerly Federico Borello, Jackson Bi, Georgios Pasisis, and Leo Bieling.

References

1. Meyer, A.: Ein versuchshaus des bauhauses in weimar (bauhausbücher 3) (1925)
2. Corbusier, L.: Œuvres Completes, p. 69 (1910)
3. Morah, N.: Humanising mega-scale habitat 67 (1925)
4. Danilovic, M., Winroth, et al.: Platform thinking in the automotive industry—managing dualism between standardization of components for large scale production and variation for market and customer (2007)

5. BrydenWood: Delivery platforms for government assets–creating a marketplace for manufactured spaces. https://www.brydenwood.co.uk/platformdesignbooks/s114123 (2017)
6. Thuesen, C., Hvam, L.: Efficient on-site construction: learning points from a German platform for housing. Constr. Innov. (2011)
7. Ulrich, K.: Fundamentals of product modularity. In: Management of Design, pp. 219–231. Springer (1994)
8. Veenstra, V.S., Halman, J.I., Voordijk, J.T.: A methodology for developing product platforms in the specific setting of the housebuilding industry. Res. Eng. Design **17**(3), 157–173 (2006)
9. Pine, J.B.: Mass Customization: The New Frontier in Business Competition. Harvard Business School Press (1993)
10. James, G., Pine, J.: The four face of mass customisation. Harv. Bus. Rev. (1997)
11. Fletcher, B.: Modern houses for industrial classes (1870)
12. Roebuck, S., Co: Modern Homes. Chicago-Newark (1936)
13. Editorial Staff Architectural Record: Guide to Home Planning. D.F. Dodge (1950)
14. Barlow, J., et al.: From craft production to mass customisation: customer-focused approaches to housebuilding. In: Proceedings IGLC. Citeseer (1998)
15. Hofman, E., Halman, J.I., Ion, R.A.: Variation in housing design: identifying customer preferences. Hous. Stud. **21**(6), 929–943 (2006)
16. Schoenwitz, M., Naim, M., Potter, A.: The nature of choice in mass customized house building. Constr. Manag. Econ. **30**(3), 203–219 (2012)
17. Hook, M.: Customer value in lean prefabrication housing considering both construction and manufacturing. pp. 25–27 (2006)
18. Nahmens, I., Mullens, M.: The impact of product choice on lean homebuilding. Constr. Innov. (2009)
19. Fisher, et al.: A common configurator framework for distributed design, collaboration and verification across the full AEC supply chain (2022)
20. Central Mortgage and Housing Corporation: Catalogue of House Building Construction Systems, pp. 302–303. Lustron (1960)
21. Radic, J., Kindij, A., Ana, M.: History of concrete application in development of concrete and hybrid arch bridges. In: Proceedings of the Chinese-Croatian Joint Colloquium on Long Arch Bridges, University of Zagreb, University of FuZhou, pp. 9–118 (2008)
22. Thomas, E.: Process of constructing concrete buildings. US Patent 1219272a. https://patents.google.com/patent/US1219272A/en (1908)
23. Lab, O.S.: The DfMA Housing Manual. An Introduction to the Principles of Design for Manufacture and Assembly for Homes. Version 1.1 (2019)
24. Jensen, P., Olofsson, T., Johnsson, H.: Configuration through the parameterization of building components. Autom. Constr. **23**, 1–8 (2012)
25. Ball, M.: Chasing a snail: innovation and housebuilding firms' strategies. Hous. Stud. **14**, 9–22 (1999)
26. Sulzer, P.: Jean Prouve Complete Works, vol. 3, pp. 1944–1954 (2008)
27. Corp, N.H.: "Custom-Line" and "Pacemaker" Houses by National Homes (1955)
28. Kendall, T.J.S.: Residential Open Building (2000)
29. Ulmer, J., Braun, S., Cheng, C.T., et al.: Human-centered gamification framework for manufacturing systems. https://doi.org/10.1016/j.procir.2020.04.076 (2020)
30. Deterding, S., Dixon, et al.: From game design elements to gamefulness: defining gamification. pp. 9–15 (2011)
31. Beck, M., Wade, J.: The Kids Are Alright: How the Gamer Generation Is Changing the Workplace (2006)
32. Squire, K.: From content to context: video game as designed experience. Educ. Res. **35**(8), 19–29 (2006)
33. Steinkuehler, C.A.: Learning in massively multiplayer online games. pp. 521–528 (2004)
34. Kapp, K.M., O'Driscoll, T.: Learning in 3D: Adding a New Dimension to Enterprise Learning and Collaboration (2009)

35. Malone, T.: Toward a theory of intrinsically motivating instruction. Cogn. Sci. **4**, 333–370 (1981)
36. Kai Chou, Y.: Actionable Gamification: Beyond Points, Badges, and Leaderboards (2014)
37. Costello, B., Edmonds, E.: A Study in Play, Pleasure and Interaction Design, pp. 76–91. ACM Press (2007)
38. Level up! a guide to game UI. https://www.toptal.com/designers/gui/game-ui (2019)
39. Sutherland, I.: Sketchpad, A Man-Machine Graphical Communication System (1963)
40. Negroponte, N.: The Man Machine: Toward a More Human Environment. MIT (1970)
41. Finn, D.: Diy urbanism: implications for cities. J. Urban.: Int. Res. Placemaking Urban Sustain. **7**(4), 381–398 (2014)
42. Playthecity. https://www.playthecity.eu/playprojects/Play-Almere%3A-Oosterwold (2012)
43. Ferri, G., Hansen, N.B., van Heerden, A., et al.: Design concepts for empowerment through urban play. In: DiGRA Conference (2018)
44. Van Straalen, F.M., Witte, P., Buitelaar, E.: Self-organisation in oosterwold, almere: challenges with public goods and externalities. Tijdschr. Econ. Soc. Geogr. **108**(4), 503–511 (1994)
45. Olofsson, J.P.T., Ronneblad, A.: Configuration and design automation of industrialized building systems (2010)
46. Suh, N.: Axiomatic design theory for systems. Res. Eng. Design **10**(4), 189–209 (1998)
47. Cao, J., Hall, D.: An overview of configurators for industrialized construction: typologies, customer requirements, and technical approaches. pp. 295–303 (2019)
48. Bogenstatter, U.: Prediction and optimization of life-cycle costs in early design. Build. Res. Inf. **28**(5–6), 376–386 (2000)
49. Wilkinson, K.P.: Social well-being and community. J. Community Dev. Soc. (1979)
50. Jansson, V.E.G., Olofsson, T.: Artistic and engineering design of platform based production systems: a study of Swedish architectural practice. **8**(2), 34 (2018)
51. Veloso, C.G.P., Scheeren, R.: From the generation of layouts to the production of construction documents: an application in the customization of apartment plans. Autom. Constr. **96**, 224–235 (2018)
52. Wikberg, O.T.F., Ekholm, A.: Design configuration with architectural objects: linking customer requirements with system capabilities in industrialized house-building platforms. Constr. Manag. Econ. **32**(1–2), 196–207 (2014)
53. Yuan, S.C.Z., Wang, Y.: Design for manufacture and assembly-oriented parametric design of prefabricated buildings. Autom. Constr. **88**, 13–22 (2018)
54. Farr, E.R.P., Piroozfar, P.A.E., Robinson, D.: BIM as a generic configurator for facilitation of customisation in the AEC industry (2014)
55. Bonev, W.M.M., Hvam, L.: Utilizing platforms in industrialized construction: a case study of a precast manufacturer. Constr. Innov. **15**(1), 84–106 (2015)
56. Said, C.T.H.M., Logan, S.: Exterior prefabricated panelized walls platform optimization (2017)
57. Archistar. https://archistar.ai (2021)
58. Buildrz. https://www.buildrz.io/ (2022)
59. Delve by sidewalk labs. https://www.sidewalklabs.com/products/delve (2022)
60. Spacemaker. https://www.spacemakerai.com/ (2022)
61. Prism-app: Prism-app.io. https://www.prism-app.io/ (2021)
62. Testfit.io: Testfit.io. https://testfit.io/ (2021)
63. ESRI: Arcgis cityengine. https://www.esri.com/en-us/arcgis/products/arcgis-cityengine/overview (2021)
64. Digital blue foam. https://www.digitalbluefoam.com/ (2022)
65. Histruct. https://www.histruct.com/ (2022)
66. Agacad. https://agacad.com/ (2022)
67. Creatomus. https://creatomus.com/ (2022)
68. Projectfrog. https://www.projectfrog.com/ (2022)
69. Cohen-Or, D., Kaufman, A.: Fundamentals of surface voxelization. Graph. Model. Image Process. **57**(6), 453–461 (1995)

70. Malmgren, L., Jensen, P., Olofsson, T.: Product modeling of configurable building systems a case study. J. Inf. Technol. Constr. (ITcon) **16**(41), 697–712 (2011)
71. March, L., Steadman, P.: The Geometry of Environment. RIBA Publications Limited (1971)
72. Nourian, P., Gonçalves, R., Zlatanova, S., et al.: Voxelization algorithms for geospatial applications: computational methods for voxelating spatial datasets of 3D city models containing 3D surface, curve and point data models. MethodsX **3**, 69–86 (2016)
73. Savov, A., Winkler, R., Tessmann, O.: Encoding architectural designs as iso-surface tile sets for participatory sculpting of massing models. In: Impact: Design With All Senses (2019)
74. Stalberg, O.: Townscaper app. www.oskarstalberg.com/Townscaper/ (2021)
75. Jensen, P., Olofsson, T., Smiding, E., Gerth, R.: Developing products in product platforms in the AEC industry. Comput. Civ. Build. Eng. 1062–1069 (2014)
76. Liu, Z., et al.: Blockchain-based customization towards decentralized consensus on product requirement, quality, and price. Manuf. Lett. (2020)
77. T-distributed stochastic neighbor embedding. https://scikit-learn.org/stable/modules/generated/sklearn.manifold.TSNE.html (2022)
78. Bhooshan, S., et al.: Collaborative design: combining computer-aided geometry design and building information modelling. Archit. Des. **87**(3), 82–89 (2017)
79. Søndergaard, A., Feringa, et al.: Robotic Hot-Blade Cutting (2016)
80. ZHACODE: zSpace. https://github.com/venumb/ZSPACE/ (2019)
81. Veltkamp, M.: Structural optimization of free form framed structures in early stages of design. In: Symposium of the International Association for Shell and Spatial Structures (50th. 2009. Valencia). Evolution and Trends in Design, Analysis and Construction of Shell and Spatial Structures: Proceedings. Editorial Universitat Politecnica de Valencia (2010)
82. Jiang, C., Tang, C., et al.: Interactive modeling of architectural freeform structures: combining geometry with fabrication and statics. In: Block, P., Knippers (eds.) (2015)
83. Tamke, M.: Aware design models. pp. 213–220 (2015)
84. Bhooshan, V., Reeves, D., Bhooshan, S., et al.: Mayavault—a mesh modelling environment for discrete funicular structures. Nexus Netw. J. **20**, 567–582 (2018)
85. Schumacher, P.: Tectonism in architecture, design and fashion: innovations in digital fabrication as stylistic drivers. Arch. Des. **87**, 106–113 (2017)
86. Topology optimization. https://www.tudelft.nl/3me/over/afdelingen/precision-and-microsystems-engineering-pme/research/structural-optimization-and-mechanics-som/som-research/topology-optimization (2022)
87. Statista: Most popular e-commerce properties in the United States as of September 2018, by average session duration. https://www.statista.com/statistics/790897/unique-visitors-average-session-durations-retail-properties-us/ (2022)
88. Møller, T., Billeskov, J.: Expanding wave function collapse with growing grids for procedural content generation (2019)
89. Rossi, A., Tessmann, O.: From voxels to parts: hierarchical discrete modeling for design and assembly. In: International Conference on Geometry and Graphics, pp. 1001–1012. Springer (2018)

From Debris to the Data Set (DEDA) a Digital Application for the Upcycling of Waste Wood Material in Post Disaster Areas

Roberto Ruggiero, Roberto Cognoli, and Pio Lorenzo Cocco

Abstract The convergence of digital and ecological transition [1] can be crucial in achieving the European Green Deal targets. In this perspective, implementing the Industry 4.0 model in the building sector acquires high value not only for the efficiency of construction processes but also for mitigating the carbon footprint and resource exploitation, traditionally related to the building industry. Considering the circular economy as a paradigm of sustainability [2], the search for synergies between "circular" and "digital" approaches in the building sector represents nowadays a strategic research sector. "Upcycling" demolition material to transform into new building components is, in particular, a topic where digital technologies can play a key role. *«Only by capturing the physical world through data»* [3] there is a real possibility to overcome the limits that have emerged to date in upcycling processes, in particular concerning the control and classification of waste materials. In this context, post-disaster areas represent a remarkable reservoir of available and potentially reusable materials: a "material bank", according to the circular economy vocabulary. DeDa (From Debris to the Data set) is a research work in progress at the University of Camerino, which focuses on reusing waste wood material in post-disaster areas. DeDa represents a new way of applying the principles of the circular economy and Industry 4.0 to debris treatment. This paper describes the aforementioned research work in its cultural and operational aspects, current limitations and future potential.

Keywords Digital innovation · Circular building · Disaster area · Data-driven design · Waste wood · Bim · Database

R. Ruggiero (✉)
Scuola Di Architettura E Design, Università Di Camerino, 62032 Camerino, Italy
e-mail: roberto.ruggiero@unicam.it

R. Ruggiero · R. Cognoli · P. L. Cocco
University of Camerino, School of Architecture and Design, Ascoli Piceno, Italy
e-mail: roberto.cognoli@unicam.it

P. L. Cocco
e-mail: lorenzopio.cocco@unicam.it

© The Author(s), under exclusive license to Springer Nature Switzerland AG 2024
M. Barberio et al. (eds.), *Architecture and Design for Industry 4.0*, Lecture Notes in Mechanical Engineering, https://doi.org/10.1007/978-3-031-36922-3_41

United Nations' Sustainable Development Goals 9. Industry, Innovation and Infrastructure · 11. Sustainable Cities and Communities · 12. Responsible Consumption and Production

1 Introduction

In the last 25 years, the advent of digital culture has touched «all aspects of existence, from everyday life to industrial manufacturing, from transport to politics, and from popular culture to the emotional relationship we form with each other» [4]. Architecture—understood as a «permanent process of transformation (…) which contributes to shape the built reality through the use of materials, (…) elements and systems»[1] [5]—has also been profoundly and progressively affected.

Nowadays in the architecture sector, digital tools are made up of a wide ecosystem of innovative technologies that concern the field of design and the industrial production of building systems and components. Parametric and computational design, CNC machines, robotics and other "enabling technologies" are—according to the Industry 4.0 vocabulary—progressively *«changing the ways»*, in many contexts, architects *«communicate and collaborate, analyze and simulate, fabricate and assemble»* [6]. Digital technologies *«not only help us to grasp very complex structures that we couldn't otherwise simulate, they also (…) allow us to gain a depth of understanding about materials, structures and building to predict performances and behaviors»* [6].

Nowadays, in what Mario Carpo called "the second digital turn" [7], digital tools represent much more than *«mere methods that can solve technical problems»*. They provide a "new possibility" *«to change the way we design, build and inhabit our world for a more sustainable future»* [8], or to conceive and produce spaces that are "ontologically" digital, a direct expression of the "digital man", able to overcome the objectives of optimization and efficiency on the one hand, and control complex forms on the other (a feature of the "first digital turn"). Whether *«the task of any architect is not about using computers to replicate or to automate what has already been thought and produced»* [9], the challenge underlying the advent of digital in Architecture is to *«give birth to a kind of architecture that is also beyond our usual capacity»* [9]. Looking at the current scenario and with specific reference to the architectural culture, nowadays we need more than ever "to give birth" to a new "architectural intelligence" that takes into account the "ecological crisis" we are experiencing, or better to say, the ecological crisis: *«the disequilibrium, which has become chronic in the Anthropocene era, between technological cycles (through which human action is expressed) and biological cycles (of nature)»* [10]. For this reason, the digital and ecological transitions can be considered, according to Luciano Floridi, "converging transitions" [1]. Following this vision, the implementation of digital culture in the context of the transformations of the built environment—i.e. the

---

[1] Original text: «permanente processo di trasformazione» che attraverso l'impiego di «materiali, (…) elementi, (…) sistemi contribuisce a dare forma alla realtà costruita» (translation by the authors).

introduction of the Industry 4.0 paradigm in such processes—can be a key factor in achieving the sustainability goals promoted by the European Green Deal.

2 Circular/Digital Approach in Building Processes

Nowadays the topic of the "circularity" of building processes has a strategic value. According to the European Union's Action Plan for the Circular Economy[2] (2020), the circularity of building processes represents a primary requirement to achieve the EU/2050 climate neutrality target. At present, the inefficiency in "environmental terms" of the technological cycles related to the transformations of the built environment is tangible.[3] Just to mention one: about 1900 million tons of construction and demolition waste (C&DW) were generated in 2018 [11]. This huge amount of material represents the largest waste stream in the EU by weight.[4] Against this backdrop, the target of 70% recycling of construction and demolition waste envisaged by 2020 by the EU Directive 2008/98/EC[5] has been largely missed.[6]

Even where recycling practices have found application, the processes of "downcycling" of crushable materials (stone, concrete, etc.) remain privileged compared to the creative reuse of waste and, in particular, of non-crushable material (wood, metal, etc.) (upcycling).[7]

---

[2] In line with the EU Green Deal goal of climate neutrality by 2050, in March 2020 the European Commission proposed the first package of measures aimed at accelerating the transition to a circular economy, as announced in the Action Plan for the Circular Economy. The proposals include the enhancement of sustainable products, consumer empowerment towards the green transition, the revision of the building materials regulation and a strategy on sustainable textiles. Cf.https://eur-lex.europa.eu/legal-content/IT/TXT/?qid=1583933814386&uri=COM%3A2020%3A98%3AFIN

[3] With reference to the European context, two data exemplify the dimension of the problem: 3.4 million companies (equivalent to 9% of the EU gross domestic product) engaged in the field of building construction produce 25-30% of all waste generated in the Union (with an expenditure of 45 billion euros for their treatment) and employ 4.3 gigatons/year of materials (of which only 12% comes from secondary sources). Source: "UNEP at 50" Report (United Nations Environment Programme).

[4] Source: European Topic Centre on Waste and Materials 2020 report: "Construction and Demolition Waste. Challenges and opportunities in a circular economy". https://www.eionet.europa.eu/etcs/etc-wmge/products/etc-wmge-reports/construction-and-demolition-waste-challenges-and-opportunities-in-a-circular-economy

[5] This EU Directive and the subsequent EU Directive 2018/851 set out measures for the reduction of impacts in waste generation and management. They represent the reference regulatory instruments for the Member States of the Union on waste matters.

[6] Source: Legambiente—https://economiacircolare.com/rifiuti-da-costruzione-e-demolizione-dovremmo-recuperarne-il-70-ma-non-sappiamo-neanche-quanti-ne-trattiamo/

[7] Paola Altamura gives a specific definition of these two terms [12]. Downcycling: *«The process of reworking a product, almost always with high energy consumption, which reduces its quality in terms of performance and/or economic value»*. Upcycling: *«The process of converting a waste material into a new material (...) characterized by better quality, which requires creativity and planning»*.

On the other hand, advanced upcycling practices are at the center of numerous research and experiments (also in the academic field) in which circular processes are triggered through advanced digital workflows ranging from raw-material digitization to the use of artificial intelligence for processing waste materials. Some research that have particularly influenced the work presented in the following pages of the paper are reported below:

- "Cyclopean Cannibalism", carried out at the Massachusetts Institute of Technology. It consists of the automated construction of a wall face using waste materials and, in particular, concrete, rubber and stone[8] [12].
- "Ashen Cabin", carried out by Hannah, an experimental design and research studio based in New York. The cabin reflects how new production methods can help to make useful what we think is wasteful. It uses waste timber through the implementation of high-precision 3D-scanning and robotic-based fabrication technologies [13].
- "Mine the scrap", developed by Tobias Nolte and Andrew Witt from Certain Measures, an international design studio. Part of this project concerns the development of a software tool that can scan scrap elements from demolished buildings and rearrange them into new architectural envelopes [15].
- The research carried out at the Institute for Advanced Architecture of Catalonia (IAAC) called "Material (data) Intelligence. Towards a Circular Building Environment". It is based on the convergence of artificial intelligence and data analysis to generate new strategies for the reuse of building materials from selective demolition [14].

These and other projects demonstrate how the "enabling technologies" underpinning the Industry 4.0 model can be decisive for achieving advanced results.[9] In many of these experiences strongly emerges «*the idea of using digital tools to exchange services and commodities, exploiting inefficiencies and redundancies, making a better use of material resources and physical assets*» [15]. The capacity of the digital to translate the sensible world into a data set constitutes, in particular, «*the technological and operational prerequisite for the application of some fundamental principles of the circular economy applied to the construction, such as "buildings as material bank" and "urban mining", which refer to the built environment as a bank of potentially reusable materials*» [10]. According to Thomas Rau and his studies about the "material passport", «*only by capturing the physical world through data is it possible to organize what is limited so that it remains available indefinitely*»

---

[8] The authors of this experimentation (launched in 2017) are Randon Clifford, Wes McGee and Johanna Lobdell, researcher at MIT, in partnership with the company Quarra Stone. Cf. http://www.matterdesignstudio.com/cyclopean-cannibalism.

[9] A recent study [18] identified ten specific enabling technologies that can be introduced in the Circular Construction processes in line with the Industry 4.0 paradigm: Additive and Robotic Manufacturing (AM/RM), Artificial Intelligence (AI), Big Data and Analytics (BDA), Blockchain Technology (BCT), Building Information Modelling (BIM), Digital Platforms, Digital Twins, Geographical Information System (GIS), Material Passports and Databanks, The Internet of Things (IoT).

[3]. Likewise, the possibility of "processing" data and transferring it, in the form of instructions, to machines capable of producing artefacts is the sequel to a digital production model that, in the construction field, is still to be explored.

3 Circular and Digital Timber Construction

Globally, many countries have set challenging targets aimed at reducing carbon emissions (towards net zero emission) by 2050. In particular, the construction sector generates about 50% of current emissions[10] attributable to the entire life cycle of buildings, including the production and construction phases. The extensive use of concrete and stone as construction materials represents a critical factor in reducing the embodied carbon footprint of buildings.

Nowadays timber is considered a key material for the ecological transition in this sector as *«it is abundant, renewable, possesses good technical characteristics, and can be converted into a host of different "engineered" wood products»* [16]. *«In the context of climate change, the main argument for using wood (…) is the fact that it is a renewable resource and that it stores CO2»* [20]. Among the many qualities of "wood" as material, there is its capacity to store CO_2 during the plant's growth process. Extending its life cycle, or postponing as much as possible the moment of release of CO_2 into the atmosphere, is a strategy of a circular approach to the use of resources [17]. It is estimated that implementing the reuse of timber as a building material could lead to a 25% reduction in emissions by 2050, especially those of CO_2 [18]. That's why timber can be considered "the" alternative to energy-intensive materials like concrete or steel.

Nevertheless, timber's availability is not unlimited. Over the past 20 years, global timber consumption raised by 1.1% per year, due to increasing urbanization and global housebuilding requirements. Global industrial roundwood timber consumption reached figures equal to 2.2 billion m^3 in 2018.[11] According to a survey carried out in 2020 by Gresham House,[12] over the next 30 years there will be an almost three-fold increase in timber consumption, from 2.2 billion m^3 consumed today to 5.8 billion m^3 in 2050.

According to Mark Hughes *«it's time to rethink how we use wood in construction»* [16], considering wood as a renewable but not unlimited resource and also taking into account the environmental costs of the engineering process of wood products. A strategy for the sustainable use of wood resources is the reuse, or "cascading",[13]

---

[10] Source: International Energy Agency.

[11] Source: FAO (The Food and Agriculture Organization of the UN); Forest Product Statistics 2018. https://www.fao.org/forestry/statistics/80938/en/.

[12] *Gresham House* is a specialist alternative asset management company (https://greshamhouse.com).

[13] The concept of cascading is comparable to downcycling: a building component is reused several times, for iteratively less demanding purposes [16].

of wooden elements embedded in existing buildings. *«Prolonging the life of wood products increases their capacity to store carbon for longer as well as avoiding the use of material with greater impacts»* [16].

In Europe, this material is often burned for energy recovery, with the consequent return of the carbon embedded into the atmosphere. This practice is also fueled by logistic, technical and economic criticalities connected to the transformation of demolition wood into another solid wood compound. Some critical issues concern the manufacturing phase, such as the need to have "clean" (and then "re-processable") wood after the removal of nails, screws or damaged parts. However, the most critical aspects concern the way in which a raw and heterogeneous material can be acquired, traced, stored and evaluated for its intrinsic qualities (firstly geometric and mechanical) and then be transformed into a set of components available for reuse.

4 Post Disaster Areas as Material Bank

Disaster causes the damage and collapse of buildings and infrastructures and consequently the production of great amounts of debris. The management of demolished parts and waste material is one of the great challenges after earthquakes. In Italy, for instance, it has been estimated [19] that after the Friuli earthquake (1976), about 188,000 m^3 of demolition waste were produced, 2.650.000 m^3 after the L'Aquila earthquake (2009) and about 364.000 m^3 in Emilia Romagna (2012). For the Central Italy earthquake (2016), according to the extraordinary reconstruction commissioner, public rubble alone is about 2.800.000 tons.

The application of the "Circular/Digital" approach in disaster areas represents a field still subject to little analysis. The treatment of debris is an aspect that often holds back reconstruction processes[14] that are usually lengthy and, especially in Italy, carried out with a low innovation content.[15] However, it has been proven that an effective debris management during the recovery and rebuilding phases has valuable social, financial and environmental impacts [20].

In the areas affected by catastrophic events it is possible to find different kinds of waste and this mainly depends on two factors: the type of event (earthquakes,

---

[14] In the case of the 2016/2017 earthquake, the reconstruction of private assets is still in a phase of substantial stalemate, especially due to the inability to dispose of the rubble of the destroyed buildings and, for the same reason, to demolish the crumbling buildings. The landfill sites are full and it is still necessary to travel more than 100 km to use a landfill (Source: Osservatorio Sisma, https://osservatoriosisma.it/gestione-macerie/).

[15] Italy is a highly seismic country where catastrophic events have followed one another, requiring almost always long and laborious reconstruction processes. The earthquakes of Belice (1968), Friuli (1976), Irpinia (1980), Marche/Umbria (1997), Puglia/Molise (2002), L'Aquila (2009) and Amatrice (which affected a large area of the sub-Apennine region of central Italy in 2016/17) represent some of the destructive seismic events occurred in Italy in the last fifty years. "Concrete" has been the reference reconstruction material in all these cases, except for L'Aquila. Here, the use of timber for the reconstruction of residential buildings has been undermined by urban planning and typological choices that are still at the center of a strong debate [23].

tsunami, etc.) and the building technologies used for the destroyed (or damaged) heritage. Wood will be the prevailing material in areas where there is a strong presence of trees (that is especially in suburban areas) or where wood was mainly used as a building material. As an example, Hurricane Andrew (Florida 1992) left an estimated 6 million tons of debris in the Greater Miami region, including downed trees and wood debris from 150.000 houses that were severely damaged or completely destroyed. Approximately 500,000 tons of wood waste from the hurricane were mulched and distributed to agricultural areas, parks and residential sites.[16] In Japan, the debris caused by the 2011 Tohoku earthquake (and the subsequent tsunami which damaged the Fukushima nuclear power plant[17]) were mainly composed of wood from the destroyed houses and trees swept away by the force of the tsunami. Following this event, Toyo Ito and other Japanese famous architects[18] realized the "Home for all" project: one of the most iconic projects of wood reuse in post-disaster reconstruction. It focused on a construction system mainly based on cedar trees (widespread in the area). The presence of seawater in the soil (resulting from the tsunami wave) "killed" a large number of plants. A team of botanical experts determined that they could still be used as pillars [21]. The local people cut and processed the cedar wood themselves and the building is surrounded by 19 local cedar pillars, seemingly erected at random.

In Italy, especially in post-earthquake contexts such as the sub-Apennine areas, there are two kinds of C&D (Construction and Demolition) wood waste: the wood coming from buildings that have been destroyed or will be demolished (floors, roofs, etc.) and the provisional structures supporting buildings awaiting restoration or demolition. All these categories constitute a vast and widespread catalogue of potentially reusable material, suitable to be implemented in a "material bank". In particular, provisional structures and temporary works to support the first emergency phases constitute a "reservoir" of wooden material destined to be discarded (in landfills or used as biomass) during the reconstruction phase, further increasing the production of C&D waste and CO_2 emissions. In particular, the shoring works [22] consisting of untreated wooden elements for structural use with homogeneous cross-sections (Table 1)[19] constitute a resource of constructive elements, potentially reusable in the reconstruction process as long as they are "processed" and acquire the characteristics identified by the legislation so that they can cease to be waste

---

[16] Source: United States Environmental Protection Agency EPA, Planning for Disaster Debris, EPA report, December 1995.

[17] With a 9.0 magnitude, the earthquake triggered a tsunami and the meltdown of a nuclear power plant, costing more than 20,000 victims and destroying more than 120,000 homes (about 1 million were damaged).

[18] Along with Toyo Ito, Riken Yamamoto, Hiroshi Naito, Kengo Kuma and Kazuyo Sejima were involved in the project.

[19] Usually solid wood elements are used for temporary works. Depending on the origin of the material (foreign or domestic) there are two different classifications given in UNI EN 338 (material of foreign origin) and UNI 11035 (material of Italian origin). See also UNI EN 460: 1996 and UNI 350: 2016 about the durability of wooden products.

Table 1 Most used cross section for shoring works [19]

Cross section (mm)	Width height (mm)	Length (mm)
100 × 100	100 × 100	6000
130 × 130	130 × 130	6000
150 × 150	150 × 150	6000
180 × 180	180 × 180	6000
200 × 200	200 × 200	6000

and become first-secondary matter.[20] In particular, provisional structures generate a large amount of "temporary object", generally destined to become waste at the end of their employment.

Always in Italy, the inspection of a wooden structure is conducted according to the criteria and procedures referred to in UNI 11,119:2004 regulation: "Cultural heritage. Wooden artefacts. Load-bearing structures of buildings—In situ inspection for the diagnosis of the elements in place", and UNI 11,035:2010 "Structural timber—Visual classification of timber according to mechanical strength".[21] These regulations also concern the assessment of waste wood.[22] The procedure is based on a visual inspection carried out by a forestry professional in order to evaluate the state of conservation and possible uses of the wooden components. It is focused on the identification of basic characteristics of materials such as wood species, geometry, biological and mechanical degradation, humidity and mechanical strength levels.

---

[20] According to the European Directive (2008/98/EC, subsequently amended by Directive 2018/851/EU), a waste is no longer such after being subjected to a recovery operation that fulfils the following conditions: (a) the substance or object is intended to be used for specific purposes; (b) there is a market or demand for that substance or object; (c) the substance or object meets the technical requirements for the specific purposes and complies with the existing legislation and standards applicable to such products; (d) the use of the substance or object will not cause overall negative impacts on the environment or human health. When all these conditions are met, the waste resulting from the recovery process is no longer "waste" and becomes a "product".

[21] Original text: UNI 11119:2004: "Beni culturali. Manufatti lignei. Strutture portanti degli edifici - Ispezione in situ per la diagnosi degli elementi in opera". UNI 11035:2010 "Legno strutturale - Classificazione a vista dei legnami secondo la resistenza meccanica".

[22] The UNI 11035 regulation only applies to Italian timber and concerns the classification of solid wood sawn timber for structural use of any size and moisture content. The material is classified in different categories (S1, S2, S3 for conifers and S for hardwoods) according to its defects. The UNI 11119 regulation sets objectives, procedures and requirements for the diagnosis of the conservation state and the evaluation of the strength and stiffness of wooden elements through in situ, non-destructive inspections and methodologies. The parameters underlying the inspection phases are: mechanical strength, geometric characteristics, biological degradation (for example: knots, inclination of the grain, density, cracks and onions), presence of chamfers and deformations, biological degradation, mechanical damage (as lesions).

This being the current situation, a question arises: can digital technologies support more effective processes for the reintegration of wood waste into a circular construction cycle? As shown in some recent research,[23] some of these surveying and inspection operations can be conducted through digital technologies related to Industry 4.0 paradigms. The application of such technologies can significantly increase process efficiency and enhance the possibilities of the reuse of waste materials as well as their field of application.

5 From Debris to Data Set (DeDa). A Digital Workflow for the Reuse of Waste Wood Material in Post-Disaster Areas

5.1 Research Goal and Field of Application

Following the earthquake that hit central Italy in 2016, the University of Camerino (UNICAM) established coordination of all its staff in order to promote research in the field of reconstruction. UNICAM boasts a long tradition in the sector of innovation technology studies for the built environment and, as it is located in a highly seismic territory, of post-earthquake reconstruction processes.

DeDa is one of the latest studies on this topic. It stands for *"From Debris to Data set. A digital workflow for the reuse of waste wood material in post-disaster areas"*. Its development began in 2021 at the School of Architecture and Design (SAAD), a UNICAM department, by a research team composed by the authors of this paper (of which the coordinator is Roberto Ruggiero). Subsequently, it has been tested during the Unicam Master on Circular Architecture titled "Circul-Ar, Shapes and methodologies of the circular architecture".[24] This master, which takes place this year for the second edition, focuses on the application of circular economy principles to the built environment through the use of innovative digital technologies. Like other European masters,[25] Circul-Ar is characterized by a strong propensity for

---

[23] As an example, the work "Matter Site. Material (data) intelligence" by Garcia et al. [14] investigates the reusability potentials from post and pre-demolitions sites using enabling technology such as Ai, Robotics and Data Analytics. The work was carried out at IaaC (Institute for Advanced Architecture of Catalonia) during the 2019-2020 edition of the Master of Robotics and Advanced Construction and subsequently developed by Driven (startup incubator) and supported by Scaled Robotics as Industrial partners.

[24] Circular-Ar is a second level master directed by Prof. Federica Ottone with the support of an international scientific committee, different companies involved in circular economy and private associations such as Symbola foundation and ANAB (Associazione Nazionale Architettura Bioecologica).

[25] At international level, the Institute for Advanced Architecture of Catalonia (IAAC) is today one of the main references for research in this field, as it uses innovative methodologies involving the integration of research and higher education, as in the case of IAAC's immersive Master in Advanced Ecological Buildings and Biocities (MAEBB).

experimenting with new ideas. For this reason, students are involved in experimenting with innovative approaches to the topic of circular construction. DeDa is a workflow based on digital tools for data acquisition and management. Its main purposes are:

- To implement a sharable cloud-based database of wooden elements, classified according to size and quality parameters: a virtual bank of raw materials;
- To demonstrate the effectiveness of this process for the reuse of waste materials.

DeDa is a digital application for the reuse of different types of waste (wood, metal, stone) in post-earthquake reconstruction. The specific background of this work is the sub-Apennine area affected by the 2016–2017 earthquake. This is a large area previously made up of small villages with a high landscape value.[26] In this area, communities are strongly linked to their building traditions and local materials (mainly stone and wood). The reconstruction works are also slowed down by the difficulty of disposal and research of materials, due to the distance from the main infrastructural network and the largest residential centers.

The first DeDa test was carried out during the 2021 edition of the Master. The material examined was the wooden material used as provisional shoring for the buildings damaged by the earthquake. In particular, the case study examined was a three-storey building located in the historic center of Santa Giusta, a small village situated in the province of Rieti. The entire main façade (12 m × 9 m) of the building was "caged" with steel tie rods and wooden uprights of homogeneous length and sections (Table 1). This kind of intervention presents recurring characteristics compared to the numerous shoring interventions carried out in the crater area.

5.2 Workflow Steps

The workflow is structured in five steps: 1. Data acquisition; 2. Data analysis and organization; 3. Database setup; 4. Database modelling of informed geometries (from Mesh to BIM); 5. Design from Database (Fig. 1).

After the first three phases, the wood components examined are "translated" into "digital objects" catalogued according to various parameters. They are brought together in a database of digital objects/materials informed, defined by qualitative and quantitative parameters and classified according to possible areas of use. According to Thomas Rau [3], this step allows to draw up a "material passport" of the wooden components from provisional structures. Through its "virtualization", this material can be "objectivized" (according to qualitative and quantitative parameters), stored in a virtual database and shared via the cloud (cloud database). The last step concerns a design hypothesis for the reuse of this (now) informed digital object through an

---

[26] In 2016 and 2017, central-Italy was affected by a destructive seismic event sequence. The crater area is about 8,000 km^2 and includes 140 municipalities and 580,000 inhabitants.

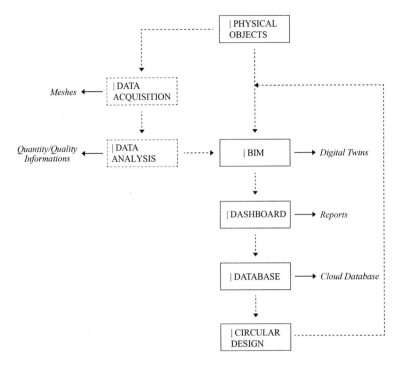

Fig. 1 The workflow is structured in five steps: 1. Data acquisition; 2. Data analysis and organization; 3. Database setup; 4. Database modelling of informed geometries (from Mesh to BIM); 5. Design from Database

approach that combines BIM and computational design (Data Driven Design).[27] A small pavilion is the subject of this step.

The development of this work will concern further steps focused on the design of new components' families as well as on their production. Regarding the latter, an automated process using a collaborative robotic arm controlled via a mobile workstation is under study (but it is not the subject of this paper).

5.3 Data Acquisition

This phase focuses on the collection of all the information (at dimensional, qualitative and quantitative level) required to create the digital twin of each previously scanned

---

[27] The term "data-driven design" refers to a process in which qualitative and quantitative data are processed to guide design decisions [24]. For a concise but effective understanding of this topic, it may be useful to consult Proving Ground, a digital design agency that enables digital transformation with creative data-driven solutions to the building industry. https://provingground.io/.

wood item. Among the different methods of digital surveying available today, "photogrammetric surveying" was used. It is based on the acquisition of metric data (shape and position of an object) through the detection and analysis in a digital environment of photograms.[28] Compared to other systems (i.e. laser scanning) this method offers, through cheaper and more flexible instrumentation, good levels of accuracy as well as the possibility of acquiring high-resolution images.[29] In particular, two photogrammetric methods were employed at this stage with a dual purpose: a. to catalogue the wooden elements; b. to acquire more accurate images and geometries:

a. Tool: drone; model: Dji MavicMini[30]; method: automatic flight parallel to the facade of the "caged" building; distance: approx. 2 m from the façade, with height increments of 1.0 m and shots every 0.5 s for a total of approx. 100 frames. The images were processed and the point cloud was extracted through Agisoft Metashape, a photogrammetric processing software which uses structure from motion algorithms. Then the point cloud was segmented using the Cloud Compare software to isolate the supporting elements protruding from the main façade by identifying the main planes of the geometries to be isolated, using "fit-plane and RANSAC algorithms" [26]. The identified portions of the cloud were exported and used for the reconstruction of the main geometries (Fig. 2).
b. Tool: digital camera Lumix DMC-lx100.[31] Once the process of partial demolition of the building and removal of the provisional structures was completed, the second photogrammetric survey was carried out on individual wooden elements (Fig. 3), using a digital camera in a controlled light environment (2 flash studio, white backdrop). This step allowed to achieve ideal and repeatable shooting conditions for each individual element. The shots were taken using a fixed camera by positioning the element on a rotating platform (one shot every 15°). The coded targets allowed to use the images for the "structure from motion" process using Agisoft Metashape software. The high-definition images with ideal lighting conditions allowed to generate both a point cloud with very high resolution and accuracy (error < 3 mm), and an accurate texture to be applied to the detected geometries.

---

[28] Photogrammetric surveying is based on "structure from motion" algorithms, i.e., a calculation method allowing the reconstruction of the three-dimensional shape of objects by the automatic collimation of points from a set of pictures [25]. This is how the procedure works: the algorithm extracts the remarkable points from the individual photos and (by cross-referencing the recognizable points on several frames) calculates the spatial coordinates of the single points. The result of this calculation is a scattered cloud of points to be, subsequently, employed to determine a dense cloud of points in space. Each point is characterized by spatial coordinates and a color.

[29] In this regard, cf. https://www.agisoft.com/ and https://www.danielgm.net/cc/.

[30] Cf. https://www.dji.com/it/mavic-mini.

[31] Cf. https://www.panasonic.com/it/consumer/fotocamere-e-videocamere/compatte-fotocamere/professionali/dmc-lx100.html.

Fig. 2 Drone automatic flight parallel to the facade of the "caged" building; distance: approx. 2 m from the façade, with height increments of 1.0 m and shots every 0.5 s for a total of approx.

Fig. 3 Studio setup (rotating platform with coded targets) for individual wood elements photogrammetry survey

Facade survey point cloud:
N.points 1701997

Fig. 4 Façade survey point cloud post-processed. N.points 1701997

5.4 Data Analysis and Organization

For the reconstruction of mesh geometries from point clouds, two methods were employed. The first is based on the use of the CockRoach, a Grasshopper3D[32] (Mcneel Rhinoceros) plugin[33]: a tool aimed at managing point clouds and reconstructing mesh geometries directly in the CAD environment. This tool is particularly useful in managing point clouds. The tool offers various options, including: 1. the reduction of the points number, the clustering and segmentation of the cloud (i.e., the definition of the main planes using RANSAC—RANdom SAmple Consensus—algorithms) (Figs. 4 and 5); 2. the management of addition and subtraction operations (Boolean) between mesh geometries. One of its current limitations is the calculation time, which is directly proportional to the number of points to be processed. For shorter calculation times, the textures obtained will be less accurate than dedicated software.

Once the outline geometries were reconstructed, it was possible to estimate (within Grasshopper3D) the volume of materials detected and the number of elements and their dimensions, providing the necessary data for the construction of an initial database (Fig. 6).

The second method used for the reconstruction of detailed mesh geometries still refers to the use of Agisoft Metashape, a software that performs photogrammetric

---

[32] https://www.grasshopper3d.com/.

[33] P. Vestartas and A. Settimi, Cockroach: "A plug-in for point cloud post-processing and meshing in Rhino environment", 2020. Cfr: https://github.com/9and3/Cockroach.

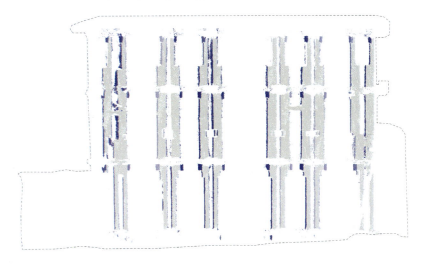

Plane detection segmentation:
N.points 311330

Fig. 5 Plane detection point cloud segmentation (RANSAC) for beams extraction from façade. N.points 311330

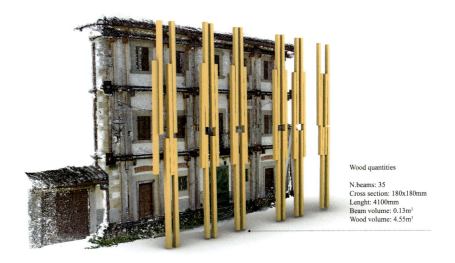

Wood quantities

N.beams: 35
Cross section: 180x180mm
Lenght: 4100mm
Beam volume: 0.13m^3
Wood volume: 4.55m^3

Fig. 6 Elements numbers, dimension and material quantity evaluation with Grasshopper 3D

processing of digital images and generates 3D spatial data to be used in a digital environment. Like the previous one, also this workflow involves the automatic construction of surfaces composed of polygons whose sides are the segments connecting the points of the cloud. This procedure allows to get surfaces consistent with the

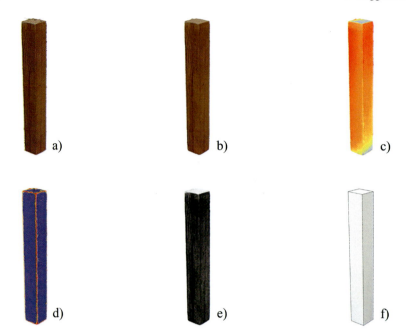

Fig. 7 Surveyed geometries and textures of the wooden elements analysis with Grasshopper 3D. **a** Point cloud; **b** Mesh geometry; **c** Deformation analysis; **d** Mesh curvature analysis; **e** Texture analysis; **f** Bim model

shape of the surveyed objects and also to construct interpolation surfaces to cover any defects or missing parts in the point cloud. The software also contains various presets and parameters able to determine the accuracy of the reconstructed geometries and manage calculation times (quality, quantity of faces, etc.). The "build texture" command allows to start of a subsequent step, obtaining a high-resolution texture (from the photographic sockets used to extract the point cloud) to be applied to the reconstructed surfaces (Fig. 7a, b).

The surveyed geometries and textures of the wooden elements are analyzed (in Grasshopper) to check their quality and performance condition and the presence of any defects or foreign bodies, by considering the parameters provided by the aforementioned UNI 11,035 and UNI 11,119 regulations (cf. note n. 21). Below are some of the main analyses undertaken:

- *deformation analysis*: aimed at quantifying the presence of excessive deformations (embossing, bowing, warping, etc.) in the surveyed elements, the method consists of evaluating the deviation between each vertex of the 3D scanned mesh and an ideal perfectly straight geometry of the building element identified by an evolutionary solver. A greater distance from the reference axis corresponds to the main deformations of the element (Fig. 7c);
- *mesh curvature analysis*: aimed at checking for structural defects (lack of material, holes, surface inhomogeneity, etc.), using the analysis of the change in the

curvature of the surfaces of the surveyed geometry. Any surface irregularities are highlighted by mapping the relative changes of direction in the perpendicular mesh faces (Fig. 7d);
- *texture analysis*: aimed at identifying any potential defects in the wood (fungi, knots, holes and, in general, inhomogeneity in the coloring of the material), manipulating the images to obtain a two-color mapping (black, inhomogeneous surface; white, homogeneous surface). By extrapolating the percentages of the two black/white values using Grasshopper 3D, it is possible to quantify the homogeneity percentages of each element (average between the values of the different faces of the element, (Fig. 7e);
- *material analysis*: aimed at obtaining a reliable indication of the wood species, average hue values, saturation and brightness by extrapolation from the detected textures and comparison with sample images of the most common wood species;
- *volume and specific weight evaluation*: aimed at assessing the presence of any defects that could alter the weight of the single element (such as the excessive presence of water or the presence of deteriorating internal parts). The digital model provides information on the volume. The multiplication of the volume thus obtained for the specific weight of the wood (variable from essence to essence) gives an indication of the expected weight of the element. Based on an empirical method, the expected weight is compared with the actual weight (measured by means of a dynamometric balance). This made it possible to obtain the first quantitative parameter to establish the material quality. For example, a ratio greater than 1 could indicate the presence of excessive humidity, and a ratio lower than 1 the internal deterioration of the material.

Following these analyzes, each element was assigned a score for each performance in order to identify and group elements with similar characteristics, drafting a classification based on possible reuses: biomass for elements with the lowest score; decorative or in any case non-structural uses for elements with average scores; structural uses for elements with higher scores according to the resistance class determined by the already mentioned standards (see note n. 21).

5.5 Database Setup

The digitization of physical elements and the systematization of the related data (at qualitative and quantitative levels) allows the creation of virtual stores. They can be shared and updated over time. Once the data collection of the shoring elements is completed (Paragraph 5.4), the DeDa workflow allows their organization into lists within Grasshopper 3D. The latter allows the simultaneous handling of heterogeneous data and their association with geometric entities (mesh). Lunch Box, a Grasshopper

Table 2 List of elements information in csv format

Element ID	Width (mm)	Height (mm)	Length (mm)	Material	Deformation factor	Quality factor
Element 00,000	180	180	4100	Pine	0.775	0.931
Element 00,001	180	180	4100	Pine	0.733	0.821
Element 00,002	180	180	4100	Pine	0.84	0.75
Element 00,003	180	180	4100	Pine	0.931	0.996
Element 00,004	180	180	4100	Pine	0.897	0.733
Element 00,005	180	180	4100	Pine	0.83	0.792
Element 00,006	180	180	4100	Pine	0.806	0.941
Element 00,007	180	180	4100	Pine	0.983	0.834
Element 00,008	180	180	4100	Pine	0.73	0.767
Element 00,009	180	180	4100	Pine	0.893	0.703
Element 00,010	180	180	4100	Pine	0.709	0.93
Element 00,011	180	180	4100	Pine	0.774	0.709
Element 00,012	180	180	4100	Pine	0.796	0.702
Element 00,013	180	180	4100	Pine	0.997	0.853

3D plugin, allows the creation of lists in.csv format[34] (Table 2), a file format that can be interpreted and edited by several commonly used applications (Excel, Open Office, Notepad). The interoperability of this format facilitates the integration of the database with data from other sources, providing the possibility of managing all information in a single software environment. This opportunity is decisive to work in the Building Information Modelling environment and to create a Material Passport dedicated to scrap materials. At the conclusion of this process, the.csv files are finally treated using Power Bi (a Microsoft software for the Data Visualization) in order to create an intelligible and more friendly interface (Fig. 8).

---

[34] A CSV (comma-separated values) file is a text file characterized by a specific format which allows to save the data in a table structured format. CSV is a common data exchange format that is widely supported by consumer, business, and scientific applications.

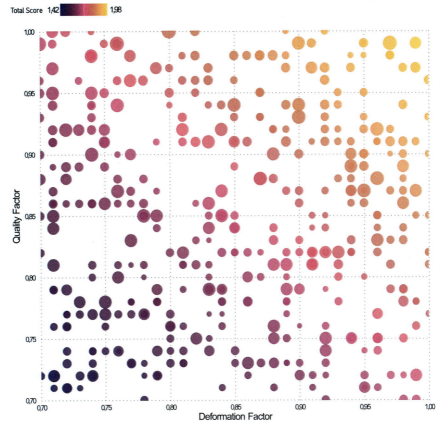

Fig. 8 Data visualization using Power Bi

5.6 Database Modelling of Informed Geometries (from Mesh to BIM)

Using current technologies, the product of a three-dimensional scan is returned in the form of a *mesh*: a network of interconnected points (vertices). The way this network of points is organized determines the "topology of the mesh" [27]. The mesh allows the digital restitution of the morphology and texture of a real object. However, the use of meshes in the BIM environment is computationally burdensome [28]. If the goal, as in the DeDa case, is the creation of the digital twin of wood waste elements and, subsequently, the organization of a material bank of all the (digitally acquired) elements, the transformation of the meshes from the scanning process into parametric objects (objects with associated information) still requires various steps using current technologies [29]. In the lack of automated (via artificial intelligence) "mesh/parametric object" conversion strategies, it is now possible to use computational software such as Grasshopper3D, able to "interface" with BIM software. Such a procedure still requires a "crafted" approach, where the choice of software and procedures strongly depends on the morphological characteristics of the object [30]. This implies a cumbersome workflow and the impossibility of defining a single method repeatable in different applications. In fact, the main BIM authoring software does not support the automatic modelling of meshes (except for special cases, such as topographic surfaces) as these are not recognized as native geometries.[35] This results in the absence of qualitative and quantitative information and parametric behaviour, typical characteristics of native BIM objects [31]. The workflow is divided into the following steps:

- *Identification of the parallelepipeds that best approximate the mesh.* Tool: Grasshopper3D. For this step, the potential of evolutionary and genetic algorithms has been exploited.[36]
- *BIM modelling* from *Grasshopper.* Tool: *Rhino.Inside.*[37] Rhino.Inside is employed to connect Rhino and Grasshopper via the API (Application Programming Interface) of the Autodesk Revit BIM authoring software.[38]
- *Conversion of modeled objects into parametric objects.*
- The following parameters are detected and attributed to the digital twin of each scanned and modeled wooden element: morphology, weight, essence, deformation, density, discontinuity, presence of foreign bodies, need for machining to eliminate unusable parts and origin concerning the original location in the building. This information was implemented and automated using Rhino.Inside Revit technology, with particular attention to data matching within Grasshopper.

The result of these three steps is a database of computerized, modelled and classified BIM components of the wooden elements taken as an object of experimentation. Despite the limitations deriving from the software technologies available today (highlighted above), this workflow actually achieved its goal thanks to the complete "interoperability" [33] of the software used, able to exchange information in real-time (Fig. 9).

---

[35] BIM software uses a B-Rep (Boundary Representation) method whereby each object can be represented by a curve governing its course (directrix) and by a curve defining its profile (generatrix) [27].

[36] These algorithms allow to find solutions to the problems described through a system of constraints and targets, expressed using a fitness function and the continuous recombination of design variables known as "genes" [32]. In this case, the algorithms control the orientation in the three dimensions of the reference planes (genes) for the construction of the parallelepipeds. They allow to minimize the difference between the volume of the starting mesh and its boundary parallelepiped (fitness function).

[37] Rhino.Inside is an open-source software that allows Rhino and Grasshopper to run within other 64-bit Windows applications (such as Revit or AutoCAD).

[38] Through direct communication with the Revit API, Rhino.Inside allows native Revit geometries (each characterized by a unique GUID—Globally Unique Identifier) to be modeled from Grasshopper's computational environment. Cf. https://www.rhino3d.com/inside/revit/beta/guides.

Fig. 9 Visualization of the set of design solutions, optimized on the basis of the objectives and consistent with the system of constraints

5.7 Design from Database

The outcome of this first experimentation is the realization of a pilot project of a wooden pavilion (Fig. 10) made with the materials contained in the virtual storage of "informed wooden components" obtained by following the previous steps. The design phase was performed using evolutionary and loop algorithms. These algorithms allowed continuous recombination of the design parameters, generating different design solutions for different design goals. The result of this process is a set of design solutions, optimized on the basis of the objectives and consistent with the system of constraints (Fig. 9). Loop procedures were used to select the most suitable elements from the database (in relation to their geometric, mechanical, conservative properties, etc.) to perform a given "task" in the pavilion construction system.

Fig. 10 Wooden pavilion made with the materials contained in the virtual storage of "informed wooden components"

This step is still in progress. It is planned to progressively increase the complexity of the design application (in order to obtain the design of a more complex object), but also to implement the database with new wooden elements. A further objective is to move from the design phase to the construction phase, i.e., to realize the artefact. This further step involves an upgrade of the database, once the hall has been assembled and subsequently dismantled. In fact, at the end of this process, some wooden components will have suffered damage or changes in some parameters due to assembly and disassembly and the operating time of the pavilion. This upgrade—which will involve the repetition of the entire workflow—sets up an infinitely circular process that, through the "virtualization" of waste wood, allows more than its simple one-time reuse and implies its re-introduction into an infinitely circular manufacturing process inspired by the "Design for Disassembly (DfD) principles" [34].

6 A Master Course as a Place for Research and Experimentation

What Mario Carpo defines as "the second digital turn" [7] represents for architecture an irreversible process meant to modify many consolidate rules in the professional practice. If there is no advancement in methodologies, tools and skills, in the next future training in architecture will prove to be inappropriate [35]. *«To create the ethical and scientific basis to allow the architect to play a leadership in a future that is announced as increasingly "digital", a radical re-thinking of the educational paths in architecture is necessary, as well as the testing of innovative methods and models based on the prefiguration of future (medium and long-term) scenarios»* [35].

As some experiences in international contexts demonstrate, the quality of educational offer requires a close relationship between training, experimentation and research. From Buckminster Fuller's experiments with geodesic structures in the 1950s with students from Black Mountain College [36] to more recent experiences such as the immersive Master in Advanced Ecological Buildings and Biocities (MAEBB) of the Institute of Advanced Architecture of Catalonia (IAAC),[39] research in architecture has often used educational environments as a place for experimentation. Mention one, the Voxel Quarantine Cabin, built during the MAEBB 2019–2020 Master's degree, represents an inspiring example of the overlap between education and advanced research.[40] In particular experiences like this are based on the use

---

[39] The MAEBB is a good example of an innovative educational format that offers interdisciplinary skills through an experimental activity focused on new categories of projects, technologies and solutions related with the design, prototyping and fabrication of ecological buildings.

[40] The project is the result of an innovative digital workflow realized with the co-partnership between the Master's students and teachers and external specialists. The Voxel is an autonomous 12 m^2 cross-laminated timber (CLT) structure clad in a parametric rainscreen, exemplifying an advanced ecological approach to architectural production. Every timber element can be traced back to its exact point of origin, and all building components have been rigorously quantified in terms of geographic

of advanced digital tools for research goals starting from the circularity of production processes. In this and other areas of research, the creation of manufacturing workshops and the offer of advanced tools such as parametric/generative software and prototyping/manufacturing machinery would become a priority for architecture departments to implement a "learning by experimenting" approach, to guarantee what Alvin Toffler already defined as an *«education in the future tense»* [37] and allow the architects of the future to meet the social, technological and, above all, environmental challenges of the coming years.

At the same time, the participation of students in research experiences can represent an added value in terms of human resources and opportunities for achieving the objectives. Especially where "digital technologies" are the basis of experimental activities, the involvement of young, open-minded, digital native people still in training could represent an added value to be achieved.

Inspired by this idea, the Master course "Circul-Ar, Shapes and Methodologies of the Circular Architecture" held at the University of Camerino represents an opportunity for connection between the educational and research sectors. Integrating multidisciplinary competencies ranging from architecture to design and engineering, the Master course proposes an innovative approach to the theme of circular architecture [15] based on the creative use of new or used, renewable, natural, biodegradable, recyclable and reusable materials. This approach is pursued through the use of advanced digital tools and the experimentation of workflows dedicated to the theme of upcycling, intended as *«creative reuse of materials»* [12]. DeDa is an experimental work developed, in its application part, within the Master's course. In particular, during the first edition of the Master (2020/2021), part of the process described so far was developed in order to experiment with different workflows, according to DeDa goals (already discussed in Chap. 5). The interaction between research and educational contexts achieved excellent results. The continuous feedback between students and teachers, the interest aroused in students by digital applications as well as the combination of social issues (such as the reconstruction in post-disaster areas), environmental issues (such as the circular building) and innovative technology created the right combination for the achievement of teaching objectives (an advanced digital training at the service of the circular economy) and research goals (the integration of digital tools to overcome the limits of software interoperability).

---

source and carbon content, taking into account any fuel or energy inputs during the entire associated life cycle. Cf. https://iaac.net/maebb-voxel-quarantine-cabin/.

7 Research Limits and Opportunities

DeDa is a methodology that could also be extended to other types of waste material. For its further development, it will be necessary to overcome some limitations that the work done so far has highlighted and which can be summarized as follows:

- *Limits of technologies*—As reported in Sect. 5.6, the transition defined "from mesh to BIM" is excessively articulated in the lack of tools enabling automatic conversion. The digital chain proposed in Chap. 5 is still improvable through either the development of a specific tool, or optimizing the procedure here proposed. Both options require the introduction of automatic design and IT (Information Technology) skills into the research group.
- *Limits of design application*—Choosing a small pavilion as design test case (with a limited number of components), helped us to have the right degree of simplification at this stage of the work. Starting in October 2022, a second test has already been planned. An existing temporary single-storey timber building will be the object of a virtual selective demolition process. It is located in the crater area of the 2016 Italian earthquake. It will be dismantled when the reconstruction of destroyed buildings will be completed. Its components will be organized in virtual storage through the DeDa workflow. With a larger quantity and quality of components than those used in the first application of DeDa, it is planned to develop a second design application focused on the production of small wooden building components to be integrated, as cladding, into façade systems.
- *Production limits of discrete systems*—For each wooden component examined, the level of degradation is recorded and any parts to be removed is precisely identified and faithfully reported in the dataset (cf. paragraph 5.6). Usually, wood waste material includes parts that are irreparably damaged. Consequently, form the original to the reusable piece a "reduction in size" (due to the removal of excessively degraded parts) is likely to be. Therefore, the reusability of these components requires, as a privileged field of application, the so-called "discrete systems". The "discretization" of building systems constitutes one of the paradigms of so-called digital fabrication [38]. It can be said that discrete systems (in building construction but also in the production of objects) count flexibility, reversibility and versatility among their advantages. At the same time, they require a great number of connections with a complexity of assembly work.

In connection with this last consideration and as a medium- and long-term goal, the research group is in parallel working on a second line of research focused on an automated process of recycled wood material. For the re-employment of the "stored components" (according to the DeDa method), a second workflow is being developed. It is focused on a mobile workstation and the use of a collaborative robotic arm. This additional work module would allow to produce, in situ, artefacts made by discrete parts from wood waste material. Arm activities range from the removal of metal elements, to repair, to the assembly of discrete parts. This second phase is not the subject of this paper (Fig. 11).

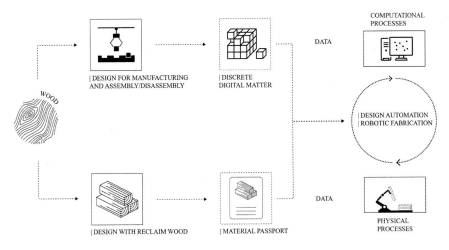

Fig. 11 Digital circular timber construction workflow

8 Conclusion

The current "environmental crisis" is an "ecological crisis". Meaning by ecology *«the exosystemic relation between something or someone and its environmental context»* [15], it can be argued that, since the first Industrial Revolution, the relationship between man and the natural environment has gone into crisis. Since that time, human actions on the environment have become more aggressive and increasingly based on linear production processes. Those processes are responsible for what we now call "climate change", which is nothing more than the progressive reaction of the natural environment (whose transformations take place through circular processes) to human actions[41] [39]. This conflict cannot be resolved by pursuing the (simplistic) idea of a "return of man to nature". According to Bruno Latour, *«the human being (…) is first and foremost a cultural being distinct from nature»* [39]: that's why any definition of ecological crisis as a "return to nature" results rhetorical and without foundation. Therefore it seems more realistic to attempt a realignment between human technological cycles and the biological cycles of nature. Nowadays, moving from a linear to a circular model of production and consumption is a condition that no longer appears questionable.

---

[41] The currently most human-driven technological cycles are still "linear". According to the Ellen McArthur Foundation, *«in our current economy, we take materials from the Earth, make products from them, and eventually throw them away as waste»* (https://ellenmacarthurfoundation.org/). In the future, the scarcity of resources will require more and more to focus on new techniques and methodologies in order to replace the principles of the linear economy with those of the circular economy, in which waste products are reused as inputs for the creation of secondary products [44]. The Ellen McArthur Foundation is a charity foundation whose core business is promoting circular economy i.e., eliminating waste and pollution and boosting the spread of circulate products and materials. Cf. https://ellenmacarthurfoundation.org/about-us/what-we-do.

This transition, as ambitious as it is necessary (given the dramatic effects of the current ecological crisis), must and will have to be *«driven by design»* [40]. Nevertheless, design needs appropriate technologies to be effective and creative as the situation requires. Digital technologies linked to the Fourth Industrial Revolution constitute, in particular, a "new opportunity" to redeem technology against the environment and help design fulfill its purpose. *«Where the danger is, also grows the saving power»* is the famous aphorism by the German poet Friedrich Hölderlin (1770–1843). If the current climate crisis has its roots in the industrial development and the first three phases of the industrial revolution [41], the advent of the Fourth Industrial Revolution (the digital revolution) gives the opportunity to face the environmental challenge with new weapons and possibilities. Technology, which has so far contributed significantly to the current environmental crisis, can therefore represent a way out in its digital form.

Some recent experiences of a "circular/digital" approach to building construction (see Paragraph n. 2) demonstrate the effectiveness of design approaches based on the interconnection of information between the physical and digital world. This approach must be placed in the context of a "digital ecosystem" that *«encompasses the technologies that enable the creation and use of digital information, following a set of data standards and interoperability. These technologies enable reality capture, computational design, data acquisition, data analytics, artificial intelligence, simulation and analysis»* [42]. However, the level of interoperability between some digital environments is still inadequate and there is a lack of "bridging tools" able to facilitate the interconnection between software [43]. These limitations have also been experienced in the application of DeDa.

DeDa is a workflow focused on the transformation of waste materials into new materials, ready to have new uses and a new life expectancy. Such a "transformative flow" requires a substantial intermediate step: the virtualization of waste material in terms of both digital twin and data set. This temporary condition of "digital object" is the prerequisite for the waste material to be framed in its performance futures. This "migration" of the physical world into a virtual world is in the middle of the current debate about "digital". In the book "Non-things: Upheaval in the Lifeworld", the South Korean philosopher Byung-Chul Han states that the "things of the world" (physical world) *«have the task of stabilizing human life by giving it a foothold»* [45]. When *«the earthly order is succeeded by the digital order»*, the latter *«dematerialize the world by computerizing it»* [45]. In Han's criticism, the ongoing process of dematerialization of the world has a negative meaning: *«it is now data and no longer concrete things that influence our lives. Non-things are taking over from the real, from facts and biology. And so, reality appears to us more and more elusive and confusing, full of stimuli that do not go beyond the surface»* [45]. In complete agreement with Han's critique, the dematerialization process at the basis of DeDa is a key requisite—*a conditio sine qua non*—so that the practices of reuse and upcycling of waste materials "come into reality" in a systematic, repeatable and scientifically controllable way. This is a paradox compared to Han's vision. In this case indeed, "computerizing" pieces of the real world in digital environments—i.e. converting scrap things into "non-thing"—is a necessary practice to redefine the "earthly order of things" called into question by the current ecological crisis.

Finally, the purpose of this work (and its future developments) is to explore a 4.0 approach to building processes from the perspective of rethinking the way we use resources in view of the current environmental crisis. If we consider post-natural disaster areas as great "material banks", the "material/data/material" approach proposed here could support new and desirable policies for the reuse of waste, transforming scrap material into a *«new raw material»* [46]. Policies and technologies need each other to achieve these goals. In the absence of effective technologies and operational chains, the ecological transition of construction processes will not take place and inefficiencies and waste of resources will continue. In the specific context of post-disaster recovery, drawing up appropriate technologies for the reuse of debris could be a key strategy to make such processes more efficient and sustainable. However, there is still a long way to go in terms of research on this topic, even if today's digital technologies open up new scenarios that were unthinkable until a few years ago.

References

1. Floridi, L.: Etica dell'intelligenza artificiale. Sviluppi, opportunità, sfide, Raffaello Cortina Editore, Milano (2022)
2. Suárez-Eiroa, B., Fernández, E., Méndez, G.: Integration of the circular economy paradigm under the just and safe operating space narrative: Twelve operational principles based on circularity, sustainability and resilience. J. Clean. Prod. **322** (2021)
3. Rau, T., Oberhuber, S.: Material matters. L'importanza della materia, Edizioni Ambiente, Milano (2016)
4. Picon, A.: The Materiality of Architecture. The University of Minnesota Press, Minneapolis (2020)
5. Gregory, P. (ed.).: Nuovo Realismo/Postmodernismo. Dibattito aperto fra architettura e filosofia, Officina, Roma (2016)
6. Tamke, M.: Foundamental changes for architecture. In: Commerel, A.H., Feireiss, K. (eds.) Craftmanship in the digital age. Architecture, value and digital fabrication, ANCB edition, Berlin (2017)
7. Carpo, M.: The Second Digital Turn. The MIT Press, Cambridge Massachusetts, Design beyond intelligence (2017)
8. Claypool, M.: The digital in architecture: then, now and in the future, "Space 10" online Journal (2020). https://space10.com/project/digital-in-architecture/
9. Morel, P.: The origins of discretism: thinking unthinkable architecture. In: G. Retsin (ed.) Discrete. Reappraising the Digital in Architecture. Wiley, Hoboken (2019)
10. Perriccioli, M., Ruggiero, R., Salka, M.: Ecology and digital technologies. Small-scale architecture as a place of connections. Agathon, Int. J. Arch., Art Des. **10** (2021)
11. Eurostat Homepage. https://ec.europa.eu/eurostat/statistics-explained/index.php?title=Waste_statistics. Last accessed 23 Jan 2022
12. Altamura, P.: Costruire a zero rifiuti. Strategie e strumenti per la prevenzione e l'Upcycling dei materiali di scarto in edilizia, Franco Angeli, Milano (2015)
13. Matter Design & Quarra Stone. (2017). Retrieved 7 Feb 2022 from http://www.matterdesignstudio.com/cyclopean-cannibalism
14. Ashen Cabin (2019). Retrieved 7/02/2022 from https://www.hannah-office.org/work/ashen

15. Mine the Scrap (2016). Retrieved 7/02/2022 from https://certainmeasures.com/MINE-THE-SCRAP
16. Batalle Garcia, A., Cebeci, I.Y., Vargas, C.R., Gordon, M.: Material (data) intelligence—towards a circular building environment. In: Globa, A., van Ameijde, J., Fingrut, A., Kim, N. (eds.) PROJECTIONS—Proceedings of the 26th CAADRIA Conferencem, vol. 1, pp. 361–370. The Chinese, University of Hong Kong and Online, Hong Kong (2021)
17. Zanotto, F.: Circular architecture. A design ideology. LetteraVentidue Edizioni, Siracusa (2020)
18. Çetin, S., De Wolf, C.E.L., Bocken, N.: Circular digital built environment: An emerging framework. Sustainability **13** (2021)
19. Hughes, M.: Cascading wood, material cycles, and sustainability. In: Hudert, M., Sven Pfeiffer, S (eds.) Rethinking Wood: Future Dimensions of Timber Assembly. Birkhauser, Basilea (2019)
20. Hudert, M., Sven Pfeiffer, S.: Rethinking Wood: Future Dimensions of Timber Assembly. Birkhauser, Basilea (2019)
21. Diymandoglu, V., Fortuna, L.M.: Deconstruction of wood-framed houses: material recovery and environmental impact. In: Resources, Conservation and Recycling, vol. 100, pp. 21–30. Elsevier (2015)
22. Bastin, J.F., Finegold, Y., Garcia, C., Mollicone, D., Rezende, M., Routh, D., Zohner, C., Crowther, T.: The global tree restoration potential. Science **365**, 76–79 (2019)
23. Faleschini, F., Zanini, M.A., Hofer, L., Pellegrino, C.: Demolition waste management after recent Italian earthquakes. In: 16th International Waste Management and Landfill Symposium SARDINIA2017. S. Margherita di Pula, Italy (2017)
24. Khanala, R., Subedib, P.U., Yadawac, R.K., Pandeyb, B.: Post-earthquake reconstruction: Managing debris and construction waste in Gorkha and Sindhupalchok Districts, Nepal. In: Progress in Disaster Science, p. 9. Elsevier (2015)
25. Saitta, P.: Fukushima. Editpress, Firenze, Concordia e altre macerie (2015)
26. Grimaz, S. (ed.).: Vademecum STOP. Schede tecniche delle opere provvisionali per la messa in sicurezza post-sisma da parte dei Vigili del Fuoco, Corpo Nazionale dei Vigili del Fuoco, Ministero dell'Interno (2010)
27. Andreassi, F.: La ricostruzione di L'Aquila. Dal modello ai progetti. Franco Angeli, Milano (2020)
28. Deutsch, R.: Data-Driven Design and Construction: 25 Strategies for Capturing, Analyzing and Applying Building Data. Wiley, Hoboken (2015)
29. Özyeşil, O., Voroninski, V., Basri, R., Singer, A.: A survey of structure from motion. Acta Numer **26**, 305–364 (2017)
30. Qian, X., Ye, C.: NCC-RANSAC: A fast plane extraction method for 3-D range data segmentation. IEEE Trans. Cybern. **44**(12) (2014)
31. Tedeschi, A.: AAD—Algorithms-Aided Design. Parametric Strategies Using Grasshopper®. Le Penseur, Brienza (2014).
32. Costantino, D., Pepe, M., Restuccia, A.G.: Scan-to-HBIM for conservation and preservation of Cultural Heritage building: the case study of San Nicola in Montedoro church (Italy). In: Applied Geomatics. Springer, Berlin (2021)
33. Qin, G., Zhou, Y., Hu, K., Han, D., Ying, C.: Automated reconstruction of parametric BIM for bridge based on terrestrial laser scanning data. In: Advances in Civil Engineering (2021)
34. Bolognesi, C., Caffi, V.: Extraction of primitives and objects from HShapes. In: The International Archives of the Photogrammetry, Remote Sensing and Spatial Information Sciences, Volume XLII-2/W9, 2019 8th Intl. Workshop 3D-ARCH "3D Virtual Reconstruction and Visualization of Complex Architectures" (2019)
35. Yang, X., Koehl, M., Grussenmeyer, P.: Mesh-To-BIM: From segmented mesh elements to BIM model with limited parameters. In: ISPRS TC II Mid-term Symposium "Towards Photogrammetry 2020" (2020)
36. De Jong, K.: An introduction to evolutionary computation and its applications. In: Reusch, B. (ed.) Fuzzy Logic. Informatik aktuell. Springer, Berlin (1994)
37. Palfrey, J., Gasser, U.: Interop: The Promise and Perils of Highly Interconnected Systems. Basic Books, New York (2012)

38. Guy B.: DfD: Design for Disassembly in the Built Environment: a Guide to Closed-loop Design and Building, Hamer center editor, State College (2008)
39. Ruggiero, R.: Learning architecture in the digital age. An advanced training experience for tomorrow's architect. Techne, J. Technol. Arch. Environ. Spec. Ser. **2** (2021)
40. Diaz, E.: The Experimenters: Chance and Design at Black Mountain College. University of Chicago Press, Chicago (2014)
41. Toffler, A.: Future shock, Bantam Edition (1971)
42. Retsin, G. (ed.) Discrete. Reappraising the Digital in Architecture. Wiley, Hoboken (2019)
43. Latour, B.: La sfida di gaia. Il nuovo regime climatico, Meltemi Editore, Sesto San Giovanni (2020)
44. Colomina, B.: Are We Human? Lars Muller Publishers, Baden, Notes on an Archaeology of Design (2016)
45. Schwab, K.: The Fourth Industrial Revolution. Portfolio Penguin, London (2017)
46. Bolpagni, M., Gavina, R., Ribeiro, D., Pérez Arnal, I.: Shaping the future of construction professionals. In: Bolpagni, M. Gavina, R. Ribeiro, D. (eds.) Industry 4.0 for the Built Environment. Methodologies, Technologies and Skills. Springer Nature (2022)

From DfMA to DfR: Exploring a Digital and Physical Technological Stack to Enable Digital Timber for SMEs

Alicia Nahmad Vazquez and Soroush Garivani

Abstract Design for Manufacturing and Assembly (DfMA) and Digital Manufacturing (DM), particularly as related to Timber construction, is expensive. DfMA is costly due to the costs related to skill acquisition—Computer-Aided Design (CAD), Building Information Modelling (BIM), knowledge about robotic production etc. Digital Manufacturing is expensive due to the initial capital costs related to digitally able machines and robots. This entails that the segment of the construction sector that is rapidly adopting these technologies to meet the productivity and ecological goals of the Construction Sector is largely restricted to large businesses—large Architectural, Engineering and Construction (AEC) firms. More importantly, it entails that the Small Medium Enterprises (SMEs) are unable/unwilling to participate in the rapidly digitizing economy. This forms the driving motivation to propose a design and fabrication paradigm based on Design for Robofacturing (DfR), a technology stack based on affordable, multiple-use machines. To enable DfR for SMEs, the research follows three main avenues: (1) the design of an easy-to-deploy micro-factory based on industrial robotic arms for timber to produce complex carpentry products; (2) a digital-management software and sensor package; (3) worker training modules and micro-credentialing. The aim is to develop a comprehensive package that brings several advantages of digitalized construction at a lower initial capital interest to SME contractors. Finally, a case study and micro-factory prototype currently being developed is presented and analyzed. An initial prototype and proof of concept of the research, is currently being developed. The researchers are taking the challenge to create a DfR-enabled digital timber factory on a remote island, to produce in situ a housing complex. The project has the particularity to offer future clients the possibility to modify and customize the design using a digital app, requiring the micro-factory to handle and produce mass-customized, geometry complex parts in a flexible yet controlled process.

A. Nahmad Vazquez (✉)
School of Architecture, Planning and Landscape, University of Calgary, Calgary, AB, Canada
e-mail: alicia.nahmadvazquez@ucalgary.ca

A. Nahmad Vazquez · S. Garivani
The Circular Factory, Roatan, Bay Islands, Honduras

Keywords Micro factory · Cyberphysical app · Democratization of digital manufacturing · Design for robofacturing · AEC entrepreneurship · Digital and robotic fabrication · Digital craft · Digital timber

United Nations' Sustainable Development Goals 9. Build resilient infrastructure, promote inclusive and sustainable industrialization and foster innovation · 11. Make cities and human settlements inclusive, safe, resilient and sustainable · 12. Ensure sustainable consumption and production patterns

1 Introduction

Timber is one of the most traditional building materials, and societies have used that for thousands of years. Contemporary construction practices suggest a growing interest in building with timber, due to its sustainable credentials and novel technologies, at a scale not previously attainable [1]. The solid timber industry continues to grow in America, Europe and globally. New production plants are opening for business all over the world, while existing plants are increasing their capacity by extending and opening new sites. Design for Manufacturing and Assembly (DfMA) has become the standard of the mass timber industry, allowing them to build with high quality, faster, efficient and more sustainable [2]. Buildings are produced offsite following a 'kit of parts' organization and then transported to their final sites. However, the advantages of this building method with timber are concentrated in large-scale manufacturing firms. SME-friendly timber technologies are behind [3, 4]. This means that SME contractors cannot participate in the digital economy, resulting in the construction industry has one of the lowest indexes of digitization [5]. SMEs represent 98.9% of all contracting firms in the Canadian construction industry and a turnover of more than 50% of the total annual business [6]. In the UK, SMEs represent 99.9% of all contracting firms and turnover more than 63% of the sector business[2]. Migrating towards offsite fabrication is causing construction skills to be lost as construction jobs shift to manufacturing jobs [7]. Young people are not entering the field with revies predicting that the industry's workforce will decline by 20–25% in the next decade [5–7]. Thus, transferring the advantages of digital timber construction to the broader industry is critical. DfMA and DM have to be made appropriate for SMEs, and benefits of offsite construction have to be democratized to all the players on the industry.

2 Design for Manufacture and Assembly (DfMA)

Digital timber manufacturing relies on using high-accuracy gantry and joinery machines such as 5-axis gantry machines and CNC joinery machines. These machines require a large initial investment for the purchase of the equipment, setup and proprietary CNC software to operate them and for the size of the facilities to host them and are better suited for high-volume production and less suited for flexibility, operation in small spaces or make-to-order manufacturing of unique pieces [1, 4]. These last three scenarios are more common to an SME contractor than a large-scale manufacturer. Additionally, SMEs are more likely to be found in a more extensive set of scenarios such as: retrofitting or upgrading existing building stock, low-volume housing, and housing in smaller towns and remote locations, which are not always accessible for volumetric components.

The use of industrial robots to manufacture wooden structures as a field has seen a significant amount of research contributions. During the last five years, roboticists from creative and architectural backgrounds have focused on several areas such as (1) the fabrication of wood panels, wood frames and spatial structures [3, 8–14]; (2) traditional timber joints and differentiated timber joints using both robots and CNC joinery machines [3, 4, 15–17].

Robots' compact size and flexibility allow for onsite or near-site manufacturing, which is impossible for joinery CNC machines [17]. Industrial robotic arms are flexible in their applications, require a lower financial commitment and allow for decentralized fabrication [18]. Their potential to feed stacks of the wood material to cutting and milling machines has also been a field explored by architects as means to increase speed and accuracy and streamline digital flows of information over manual CNC production workflows [17, 19, 20]. These efforts have proven great potential for the robot as a mediator between the digital files and the machines used to cut and process the wooden elements. However, their focus has been mainly on processing long wood elements picked from a loaded pallet and translated back and forth for processing.

3 Design for Robofacturing (DfR)

Design for Robofacturing (DfR) is a fabrication paradigm in which flexible robotic cells can robofacture a range of high-quality timber products by integrating different fabrication processes [21]. DfR represents a shift towards localized, configurable spaces equipped with a broad range of flexible digital fabrication processes enabled by low-cost machines such as industrial robotic arms. The research presented in this paper aims to capitalize on industrial robotic arms' characteristics, flexibility and potential to extend their capabilities for integrated timber production. It proposes the Design for Robofacturing (DfR) paradigm to develop a specialized timber microfactory with its corresponding hardware, software integration, communication protocols and micro-credentialing program. The research involves both the creation and adoption of bespoke timber-based DfMA software to produce and adapt architectural designs for DfR.

The developed DfR software toolbox consists of a set of principles to produce complex architectural timber products with their accompanying custom-designed physical robotic fabrication facility. The DfR communication platform connects different fabrication devices, including human tasks required throughout the fabrication process. The DfR platform also works as a communication system between the various stakeholders on the project, from designers, shop managers and fabricators to the trades and product finishing the pieces. It enables providing feedback on time, material, production times and costs to the designers. Finally, the UI is specifically developed to be used by trades allowing them to participate in the digital economy and learn novel fabrication methods. DfR positions itself within the paradigm of digital craftsmanship [22]. It integrates traditional construction trades, crafts and skills with digital manufacturing and digitally produced materials. DfR considerations, in this case, are focused on onsite and near-site fabrication utilizing local resources. The design keeps a low-entry point on space, cost and software skills. It aims to make the hardware and software accessible for SMEs, establishes hybrid human–machine work models and ultimately allows them to understand the benefits that adopting novel technologies can bring to their workflows [23].

4 Development of the Research Project

For construction to work as mobile manufacturing and evolve as a democratic, accessible solution for SMEs and all the players in the construction industry, three avenues of research are explored: Physical setup, digital communication platform and a micro-credentialing program.

4.1 Physical Setup

For the design of the fabrication facility, five predefined design principles with their quantifying and guiding parameters were researched and established:

> **Flexibility**: measured by using generic, multifunctional tools instead of specialized single-purpose equipment in the machines. Allow interchangeability of tools between machines. The floor plan should remain as open as possible, and machines to be mobile to encourage reconfigurability.
> **Efficiency**: focus on using local materials and minimizing waste. Design the machine configurations to allow the maximum variety and range of processes to be done utilizing the digital fabrication machines and tools. Keep the remaining manual works to those requiring higher skills while looking for a high level of integration between the DfR facility and the SME workflows.
> **Legibility**: SMEs require to be doing many similar yet different parts with the same tools and machines. The system should be legible and provide information to workers, managers and stakeholders as directly and accessible as possible.

Safety: the facility should minimize risks to the workers whilst maximizing their comfort (i.e. avoid heavy lifting and repetitive movements, allow swapping between tasks and avoid monotonous jobs; maximize visibility and keep the environment as open as possible)

Robustness: keep a low maintenance and service workflow while providing redundancy that allows swapping between machines and amongst workers.

After defining the design criteria, a timber housing project by Zaha Hadid Architects was analyzed and benchmarked for the design of the fabrication facility (Fig. 1). The project was selected due to the complex geometry, its requirement for customized components and its remote location near to natural resources and SME cotractors.

A basic unit of the ZHA project was analyzed to identify how a DfR approach could be implemented and target the capabilities and dimensions of the proposed facility. The unit was broken down into its constituent parts. The complex carpentry components will be the target products for the fabrication and assembly activities in the proposed fabrication facility. The unit was split into different components: First the "cassette" -style timber frame modules for floors, walls and roofs, containing structural and functional components that can be made using traditional carpentry tools and machines (Fig. 2).

Second, the parts that contain single and double curvature, which require specialized processes and robot end effectors (Fig. 3). The units also have vault-style roofs, requiring specialized fabrication processes. To start defining the machine and human processes that transform raw material into these desired products, specific production tracks were defined (i.e. a material and processing sequence).

After identifying all the processes that could be performed using robot end effectors and the processes that require additional machines, the machinic processing tracks were transformed into a functional spatial configuration with maximum flexibility and minimal footprint. The different activities required, materials and machinic sequences—including human and material—were laid out over time and in several states to define the spatial distribution (Fig. 4). The main areas are stocking,

Fig. 1 Beyabu project by Zaha Hadid Architects (© Zaha Hadid Architects)

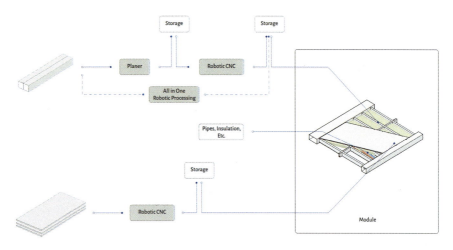

Fig. 2 Production track for a planar module (©CF in collaboration with IAAC)

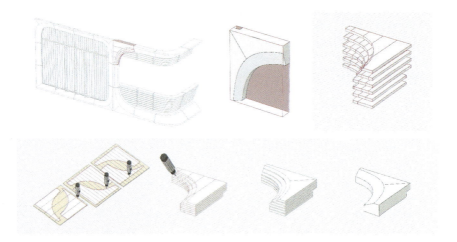

Fig. 3 Production track for a curved component. (1) design received from the architects; (2) Principal Component Analysis (PCA) for orientation on the stock material; (3) lamination strategy based on the properties and dimensions of the local available material; (4) Nesting for milling of the laminations; (5) stacking of the laminations; (6) layered stack and bandsawing; (7) final piece (© CF)

robotic fabrication, assembly (human-based), finishing area and storage before the components go out to the site.

To enable a system of low capital investment with maximum flexibility, the equipment of the fabrication facility and machine setup are critical decisions. Production scenarios based on industrial robotic arms were studied and analyzed. Machining operations such as routing, drilling and cutting require rigid and precise devices that

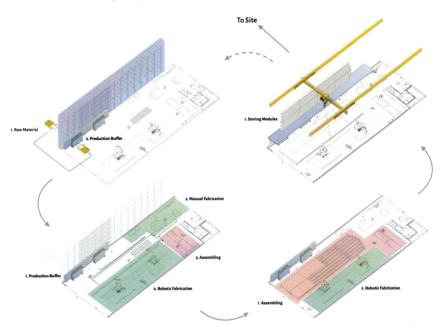

Fig. 4 Overview of the different processes (©CF in collaboration with IAAC)

provide the stiffness needed for vibration-free and dynamic processes. Additionally, they require high repeatability to ensure dimensional accuracy and high-quality finish of the machined parts. CNC machines have been developed specifically for these functions, making them the industry standard for timber operations. However, specialized woodworking and milling CNC machines are more expensive than industrial robotic arms. The technically ideal 5-axis gantry machines require an even higher investment cost. Hence, the focus is on robot-based setups that allow for the same workability at a comparable scale. The robots provide greater flexibility at a lower investment cost. However, their inherent kinematic disadvantages for machining applications need to be compensated using oversized robots [17].

The adaptability of the industrial robot for different applications makes them an extremely versatile, generic tool but also demands additional effort to set up and integrate all the components needed for specific machining applications. Purpose-built tools like piece holders, end effectors, tool changers, safety light guards, tool and workpiece probes and additional linear axis must be designed and considered.

Four different setup scenarios are designed that satisfy all the considerations: (1) 2 robots on a shared, floor-mounted linear track (Fig. 5); (2) 2 robots on a shared elevated linear track (Fig. 6); (3) 2 robots on individual rails (Fig. 7); (4) 1 stationary robot and a 5-axis CNC machine (Fig. 8).

Setup 1, which considers two identical ABB robots in a shared linear track, allows creation of a large elongated combined working envelope. This setup would allow the processing of the entire length of the largest edge beams. Since they share the same

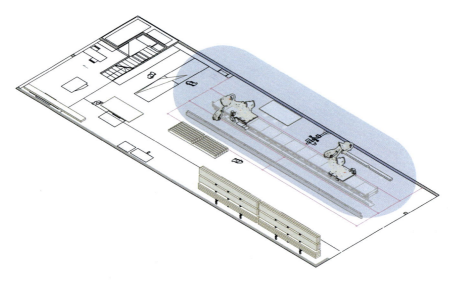

Fig. 5 Two identical ABB IRB6700 robots share a linear track (©CF in collaboration with IAAC)

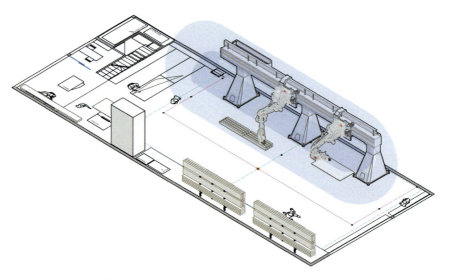

Fig. 6 An elevated GÜDEL linear track of 13 m in length is shared by two ABB IRB6700 robots (©CF in collaboration with IAAC)

rail, applications in which their working envelopes need to intersect are possible, creating the flexibility to share handling and machining tasks between both of them, but also to act as 2 independent robots that are working in parallel or on one large workpiece simultaneously.

From DfMA to DfR: Exploring a Digital and Physical Technological ... 845

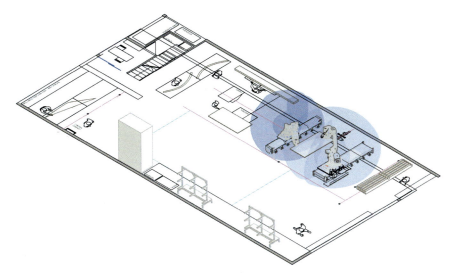

Fig. 7 Two robots in individual rails (©CF in collaboration with IAAC)

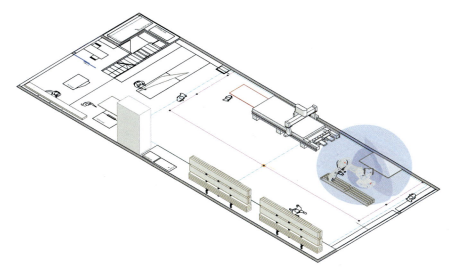

Fig. 8 One floor-mounted robot with a 5-axis CNC machine (©CF in collaboration with IAAC)

Setup 2, an elevated GUDEL track with two ABB robots on an inverse mount, also provides the capability of processing long beams like the previous one, but its floor mounted. This means that support columns occupy a smaller area of the factory floor than the floor-mounted track. The sideways-hanging configuration also allows for better utilization of the robots' theoretical maximum working envelope, as it is not intersecting with the building floor and walls. The considerable upcharge, the introduction of an additional supplier and extra integration effort (tracks being produced in Switzerland by GÜDEL) make this option less attractive from an economic standpoint. If large format 3d-milling of freeform surfaces is really needed, it might still be a viable option but will only be justifiable if the requirements actually match the theoretical workspace of this setup.

Setup 3 considers one ABB IRB6700 for machining and one ABB IRB6620 as a handling robot mounted on individual linear rails, facing each other with partially overlapping working envelopes. Here, the robots have a limitation in the task assignment. The ABB6620, while cheaper, is not appropriate for milling applications (ABB). While providing a small advantage in investment costs, this comes at the expense of flexibility and robustness, as the most critical tools cannot be shared between the machines. Furthermore, the more equally-sided footprint of this setup suggests a less elongated floorplan, which is less beneficial for the production of the long elements on this project.

Setup 04 considers one floor mounted ABB IRB6700 accompanied by a Laguna 5X CNC router. The router can be custom built to have a large table (4,27 m) for a limited surcharge, potentially enabling it to machine both ends of the longest timber elements used in the current design.

In addition to these four robot scenarios, a system for the clamping and registration of curved elements was studied. To allow precise, flexible clamping, it is important to set up a reliable method that obtains the actual piece configuration and feeds it back to the system in an adaptive way. Four possible solutions (Fig. 9) are analyzed to obtain the workpiece configuration: (1) a mechanical, robot-mounted workpiece probe; (2) using motion-tracking sensors and targets; (3) a robot-mounted depth camera; (4) a laser probe on a linear rail.

An assembly and multifunction space were considered for the pieces' assembly, drying and finishing. These areas will be used mainly by humans doing those activities

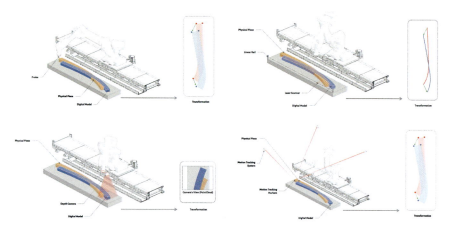

Fig. 9 Registration methods for machining of curved beams (©CF in collaboration with IAAC)

and configured with assembly tables. Finally, three internal storage areas with the ability to fit a forklift are considered. A production buffer that provides dynamic, hand-accessible storage space during production. Small-parts storage in the vicinity of the pre-assembly area and a high-load cantilever shelf for finished modules, located above the production buffer.

4.2 Digital Workflow and Communication Platform

Digital workflows from design to fabrication are segregated, as they are composed of multiple actors and stakeholders that work in isolation in various software platforms to complete different tasks [24]. Current product development cycles for architectural elements go through the design team, DfMA engineers, and constructors in separate packages (Fig. 10). While this entire team is part of one unified system, each member is in a different location, using different tools and file formats [25, 26].

For DfR to be effective, a communication system is a key to connecting the different machines of the production process into a network. The digital platform aims to provide a unified and transparent overview of the process, leading to a more productive design to fabrication approach. The developed web-based communication workflow is built mostly using existing open-source tools, which can enhance the

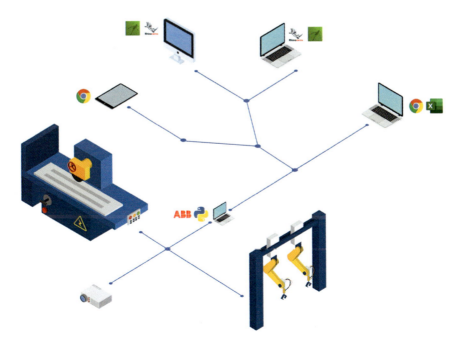

Fig. 10 Data connections between different Machines in the fabrication process

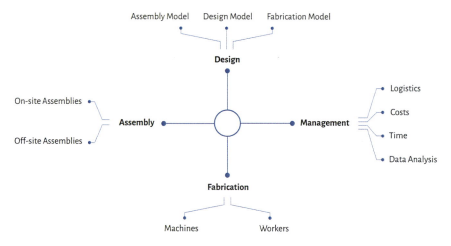

Fig. 11 Data connections between different Stakeholders in the fabrication process

project's overall management and better understanding and data exchanges between different stakeholders (Fig. 11).

The information management system is based on a database that contains rich data embedded geometry and the files required for different processes such as fabrication, visualization, management, etc. In this approach various stakeholders of the process build together a holistic model. The database allows users to query and access only the required information. Therefore, this information will be easily accessible for the various stakeholders' requirements during the fabrication process. Some of the key features of this communication platform are listed below [27] (Fig. 12).

4.2.1 Parts Information

The metadata stored in each part consists of 6 sections (Fig. 13).

- Identification: includes the name and description of the part to help identify the piece.
- Geometry: includes important geometrical data which facilitates further processes during fabrication and planning, some of this information include dimensions, curvature, and important features such as axis line, holes, important curves, etc.
- Fabrication: includes fabrication related information for the piece; The sequence of processes that needs to be performed on the material, as well as instructions and simulation files for these processes.
- Assembly: includes assembly instructions, assembly order and other information regarding the assembly process.
- Logistics: includes information that can help the production manager to optimize the fabrication process. This includes time estimations for fabrication and assembly, raw materials needed for the production of the piece, etc.

From DfMA to DfR: Exploring a Digital and Physical Technological ...

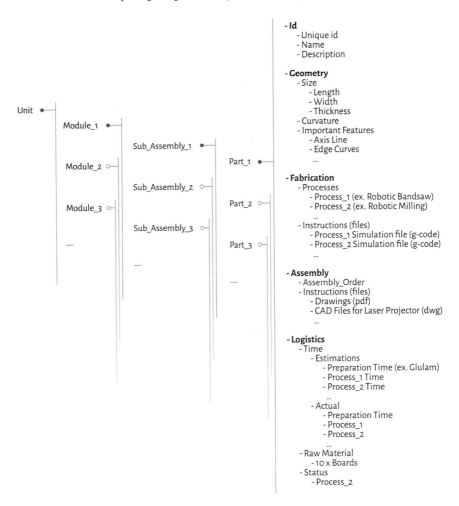

Fig. 12 Metadata stored for each part

- Quality Control: includes as-built information after the fabrication. Which can be used for further documentation of the construction process. It can contain images, and 3d point cloud data.

4.2.2 Fabrication Schedules

For each piece, there are several processes required in the production pipeline to manufacture the part. DfMA engineers rationalize the design and define the fabrication details, such as joineries and other details. Then, fabrication experts define the processes required to fabricate the piece with such details. For each piece to be fabricated, several processes happen on manual and digital fabrication machines, from

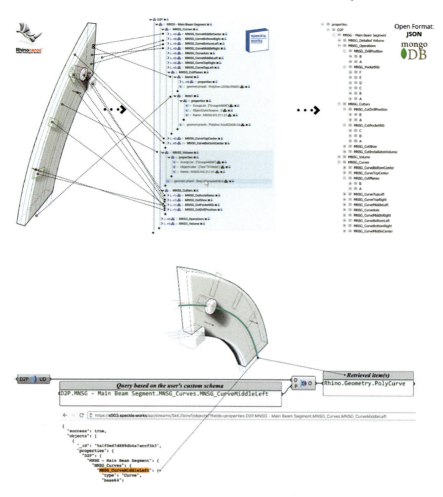

Fig. 13 Meta-data stored in the database for each part: A "MainBeam" modeled by Design-to-Production containing a hierarchy of different objects and sub-objects translated successively into a UserDictionary and a JSON string. (top) Querying process for specific information from the database (bottom) [24]

planar and table saw to robotic milling cells. Each process needs to be scheduled and communicated to the specified operator in the production workshop (Fig. 14).

The scheduling process is based on the *job shop* scheduling problem, in which multiple jobs are processed on several machines. Each job consists of a sequence of tasks, which must be performed in a given order, and each task must be processed on a specific machine. The problem is to schedule the tasks on the machines to *minimize the total length of fabrication*—the time it takes for all the jobs to be completed. There are several constraints for the job shop problem:

- No task for a job can be started until the previous task for that job is completed.

From DfMA to DfR: Exploring a Digital and Physical Technological ... 851

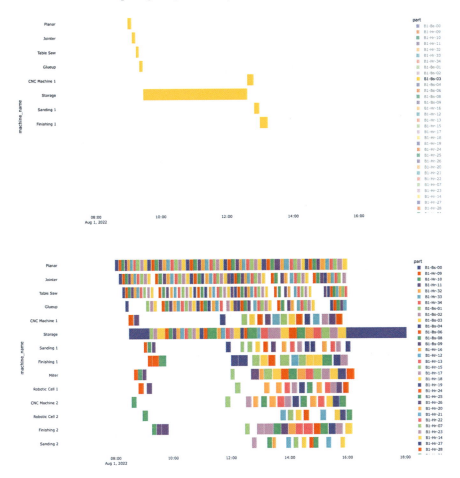

Fig. 14 Schedules: daily plans for the fabrication of pieces

- A machine can only work on one task at a time.
- A task, once started, must run to completion.

We leverage Google's OR-Tools library, open-source software for *combinatorial optimization,* which seeks to find the best solution to a problem out of a very large set of possible solutions. Below is a snapshot of a daily schedule for production and the processes for one part within that day.

By defining the schedules, the manufacturer can easily input their product data into the network database and quickly download the appropriate work materials, tools, machine tools, and fixtures based on the fabrication requirements.

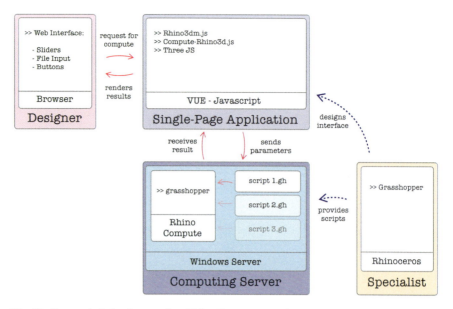

Fig. 15 Geometrical cloud computing (Rhino Compute) Interface between designer and engineer

4.2.3 Feedback for Product Development

Integration of this platform with server-side computing enables the specialist construction team to provide an interface/tool for the designers to assess their design in real-time using the scripts provided by the specialist.

For example, in geometry calculation and analysis, solving a process on a cloud-based Rhino Compute instance; Users can input a local design file and process analysis (CAA) and CAM workflows for the input, getting estimates of fabrication time, cost, waste, and materials needed. Other use-cases can be regenerating a CNC tool path or updating the fabrication schedules based on onsite modifications needed during the construction (Fig. 15).

Providing a web-based interface for solving different CAM and CAA workflows enables users with little or no knowledge of scripting to use a complex script in a fast and simple manner.

4.2.4 Interoperability

Leveraging the Speckle database system and cloud computing services like Rhino-Compute, the platform can connect with most of the software used in the AEC industry, such as Unreal Engine, Rhinoceros, etc. It can also integrate with database systems such as graphQl, or blockchain databases (Fig. 16). The interface is accessible through CAD software and programming languages for developers and Web-based access (Browser) for less technical users.

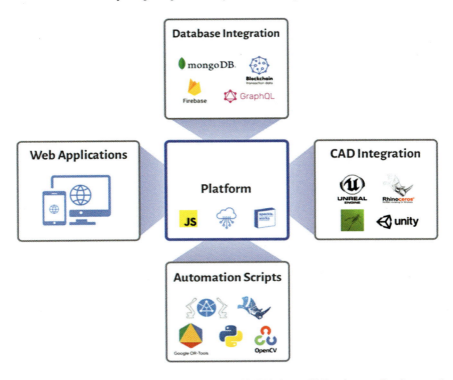

Fig. 16 Interoperability of the platform interface with Web Apps, CAD softwares, Databases and other Scripting libraries

4.2.5 Activity Logs Database

While this communication platform provides tools for communication and real-time computing from different design to construction stages, it also enables the SME's team to log different interactions and information about the design to the fabrication process in a database. Some of the information includes the stock of materials needed, the actual time for fabrication, usage of tools, the performance of the machines and construction team, quality prediction and control, etc. This database can be used for further studies on the process to reduce the gap between estimations and simulations with reality in terms of material use, fabrication time, etc.

4.3 Upskilling and Microcredentials

A DfR paradigm requires promoting and developing a trained and well-informed workforce that can take advantage of the digital processes while maintaining traditional skills and crafts required by the construction industry (Fig. 17). The objective of the micro-credentialing program is twofold: (1) to keep the construction skills and jobs within the sector by introducing digital fabrication skills, and avoiding construction jobs going to manufacturing professionals. (2) to attract new and young talent to the industry through a digitized onsite/near-site offering.

Modern Methods of Construction (MMC) traditionally refers to offsite prefabrication tools and techniques. In the context of DfR MMC is extended to include onsite and near-site digital fabrication techniques. The micro credentialing course is intended to address specific training needs in digital carpentry and focuses on the fabrication of curved façade components. The pieces selected for training require from material preparation, lamination, and several robotic processes such as routing, bandsawing and milling. Participants are also introduced to the digital workflow and communication platforms for preparing a design for DfR. Based on the research of various micro-courses on the construction sector [28, 29] two courses are proposed:

(1) A full-time 9-week course combining practical and classroom activities is envisioned for technicians with construction experience. The program will teach technical skills (knowledge of tools, equipment and processes). Safety training and essential skills are embedded throughout the program, including using 3D models, measurements, calibration, communication protocols, physical dexterity and working with others. This course is designed for technicians with construction experience.
(2) A two-part full-time 9-week course (total of 18 weeks) for technicians willing to train, do a career change and those who don't have experience in the construction industry. The first part of this course focuses on identifying components, processes and procedures to demonstrate an awareness of best practice standards before moving to the second course.

The first iteration of both curriculums will be deployed after the micro-factory completion. The construction of the technology demonstrator is envisioned using local trades that undergo the training curriculum.

Fig. 17 Laminated timber prototypes produced using hybrid methods

5 Discussion

The digitization of the construction industry faces many challenges—the digital fabrication machines' financial, skill and space requirements limit the advances to large companies able to afford them. Through applying innovation in computational design within a DfR workflow, the research aims to enable SME contractors access to digital machines and tools, upskilling them and making them active participants both onsite and near-site in the digitization of the construction industry. In the current stage, local contractors have expressed interest in going through the training and micro-credentialing program. They also see the potential for renting or buying the facility. The engagement of the local people and results will be analyzed. However, this is a step towards sustainable digital craftmanship [22] in the construction trades. DfR for timber advocates for the democratization of digital manufacturing tools and techniques. It aims to upskill local labour by disseminating the use of digital fabrication machines with their efficient and sustainable credentials.

Different stakeholders in the design to fabrication process include the design team, the engineering team and the fabrication team. The fabrication team involves from fabrication experts and production managers to the workers in the factory and the machines processing the materials. The DfR platform presented in this research allows for communication between stakeholders. Connecting the design process to fabrication parameters that include material layout, machine time, human time, laminations and ultimately GCode generation. This interoperability would allow SMEs to be more precise on their estimates but, more importantly, designers and other stakeholders to have a higher degree of transparency, similar to the one that offsite manufacturers offer.

Furthermore, the feedback provided to designers and stakeholders would allow them to understand the implications of their decisions to each of the different parameters (i.e. machine time, human time, bill of materials, etc.) and make informed decisions with confidence using local trades. The platform collects information once the work starts to identify discrepancies between time estimates and actual working times and iteratively refine its calculations. The data collected from the working pieces will help the designers and stakeholders understand limitations and bottlenecks and better plan their designs and schedules using quantifiable information. The lack of For the future development of the software platform, integration with blockchain and smart contracts is considered. Blockchain can enable SMEs to get paid as processes are finished adding more transparency to trades and processes often considered opaque on their inner workings, cost calculation and timeframe delivery. Another consi.

The developed digital and physical DfR framework contributes to promoting inclusivity by enabling a sustainable approach to industrialization. It makes advanced digital fabrication available to a sector often overseen by the construction industry, SMEs (UNsdg 9). The technology demonstrator is purposely built in a country under development. The financial requirements to acquire the machines and the corresponding digital fabrication process were iterative research during the design of the micro-factory. Packing as many processes as possible within a DfR workflow enables a financial model in which SMEs can engage. The technical platform aims to encourage 'building sustainable and resilient buildings utilizing local materials' (UNsdg 11). The DfR cyber-physical platform is a step toward supporting developing countries to strengthen their scientific and technological capacity. Skilled workers must have access to resources in their toolkit that can move them towards more sustainable production patterns (UNsdg 12), such as access to digital fabrication tools.

6 Conclusion

The paper presents the author's entrepreneurship and efforts to construct a DfR fabrication facility in a developing, remote location—rich in resources. It offers a digital and physical strategy to upskill and digitize the local trades. The authors will produce a housing complex using the hardware, software, and training program in situ as an initial project. The project has the particularity to offer future clients the possibility to modify and customize the design using a digital app, requiring the DfR workflow to handle and produce mass-customized parts in a flexible yet controlled process. The paper presents the development of the corresponding cyber-physical DfR fabrication platform for complex timber products with its upskilling and training program. The micro-factory facility can subsequently be acquired by the local trades trained in the use of the machines and software. More construction companies are moving towards adopting manufacturing processes in a bid for digitization and in response to problems attracting more workers onsite [5]. This research envisions democratic digitization of the construction industry by providing existing trades with a digital toolset, new talent will get interested and onsite/near-site skills, and construction knowledge will be continued and augmented.

The DfR framework, with its SME-friendly components, aims to enable them to: reduce construction costs by being able to engage with material-reducing design technologies; reduce total construction times by using a low-capital investment, flexible digital micro factory; enabling more transparent processes on time and costs between stakeholders that work with SMEs; reduce the ecological impact by using local materials and processing elements on/near site as opposed to transporting large volumetric materials. Finally, it aims to contribute to the democratic digitization of the supply chain in the construction industry.

Acknowledgements Phae 1 of the research presented in this paper was developed in collaboration with the Insitute of Advanced Architecture of Catalunya (IAAC) and their external consultants with expertise in wood manufacturing and architecture, robotic fabrication and fabrication laboratory setup. In particular Raimund Krenmueller, Alexandre Dubor, Shyam Zonca, Lana Awad, Tom Svilans and Daniela Figueroa. Their contributions and work during phase 1 of this project shaped the initial configuration of the fabrication facility and the skeleton of the digital and communications platform. Additionally, Jean-Nicola Dackiw (CF) is part of the team working with the authors to refine the digital workflow related to the physical fabrication processes. He has been instrumental in refining the factory layout from initial studies to its final configuration—currently under construction- and towards its final physical setup.

References

1. Ramage, M.H., Burridge, H.,Busse-Wicher, M., Fereday, G., Reynolds, T., Shah, D.U., Wu, G., Yu, L., Fleming, P., Densley-Tingley, D., Allwood, J., Dupree, P., Linden, P.F., Scherman, O.: The wood from the trees: The use of timber in construction. Renew. Sustain. Energy Rev. **68**, 333–359 (2017). https://doi.org/10.1016/j.rser.2016.09.107
2. Pryke, A.: What is Design for Manufacture and Assembly (DfMA)? https://www.bam.co.uk/what-we-do/dfma
3. Robeller, C., Weinand, Y.: Design and fabrication of robot-manufactured joints for a curved-folded thin-shell structure made from CLT. Robot. Fabr. Archit. Art Des. (2014). https://doi.org/10.1007/978-3-319-04663-1
4. González Böhme, L.F., Quitral Zapata, F., Maino Ansaldo, S.: Roboticus tignarius: robotic reproduction of traditional timber joints for the reconstruction of the architectural heritage of Valparaíso. Constr. Robot. **1**, 61–68 (2017). https://doi.org/10.1007/s41693-017-0002-6
5. Manyika, J., Ramaswamy, S., Khanna, S., Sarrazin, H., Pinkus, G., Sethupathy, G., Yaffe, A.: Executive Summary Digital America: A Tale of the Haves and Have-Mores. McKinsey & Company (2015)
6. Statistics Canada: Key Small Business Statistics. Gov. Canada. 1 (2019)
7. Southgate, M.: Young People and the Future of Homes. https://www.mobie.org.uk/mobie-blog/2021/08/19/make-the-future-yours
8. Schwinn, T., Krieg, O.D., Menges, A.: Robotically Fabricated Wood Plate Morphologies. Presented at the (2013)
9. Johns, R.L., Foley, N.: Bandsawn bands feature-based design and fabrication of nested freeform surfaces in wood. In: McGee, W., de Leon, M.P. (eds.) Robotic Fabrication in Architecture, Art and Design, pp. 17–32. Springer, Berlin. https://doi.org/10.1007/978-3-319-04663-1
10. Robeller, C., Weinand, Y.: Fabrication-aware design of timber folded plate shells with double through tenon joints. Robot. Fabr. Archit. Art Des. **2016**, 166–177 (2016). https://doi.org/10.1007/978-3-319-26378-6_12
11. Sondergaard, A., Feringa, J.: Scaling architectural robotics: construction of the kirk kapital headquarters. In: Menges, A., Sheil, B., Glynn, R., Skavara, M. (eds.) Fabricate: Rethinking Design and Construction, pp. 264–271. UCL Press, Stuttgart (2017)
12. Williams, N., Cherrey, J.: Crafting Robustness: Rapidly Fabricating Ruled Surface Acoustic Panels BT—Robotic Fabrication in Architecture, Art and Design 2016. Presented at the (2016). https://doi.org/10.1007/978-3-319-26378-6_23
13. Satterfield, B., Preiss, A., Mavis, D., Entwistle, G.: Bending the line zippered wood creating non-orthogonal architectural assemblies using the most common linear building component (The 2X4). In: Burry, J., Sabin, J., Sheil, B., Skavara, M. (eds.) Fabricate: Making Resilient Architecture, pp. 58–65 (2020)
14. Wood, D., Grönquist, P., Bechert, S., Aldinger, L., Riggenbach, D., Lehmann, K., Rüggeberg, M., Burgert, I., Knippers, J., Menges, A.: From machine control to material programming: self-shaping wood manufacturing of a high performance curved CLT structure—Urbach tower. In: Burry, J., Sabin, J., Sheil, B., Skavara, M. (eds.) Fabricate: Making Resilient Architecture, pp. 50–57. UCL Press (2020). https://doi.org/10.14324/111.9781787358119
15. Tamke, M., Thomsen, M.R.: Designing parametric timber. eCAADe **26**, 609–616 (2008)
16. Tamke, M., Thomsen, M.R.: Digital wood craft. In: Joining Languages, Cultures and Visions—CAADFutures 2009, Proceedings of the 13th International CAAD Futures Conference, pp. 673–686 (2009)

17. Søndergaard, A., Becus, R., Rossi, G., Vansice, K., Attraya, R., Devin, A., Owings, S., Llp, M.: A Factory on the Fly Exploring the Structural Potential of Cyber Physical Construction (2019)
18. Arrival: Arrival—why the microfactory. https://arrival.com/world/en/card/why-arrival-microfactory
19. Eversmann, P., Gramazio, F., Kohler, M.: Robotic prefabrication of timber structures: towards automated large-scale spatial assembly. Constr. Robot. **1**, 49–60 (2017). https://doi.org/10.1007/s41693-017-0006-2
20. Eversmann, P.: Concepts for timber joints in robotic building processes. Rethink. Wood Futur. Dimens. Timber Assem. (2019)
21. Siciliano, B., Shishkov, R., Smagt, P. van der.: Robofacturing. https://en.everybodywiki.com/Robofacturing
22. Klein, T.: Digital craftsmanship. Lect. Notes Comput. Sci. **9187** (2015)
23. Brosque, C., Fischer, M.: A robot evaluation framework comparing on-site robots with traditional construction methods. Constr. Robot. (2022). https://doi.org/10.1007/s41693-022-00073-4
24. Poinet, P.: Enhancing Collaborative Practices in Architecture, Engineering and Construction through Multi-Scalar Modelling Methodologies (2020). https://doi.org/10.13140/RG.2.2.25478.73280
25. Daniel, D.: Modelled on Software Engineering: Flexible Parametric Models in the Practice of Architecture (2013)
26. Svilans, T.: Integrated Material Practice in Free-Form Timber Structures (2020)
27. Scheurer, F., Stehling, H., Tschumperlin, F., Antemannn, M.: Design for assembly—digital prefabrication of complex timber structures. In: Proceedings of IASS Annual Symposia, IASS 2013 Wroclaw: 'Beyond the Limits of Man'–Timber Spatial Structures, pp. 1–7. International Association for Shell and Spatial Structures (IASS) (2013)
28. SAIT: Centre for Continuing Education and Professional Studies. https://coned.sait.ca/search/publicCourseSearchDetails.do?method=load&courseId=1027416
29. Council, M.C.S.: MCSC Micro Courses. https://mbcsc.com/mcsc-micro-courses/

Spatial Curved Laminated Timber Structures

Vishu Bhooshan, Alicia Nahmad, Philip Singer, Taizhong Chen, Ling Mao, Henry David Louth, and Shajay Bhooshan

Abstract The paper describes the physical realisation of a demonstration prototype produced by mouldless wood bending of discrete laminated timber elements which are interconnected to create a predominantly compression only spatial structure. Integrated design to production pipelines is increasingly valued in Architecture, Engineerinng and Construction, as it has contributed to developing methods of generation of the so-called architectural geometry and in bringing the various disciplines in the industry closer together. The research presented is motivated by the application and use of timber in such a realm. It details a design to production toolkit along with development of custom actuator-based tool to deliver sustainable benefits of reduced material usage and wastage in addition to efficient production of bent wood structures. Furthermore, the paper proposes an alternative procedure for polyhedral

---

V. Bhooshan (✉) · P. Singer · T. Chen · L. Mao · H. D. Louth · S. Bhooshan
Computation & Design Group (CODE), Zaha Hadid Architects, London, UK
e-mail: vishu.bhooshan@zaha-hadid.com

P. Singer
e-mail: philip.singer@zaha-hadid.com

T. Chen
e-mail: taizhong.chen@zaha-hadid.com

L. Mao
e-mail: Ling.Mao@zaha-hadid.com

H. D. Louth
e-mail: henry.louth@zaha-hadid.com

S. Bhooshan
e-mail: shajay.bhooshan@zaha-hadid.com

A. Nahmad (✉)
Laboratory for Integrative Design (LID), School of Architecture, Planning & Landscape, University of Calgary, Calgary, AB, Canada
e-mail: alicia.nahmadvazquez@ucalgary.ca

V. Bhooshan
Architectural Computation, School of Architecture, University College of London (UCL), London, UK

reconstruction of disjointed force polyhedrons from an input graph, which enables the creation of spatial structures in static equilibrium.

Keywords Digital timber · Laminated timber · Computational parametric design · 3D graphic statics · Fabrication aware design · Digital robotic fabrication · Mesh modeling environments

1 Introduction

The research presented in this paper is motivated by the resurgence of timber as a construction material and the current lack of publicly available integrated design, structure and fabrication aware toolkits especially in the domain of timber spatial structures. The paper details the development of such a tool chain that enabled the production, fabrication and assembly of a demonstrator spatial structure made up of 24 nodes of curved laminated timber (Figs. 1 and 2). The objective was to create a toolkit emphasising on knowledge capture and reuse instead of an individual stand-alone solution [1].

1.1 Digital Timber

Timber is one of our most traditional construction materials, but its resurgence has come out of recent technological advances in digital production techniques, processing technologies, digital design tools etc.; as well as a ecological factor of timber being environmentally friendly as compared to other construction material

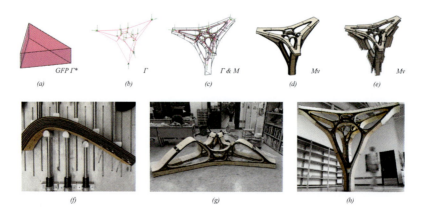

Fig. 1 Design to production tool chain and physical demonstrator. **a–e** procedure to design a spatial curved laminated timber structure, **f** pneumatic bending multi-layer laminates, **g** on site assembly, and **h** finished structure

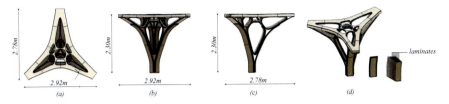

Fig. 2 Schematic drawings—**a** top view, **b** sections, **c** front elevation, **d** right-side elevation, and **e** perspective

such as concrete or steel. Timber is believed to have a key role to play on creating a net zero built environment [2–5].

The recent and rapid advancements in robotic and digital fabrication systems (RDF), have become extremely useful as they are not only leading to considerable time saving but can directly transfer computational design data directly to manufacturing and assembly operations which are enabling construction of non-standard timber structures. These advancements have also brought about bidirectional feedback of production back to the design conception—the so-called design for manufacture and assembly paradigm (DFMA) [6–9]. The robotic assembly of spatial structures with timber elements have also been researched and documented [10–12].

The benefits of aligning of geometry to structural principals and static equilibrium shapes have been well established especially in the domain of large span timber shells [13–15] etc. Furthermore, albeit not specifically on timber as a material, the benefits of structure aligned geometry paradigm to improve recyclability, and the repair and reuse of material and structural components due to dry assembly have also been recently highlighted [16–19].

The delivery of timber architecture, once endowed to the architect-builder or "master craftsman" must evolve into a collaborative model of specialised skill sets in various trades and "digital craftsmanship" to build the promise of environmentally friendly, sustainable built environment [20–22].

1.2 Key Contributions

The main contribution of the paper is the development of a design to production toolchain that enabled the design and build of a spatial structure in timber. Specifically, it enabled:

- to reduce the amount of material required by precisely computing the number of cross section laminates required based on the compressive and tensile forces acting on the structure;
- to reduce complexity of node connection, a typical issue for high valence spatial structures, by having the timber joinery away from node;

- the development of an actuator-based machine for bending of glulams which used electric linear motion over pneumatic actuation, allowing for less waste and tighter tolerances in comparison to its precedents;
- to register real time and efficiently bend the multiple varying curvature elements in a short period of time via a physical and digital configuration of linear actuators on a bed-like table.

In addition, the paper proposes an alternative procedure to the *extended Gaussian image-based* method detailed in [23], to create a polyhedral force diagram from an input graph—which are amenable for quick design manipulation in commercially available Mesh Modeling and CAD Environment's.

2 Prior Work

The prior work stems from domains which are relevant to the main contributions of the paper, which are expanded in the sections below.

2.1 Geometry Representation and Architectural Geometry

Discrete geometry representations—graphs, meshes, volumetric meshes etc. aided by rapid development in applying discrete differential geometry to architectural problems such as planarisation, developability, cylinder fitting of panels etc., have become the preferred geometric representation for both the conceptual design and downstream production stages of the design workflow [24–27].

Coupled with an interactive mesh modeling environment (MME) as detailed in [28], it presents a significant opportunity for a holistic mode of design exploration. They are increasingly valued as it facilitates.

- use of contemporary paradigm of edit and observe interactive modelling [29–31];
- exploration of static equilibrium shape design [31–36]
- greater design control in the production and delivery stages [8, 11, 16, 34, 37, 38]
- and provides a feedback loop between the various stages of the design workflow [8, 11, 16, 20, 28, 37, 39].

Such workflows have contributed to developing methods of generation of the so-called *architectural geometry* [40] and bringing disparate disciplines within Architecture, Engineering and Construction (AEC) industries closer together [37]. For further readings on the use of discrete representation for architectural applications we refer the reader to [16, 31, 35, 36, 38, 39, 41–47] etc.

2.2 Polyhedral 3D Graphical Statics

Polyhedral 3D graphical statics (3DGS) is a recent development of graphic statics in three dimensions based on a historical proposition by [48, 49]. The construction of the dual and reciprocal form and force diagrams, the design exploration and structural form-finding via manipulation of the diagrams have been thoroughly explicated by [23, 32, 50–53].

The procedures have been fully extended to computational frameworks and made into publicly available toolsets for CAD application by [54, 55] etc.

Various materialisation strategies have been explored for the realisation of 3DGS structures—concrete, mycelium, glass etc. [18, 19, 56]; However, its exploration in timber is very recent and in early-stages of research and development [57–60].

2.3 Wood Bending

Wood bending methods date back to antiquity. Bending started with the use of fresh twigs, which can be easily bent into almost any shape. Dried wood is more difficult to bend as it becomes rigid and breakable and hence relies on the use of heat and water. They were specially developed in the shipbuilding industry in which lamella were turned into boat hulls and bark ribs into canoes [61–63]. These mechanisms relied on using nature found (i.e., tying the wood around a tree after pouring boiling water over it) and purpose-made moulds (Fig. 3a) around which wet, thick sections of wood would be tied and dried to hold the curvature—predominantly single curvature.

In addition to the different steam bending methods for dry wood, other bending practices have emerged and evolved. Glue-laminated timber—abbreviated to *glulam*—the technique of gluing together a slack of thin rectangular-section wood elements to create a larger section element that is typically used as a column or beam. This along with development of engineered timber products with glue and resins have made possible to have mechanically fastened compound sections amenable for large spans, improved element consistency, load capacity etc. Glue-lamination also presents the opportunity of creating curvatures in large timber elements by individually curving the component lamella (which are flexible) during the glue-up. The

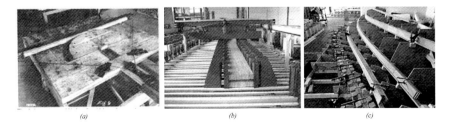

Fig. 3 Precedents for wood bending machines

bending of lamella demands special systems of glue-up jigs and clamping, which limits the flexibility and speed at which different curvatures can be achieved (Fig. 3b, c). Typically, as a rule of thumb, the radius of curvature achievable is 200 times the lamella thickness [64]. Digital methods have been explored as a flexible alternative for glulam bending, suggesting how wood lamination curvature could inherently be programmed or predetermined [65–69].

Contemporary pavilions and architectural applications continue to expand the wood bending technology beyond furnishing scale [70] tending towards the design and production of elastically bent wood sheets amenable to robotic fabrication and assembly [71]. This paper describes the use of ubiquitous two axis Computer Numerical Controlled (CNC) technology to create bent wood skeletal structures (Fig. 1). For further understanding of the economic benefits of using such a technology we refer the reader to [38, 47, 59, 72].

2.4 Software Add-Ins and File Formats

We implemented the toolchain detailed in Sect. 3 including the polyhedral 3DGS procedures detailed in [50, 51] using a C++ based framework—ZSPACE [73].

The collaborative design to production (DTP) toolchain described below is based on a discrete geometry-processing paradigm—graphs, meshes and volume meshes. This allows for lightweight transmission and reconstruction of information by various participating tools in DTP. The DTP toolchain is supported by a custom file format which uses JavaScript Object Notation (JSON) and half-edge and half-face data structures that are common to this paradigm, to transmit and process 3D model information [74].

The authors have previously developed add-ins to incorporate form finding methods such as Force Density Method (FDM) [75] and Thrust Network Analysis (TNA) [33] within the MME of Autodesk Maya [28, 57]. Further the authors have made previous investigations to use *dynamic relaxation* (DR) techniques on meshes to address fabrication related constraints [38, 42, 44].

3 Design-To-Production Toolchain

The collaborative, multi-author DTP process that we developed to physically realise the demonstrator spatial structure, is composed of the following steps:

- **Shape-design**: designer-guided shape-design exploration of a medial spatial graph and form finding using polyhedral 3DGS (Fig. 4a).
- **Architectural geometry**: Generation and thickening of geometry of the spatial graph using the structural and polyhedral information from the previous step and decomposition of it into fabricable parts per node; (Fig. 4b).

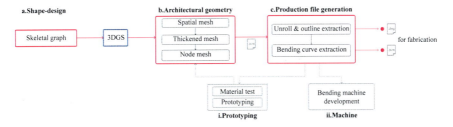

Fig. 4 DTP Tool chain unrolled into threads. **a, b, c** serial design thread, (i) material testing and prototyping thread, and (ii) bending machine development thread

- **Production information generation**: fabrication aware optimisation of the geometry and generation of the production information per node of the spatial structure (Fig. 4c).

In addition to these, there are two other steps of material testing and design and creation of actuator-based machine for bending of the glulams. These steps informed the material thickness, size and permissible bending radius in the optimisation process of AG.

3.1 Shape Design

The global shape design and manipulation of the spatial structure was carried out using the framework of polyhedral 3DGS. In a typical polyhedral 3DGS application scenario, the process begins by defining the force (Γ^*) and form (Γ) diagrams which are made up of polyhedral cells. (Γ^*) can be decomposed into a *global force polyhedron* (GFP)—representing the static equilibrium of external forces—and *nodal force polyhedrons* (NFP)—the rest of the cells inside the GFP, representing the equilibrium of forces coming together at that node in Γ [51]. In the initial step of the design a GFP was defined (Fig. 5a), which were subsequently subdivided using the schemes described in [76] to create the (Γ^*) (Fig. 5b).

Next an initial Γ is constructed from (Γ^*) using the reciprocation algorithm [77] (Fig. 5c, d). The equilibrium of the external forces of the ith node $\mathbf{v}_i$ of Γ (Fig. 5g) is represented by a closed polyhedron, $\mathbf{c}_i^*$ (Fig. 5b) of (Γ^*). For each $\mathbf{c}_i^*$, the normal $\mathbf{n}_{i,j}$ and the area $\mathbf{A}_{i,j}$ represent the direction and magnitude of the axial force $\mathbf{f}_{i,j}$ in the corresponding edge $\mathbf{e}_{i,j}$ of Γ [23]. This equilibrium is achieved by updating the vertices of Γ and/or (Γ^*) using a so-called perturbation procedure detailed in [77]. Figure 5e shows the equilibrium state of Γ.

This sequence of defining the GFP first could be counter-intuitive for designers as the shape that they would like to exercise control on—the Γ—is generated as topological dual only in the second step. The representation Γ as a collection of polyhedral cells would lack ease of manipulation in a interactive MME, as such commercially available interfaces like Autodesk Maya, generally doesn't ship with the volume

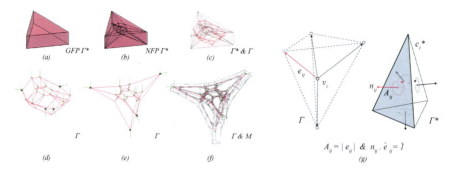

Fig. 5 Shape design process—**a** global force polyhedron, **b** nodal force polyhedrons, **c** generation of initial form diagram from force diagram via reciprocation, **d** initial state of form diagram before equilibrium, **e** equilibrium state of form diagram, **f** correlation of form diagram and spatial mesh (see Sect. 2.2), and **g** principle of equilibrium as detailed in [77]

mesh data-structure [78] required for efficient manipulation of such geometries. An alternative procedure using *disjointed force polyhedron* [23] which represents Γ as 3D graph input is presented next.

3.1.1 Geometric Procedure for Polyhedral Reconstruction

In polyhedral 3DGS, as noted previously, a Γ^* is typically represented as consolidated polyhedron where all pairs of adjacent cells have co-planar and co-incidental contact faces. McRobie [79], Lee et al. [23] displayed the concept of the disjointed force polyhedron (Ψ^*), where neighbouring cells could be made of dissimilar geometry and non-coincidental contact faces if they are equal in area and have equal but opposing face normals. The corresponding form diagram (Ψ) can be in static equilibrium but is not a topological dual of $\Psi*$ and not necessarily polyhedral in geometry. In such a representation Ψ^* is a collection of disjointed NFP (Fig. 3), with reciprocal dual diagrams Γ & Γ^* being a special instance of Ψ & Ψ^* respectively.

Like all graphic static scenarios, the polyhedral reconstruction of the Ψ^* is a crucial step. Developing on the Minkowski's theorem, the best suited method for reconstruction for 3DGS would be the reconstruction from face normals and areas [23, 80]. It displayed an iterative minimisation solver using the *extended Gaussian image* (EGI) [81] and *area pursuit algorithm* to reconstruct Ψ^*. We present an alternative geometric procedure for EGI using the convex hull algorithm [82].

Given a set of face normal vectors $\mathbf{f}_{i,j}$ with length of the vectors representing the face area $\mathbf{A}_{i,j}$ (Fig. 6a)—either known or as target—the force polyhedron (Fig. 6f) is derived through the following sequence of geometrical operations:

1. compute a convex hull η, from the set of points $\mathbf{P}_{i,j}$ at the head of the unitised face normals (Fig. 6b);
2. compute the topological dual η^* (Fig. 6c); By principle of reciprocity the number of faces of η^* will be equal to number of vertices of η, which in turn would be

equal to the number of input face normal vectors. It should be noted it is not necessary to have the face normals $\mathbf{n}_{i,j}$ and area $\mathbf{a}_{i,j}$ of η^* to have the correct orientation and required area respectively by construction (Fig. 6d);

3. An iterative projection solver is used to perturb the vertices of η^* to meet the dual objectives (Fig. 6e).

- $\hat{n}^*_{i,j} \cdot \hat{\mathbf{f}}_{i,j} = 1$, where $\cdot$ is the vector dot product. The projection force per vertex is computed using the procedure described in [38, 83].
- $\mathbf{a}_{i,j} = \mathbf{A}_{i,j}$ using the area pursuit method described in [23].

In exceptional cases, such as a 2D node where all members at that node are coplanar or an open node (Fig. 7), like [23] we would be required to introduce virtual force vectors to help facilitate the construction of η & η^*. As these are only to facilitate construction of the diagram, the corresponding virtual faces in η^* for each input virtual force vectors would not have a target area or orientation during step 3 of the above procedure and will have no corresponding member in Ψ.

It can be noted, the special case of adding *zero face*, as in [23], is not required as by construction the number of faces in η^* will be equal to number of input face normal vectors $\mathbf{f}_{i,j}$ (Fig. 6).

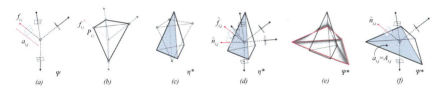

Fig. 6 Polyhedral reconstruction—**a** force vectors extracted from input form graph, **b** normalised force vectors, **c** convex hull generated from points $P_{i,j}$, **d** topological dual of convex hull, **e** iteration steps of the equilibrium projection solver and **f** force polyhedron

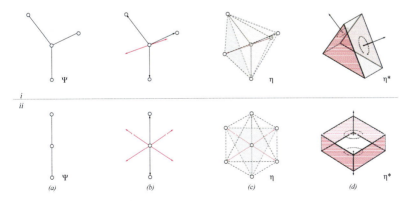

Fig. 7 a Node condition, **b** added virtual vectors, **c** initial convex hull and **d** topological dual of convex hull with virtual faces highlighted for specials of (i) a 2D co-planar node and (ii) an open node

The procedure is easily adaptable to work with multi-node initial input of Ψ. In this case for each node v_i of Ψ, the outgoing edges $e_{i,j}$ and target face area $A_{i,j}$ would be used as inputs for the geometrical construction of Ψ^* (Fig. 7a). It can be noted the procedure after perturbation, each unitised face normal of Ψ^* would meet the constraint of

$$\hat{n}^*_{i,j} = \hat{e}_{i,j}$$

The equilibrium procedure, like in [34], can furthermore be weighted by using the weighting factor $\gamma = 0, ..., 1$. This factor increases or decreases the influence of the Ψ or Ψ^*, on its counterpart during the iterative process. To achieve this, a target vector $\mathbf{t}_{i,j}$ for each pair of corresponding normalised vectors $\hat{n}^*_{i,j}$ and $\hat{e}_{i,j}$ is defined as follows:

$$\mathbf{t}_{i,j} = \gamma \hat{e}_{i,j} + (1-\gamma)\hat{n}^*_{i,j}$$

This weighting factor allows to define the degree each diagram stays fixed during the iterative process of finding equilibrium. If $\gamma = 1$, only the vertices of the Ψ^* are affected, and if $\gamma = 0$, only those of Ψ are affected. Figure 8, displays the iterative steps for both Ψ & Ψ^* when $\gamma = 0.5$.

In addition to the above control, the weighted iterative perturbation procedure, makes it amenable to incorporate additional design-based constraints on Ψ such as fixing certain vertices as per contextual conditions or incorporating certain minimum and maximum edge lengths along with other projection-based constraints detailed in [34, 50, 84, 85] etc. The detailed description of projection forces for each of the above constraints is out of scope of this paper. Figure 9, showcases the steps of polyhedral reconstruction on simple multi-nodal setup, while Fig. 10 displays it on the graph of the demonstrator spatial structure.

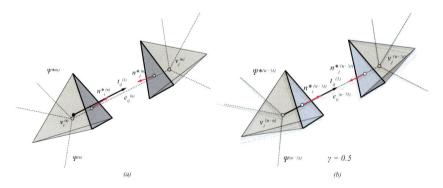

Fig. 8 Bidirectional equilibrium in a two node situation using a weighting factor γ—**a** initial state at time u **b** converged equilibrium state at time u + Δ t along with the intermediate states

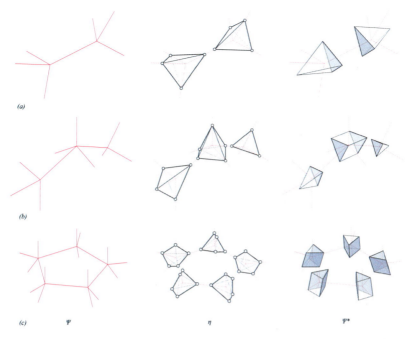

Fig. 9 Polyhedral reconstruction of disjointed force polyhedrons from input graphs of varying nodal conditions—**a** 2 Node, **b** 3 Node and **c** 5 Node

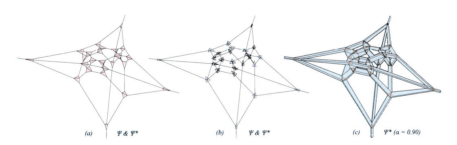

Fig. 10 Polyhedral reconstruction of disjointed force polyhedrons from a complex input graph (Ψ) of the demonstrator spatial structure showcasing—**a**, **b** initial and equilibrium state of $\Psi*$ and **c** unified diagram $\Psi*(\alpha)$

3.2 Architectural Geometry

First, a spatial structure mesh **M** is derived by adapting the procedure described in [59] to work with disjointed force polyhedrons.

Given Ψ & Ψ^*, **M** is created by the following sequence of geometric operations (Fig. 11):

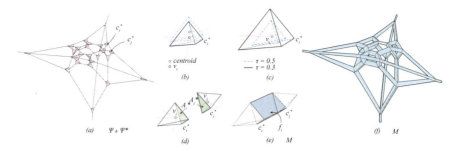

Fig. 11 Generation of Architectural geometry—**a** Input Ψ and $\Psi*$, **b** checking and aligning position of cell $c*i$ such that the centroid of cell matches with the corresponding node vi of Ψ, **c** scaling of cells using a factor τ, **d**, **e** bridging between face- pairs of $\Psi*$ to create a faces of M and **f** resulting spatial mesh M

1. position each cell c_i^*, such that the centroid of the cell lies on the corresponding node v_i of Ψ;
2. scale each cell towards/away from its centroid by some factor τ. This value can vary per cell and is used to control the sizing of the cross section and edge length at each node;
3. bridge between the face-pairs of Ψ^* to create faces f_i of **M**. For each face of **M**, the corresponding area A_i^* of the face-pairs of Ψ^* is stored as an attribute. This attribute would be used to compute the thickness of **M**, thereby making the thickness a function of the force acting on it.

The resulting **M** (Fig. 11f) will be a 2-manifold mesh with exclusively planar quadrilateral faces (PQ Mesh) based on generalization of Varignon's theorem [86]. For further understanding of it in the context of polyhedral cells, we point the reader to [59].

An additional geometrical bevel operator was introduced to the procedure to control the bend radius of the timber laminates on higher subdivision of **M** produced using the adapted Catmull-Clark subdivision scheme [87] as described in [57] (Figs. 12 and 13). It can be noted that even with these operators the property of PQ mesh is maintained thus ensuring that **M** is developable and can be unrolled into a plane without intersection [24].

Next, **M** is decomposed into patches of faces around each node (**Mv**$_i$) (Figs. 1 and 17) based on the fabrication bed constraints of the laser cutter or CNC machine (see Sect. 4). Each strip of **Mv**$_i$ is offset to create layers of timber laminates in proportion to the corresponding face areas in c_i^*, which as noted previously is stored as a face attribute of **M**. This entails that the material thickness is proportional to the force acting on it (Fig. 14). For the demonstrator, based on the forces acting and material prototyping tests (see Sect. 4) the number of layers varied between 3 and 9 layers, which was composed of combination of 1/4 and 1/8 inch laminates. 1/4 inch laminates were used for the first 5 layers and the subsequent layers were of 1/8 inch laminates. The color variation in the laminates were used to capture and highlight this tectonic variation of material thickness based on structural performance (Fig. 1).

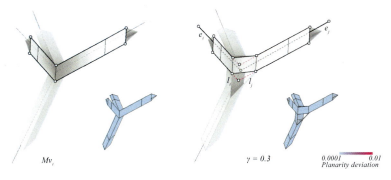

Fig. 12 Bevel operator driven by an user specified factor $\gamma = \text{li}/\text{ei} = \text{lj}/\text{ej}$, $\gamma \in (0, 1)$. The inset colored geometry showcases the planarity deviation

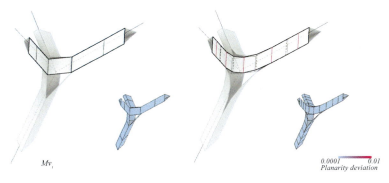

Fig. 13 Subdivision operator using an adapted Catmull-Clark subdivision scheme [87] as described in [57]. The inset colored geometry showcases the planarity deviation

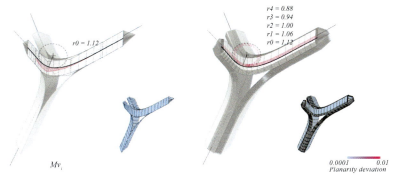

Fig. 14 Offset operator to create layers of timber laminates in proportion to forces acting on it by generating offsets along the normal. The inset colored geometry showcases the planarity deviation

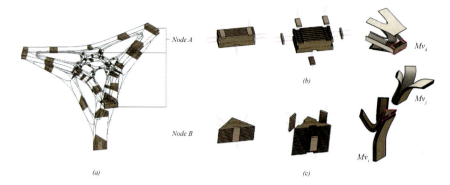

Fig. 15 Two types of nodal connectors **a** an end connector positioned at the end of a single node, and **b** an internal connector positioned in between two nodes

The connector detail between any two adjacent node mesh $\mathbf{Mv}_i$ and $\mathbf{Mv}_j$ was made as a block which fits into the void of the spatial node mesh. The strips of the node mesh were then screwed to the connector block (Fig. 15). This connection detail, used due to the time constraint of making the demonstrator within a 10-day design to fabrication workshop, could be improved to create a more robust and interlocking detail using the timber laminates (see Sect. 6).

3.3 Production Information Generation

The structurally verified and thickened $\mathbf{Mv}_i$ serve as the input to create the production information for fabrication. All the layers of $\mathbf{Mv}_i$, as noted previously, are a PQ Mesh. Such PQ meshes are amenable to fabrication methodologies such as 2D cutting via CNC technology [38, 57, 59], robotic hot wire cutting [88], curve creased folding [43] etc. For the demonstrator, the 2D laser cutting was used and the outline of the unrolled strips in plane (Fig. 16) were used as the production information.

The production information for the custom bending machine were also generated using $\mathbf{Mv}_i$ (Fig. 17). The center-line graph of the top and bottom layer of each node strip is extracted and is inserted into the JSON file format and is subsequently parsed in the timber bending thread of the DTP (see Sect. 4.3).

3.4 Implementation

The first 2 steps of the DTP toolchain, including 3DGS routines have been implemented as a software add-in to the MME of Autodesk Maya for interactive design explorations. We use the inbuilt features in the MME to create and manipulate the

Spatial Curved Laminated Timber Structures

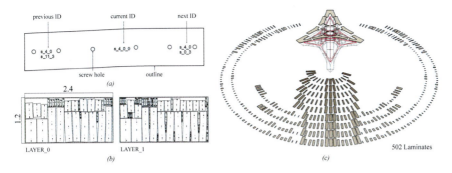

Fig. 16 Unrolled laminates **a** programmed laminate ID and its connecting part ID, **b** compact unrolled outlines for laser cutting on a 2.4 m by 1.2 m plywood sheet, and **c** laminates in a total number of 502

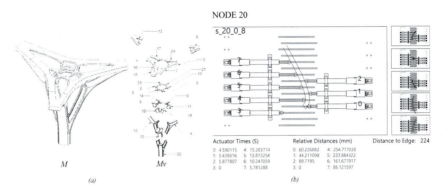

Fig. 17 Demonstrator of the machine data in a bending process **a** node assemblies of the structure, **b** actual operation distance and time on node 20, and **c** fabrication result

spatial graph Ψ or polyhedral mesh Ψ^*. The user can carry out topological changes on the diagrams using Conway operators [89], addition/deletion of cells of Ψ^* etc. Given an initial Ψ^* or Ψ, the plugin generates counterpart and form find using the weighted perturbation method. After the equilibrium step, the user can use the GUI to generate the AG with parameter sliders to control the number of laminates, the offset layers, thickness, and the feature to export the final state of AG into the JSON file. An overview of the results using such a workflow in an MME is presented in Fig. 18. Step 3 of the DTP is carried out in the CAD environment of McNeel Rhinoceros with the usage of Grasshopper to parse the JSON, unroll the timber strips and generate the necessary production information for fabrication.

The above plugins were successfully used in teaching and design explorations in couple of 5 day computational design workshops. Participants with minimal knowledge of 3DGS, static equilibrium and design for timber were able to intuitively understand the action of forces, the weighting of the form and force diagrams and how structural and fabrication aware geometries could be generated. Apart from learning

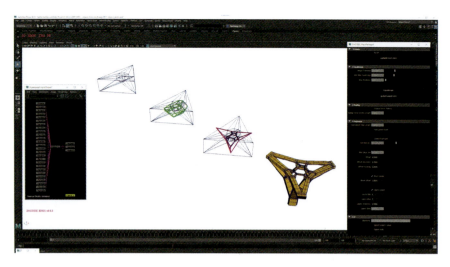

Fig. 18 Custom add-in within the MME of Autodesk Maya highlighting the various diagrams and geometries in the DTP, custom add-in UI, incorporating add-in node in the hypergraph environment of Maya to maintain dependency graph structure

the toolchain, the participants were able to design structures such as bridges, residential units, and columns within the span of a short workshop. Some the participant result from the workshop which use the proposed DTP are displayed in Fig. 19.

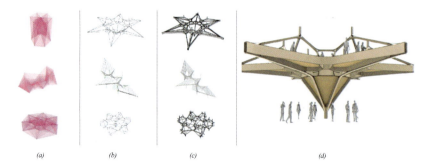

(a) (b) (c) (d)

Fig. 19 Some of the participant work from the computational design workshops using the custom add-ins showcasing **a** force polyhedrons, **b** 3D graphic statics, **c** 3D geometry, and **d** spatial combinatorics

4 Machine Design for Bending Timber

To achieve the design's different radii and cross-sectional condition of the laminates required the development of a customized tool based on re-configurable electric linear actuators on a table (Fig. 20). Producing bespoke, programmed wood curvature based on actuator operation that applies bending forces based on selected control points has been explored through machine prototyping [69]. Wood can be formed in unique, customized, digitally driven, programmable curvature, which is determined through the automated process of bumper placement and pneumatic actuation against them. Differently from previous work, the bending machine used on this prototype was designed using electric linear motion over pneumatic actuation. Electric actuation allows for less waste and tighter tolerances [90].

Additionally, we can know how hard the actuator is pushing and coordinate the movement with force and position through force feedback. Furthermore, bending tables using pneumatic actuators rely on using a stopper against which the pneumatic actuators press, they incorporate adjustable bending blocks which are lined up with a projected image of the desired shape. Lamella are held in place while the pneumatic actuators provide the pressure to bend the wood [69]. Through different iterations, it became clear that linear actuators allow achieving the desired positions by controlling the actuator device alone and without additional machine assemblies.

4.1 Bending Machine Components

The bending table, in order to use linear motion—travel distance divided by speed, is composed from the following equipment (Fig. 21).

- Seven PA-17 Heavy-Duty 14-inch linear actuators of 12 V with a 14″ stroke—and 850 lbs. of force are oriented in opposite directions to reach the maximum contact area on any given curvature. The opposing forces between the actuators -at each side of the table-aid in keeping the curve in place. Clamps are then positioned

(a)   (b)

Fig. 20 Iterations of bending machine development **a** physical prototype on multi-layered laminates (~0.125 m), **b** digital prototype on multi-directional laminates (~0.5 m), and **c** dual robotic bending (~3.0 m)

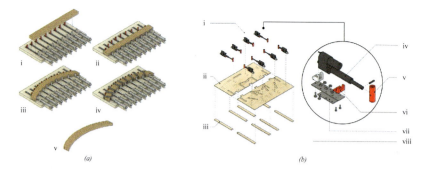

Fig. 21 Bending process—(i) neutral table/linear lams with glue (ii) computationally adjust stoppers (iii) activate pneumatic cylinders (iv) add clamping along glulam (v) final curved glulam member; (b) Bending machine (i) linear actuators (ii) CNC-milled plywood table (iii) CNC-milled table supports (iv) heavy duty electric linear actuator (v) 3D-printed bumper (vi) 3D-printed actuator seat (vii) aluminum mounting plates (viii) heavy-duty mounting bracket

along the curved piece. Clamps—although not required- allow removing the part from the table, freeing it for the next bend.
- Seven heavy-duty mounting brackets are used to hold each actuator to the bending table
- Seven custom-designed 3D-printed bumper heads are attached to the pistons at the actuator's end. The bumpers are designed to maximize the contact area of the actuator against the laminations.

The base to hold all the above has two 12.7 mm (1/2inch) plywood sheets. Slots were milled on the plywood sheets to provide the bumper heads with and stabilize their motion during actuation. The position of the slots was based on the analysis of the curvatures. The aim was to maximize the relationship between the curves' control points and the actuators' positions. Finally, wooden pallets were used as a substructure to the plywood sheets due to their inherent robustness. Digitally, the actuators were controlled through seven high-current motor drivers, four Arduino UNO control boards (each Arduino controlling two actuators) and multiple DC power supplies.

4.2 Material Testing

Various material tests were performed to refine the thickness of the lamellas, the stepping between them and define the curvature radii that were achievable. Initially, thin lamellas of Douglas Fir were tested. However, the lamellas required thinner sections to achieve the tight curvature on the upper part of the demonstration prototype. Veneer-core plywood, a material commonly used in furniture making, formed by two layers of plywood with a central core made of veneer was used due to its high degree of flexibility. Veneer-core plywood is traditionally not used for glulams.

Nevertheless, its behaviour was more flexible than the thinnest available lamella that could be made from other wood whilst achieving a high degree of stiffness after lamination. Tests were made using veneer-core plywood in 1/4 and 1/8 of an inch thickness. Furthermore, experiments mixing both thicknesses on the same lamination were successful. An important consideration when using this material over wood or plywood lamellas is the grain direction of the central veneer. The material is very flexible in the direction of the veneer grain, but it does not bend in the opposite direction. This characteristic defined the maximum length of each component on the final piece. It also characterized how the laminations were laid out on the sheets for 2D cutting.

4.3 Digital Workflow

The orientation of the curved members on the bending table is iteratively adjusted based on the pre-defined position of the bumpers and the control points of the desired curve/graph—parsed from the JSON file created in Sect. 3.4. The aim is to achieve the maximum contact area between the bumpers and the curve. Additionally, the contact should be on the strategic positions that define the shape—such as those defined by the control points. Another constraint on the definition was to ensure that all the actuator distances were achievable. Finally, the thickness, translated as the distance between the actuators at each side of the table, is considered. Once these three conditions are satisfied, the distance values of each actuator and the position of the lamellas on the bending table—measured from the edge—are extracted. The distance values extracted from the grasshopper model are then translated to the Arduino IDE interface [91] (Fig. 22). The linear actuators are driven to the correct distances by stacked Arduino UNO circuit boards, motor drivers and their corresponding Arduino proprietary programming.

Fig. 22 Computational simulation of bending process **a** digital machine sets **b** relative distance per actuator and period of holding position

5 Fabrication and Assembly

All constituent blocks of the demonstrator were produced using the DTP workflow described in the previous sections and the resultant output files. In total, 24 nodes consisting of 502 timber laminate strips in total were laser cut, glued to create 129 glulams of which 114 glulams were bend into shape using the custom actuator tool in 10 hours.

A digital workflow with corresponding physical tools was developed to bend the unique 114 pieces. The lamella for the different laminations are realized via a two-axis laser cutting machine and a CNC router. The 2D curvature of the glulams is registered and held in shape via a custom build actuator- tool. The bending machine uses linear motion. Linear actuators are driven by a custom-developed McNeel Rhinoceros—Grasshopper script that iteratively positions each element on the bending table, aiming to maximize the contact area given the curvature of the piece and the actuators on the table. The script then adjusts the travel distance based on the thickness—number of laminations—the part will require. It is important to note that laminations changed in number and thickness in all the parts as they were driven by structural requirement.

The assembly sequence (Fig. 23), given the small scale of the demonstrator although not a scalable solution for large scale application—was carried out upside down for ease of handling of the node blocks and minimisation of false work. The lightweight prototype was assembled in 12 h and flipped to its final orientation (Fig. 24). The expected external forces at the top corner of the spatial structure as seen in the initial GFP (Fig. 5) were resolved via the use of tension cables.

It should be noted that the use of individual laminates was a proxy material for ply/dimensional lumber elements and the use of laser cutter was a proxy fabrication for lam shape profiles to produce customised flat part profiles given the short workshop timeline.

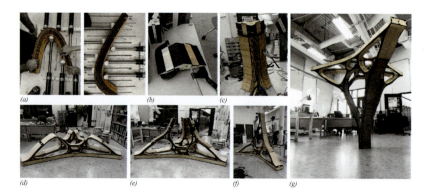

Fig. 23 Assembly process **a** bent laminates **b** glued connectors **c** assembly a connector on a node **d** phase 1 assemblage **e** phase 2 assemblage **f** phase 3 assemblage **g** final product

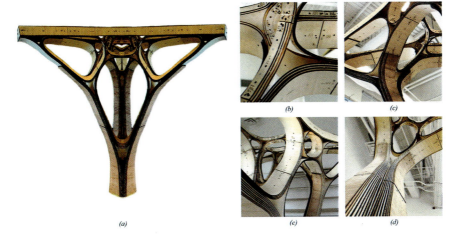

Fig. 24 Shape profile and overall detail

6 Outlook and Conclusion

The proof-of-concept spatial structure and the associated DTP toolchain together demonstrate the viability for it to be developed further for large scale applications. The following areas of work could be improved to help streamline the process for full-scale application:

- **Digital design interface**—further development and improvement of the designer friendly interface in MME's for more direct and willful manipulation of spatial structures.
- **Closed node geometries**—development of geometric procedure to compute intersection for closure of node strips for application with multi layered laminated system.
- **Cross section optimisation**—development of a fast, optimisation method to optimise the cross-section size via the scaling factor (see Sect. 3.2) per node would help alleviate manual intervention / usage of global scaling factor.
- **Assembly sequencing**—development of a more robust and scalable assembly sequence with minimal false work and potentially with self-registering interlocking node block strips.
- **Interface detail**—development of male–female interface details to avoid misalignment during assembly. The design of the interface can also consider registration points that will aid the assembly process.
- **Feedback incorporation**—development of a more robust geometrical method to incorporate the constraints of the bending machine within DTP could allow averaging the control points of each curve so that the positioning of the linear actuators better corresponds to most of the curves. The position of the actuators

can be modified, and the base tables rationalized into 4 or 5 tables that could cover all the different curves.
- **Machine development**—engaging multiple actuators simultaneously is a crucial factor. Controlling multiple actuators is complex; an inter-integrated circuit (I2C) with its highly modular approach that can be used to control various Arduino boards from a single master could be tested in the next iteration of the machine. This would allow the actuators to operate simultaneously when information is relayed from the DTP thread to a master Arduino board instead of having to divide it into multiple Arduino boards. It will simplify the digital operation and make for a more friendly interface.
- **Scale Up**—adapt and develop the current DTP to work on a larger scale architectural and industrial construction scenario.

In conclusion, the proposed DTP for spatial timber structures showcased procedures in computational design and fabrication to reduce the material usage for spatial structure by aligning geometries to structural principles, timber processing technique to efficiently achieve the varying structural thickness using ubiquitous 2 axis CNC technology, reduce the complexity of node connection of spatial structure for ease of assembly and time-efficient ways to achieve bent timber using an actuator tool which reduces wastage. All the above features are in sync with some of the sustainability goals set out by the United Nations, especially in the domain of consumption and production patterns (UNSDG 12.5 & 12.8).

Further, the paper showed that embedding such a workflow in an MME would enable the designer to manipulate and explore novel spatial design tectonics and highlighted the benefits of using discrete geometry based workflows for downstream production. In addition, the paper displayed the didactic value and knowledge transfer of 3DGS, equilibrium and static aware geometry creation by using the DTP in an interactive MME (UNSDG 4.4 & 4.8). We also described the design and production of custom machine tools and the fabrication and assembly process of a demonstrator timber spatial structure and the key learning thereof, which opens several trajectories for further investigation [76].

Acknowledgements We would like to thank all the student participants of our free open to all online Timber Tectonics workshop at DigitalFUTURES 2021 and CAAD Futures 2021. Some of the results produced during the workshops are shown in Fig. 18.

The research paper focused on describing the integrated DTP toolchain for the demonstrator prototype. The project team for the physical demonstrator comprised many more contributors. The full project credits are listed below.

Design
ZHA CODE: Vishu Bhooshan, Henry Louth, Shajay Bhooshan.
Fabrication Design
ZHA CODE: Taizhong Chen, Vishu Bhooshan, Henry Louth.
LID: Alicia Nahmad.
Fabrication & Assembly
LID: Guy Gardner, Matt Walker, Youness Yousefi, Jo-Lynn Yen, Anagha Patil, Alicia Nahmad.
Documentation
ZHA CODE: Ling Mao, Taizhong Chen, Philip Singer, Vishu Bhooshan, Henry Louth.
LID: Matt Walker, Youness Yousefi, Alicia Nahmad.

with support from
ZHA CODE: Jianfei Chu, Edward Meyers.
AA: Jean-Nicola Dackiw.

References

1. Verhagen, W.J., Bermell-Garcia, P., Van Dijk, R.E., et al.: A critical review of knowledge-based engineering: An identification of research challenges. Adv. Eng. Inf. **26**(1), 5–15 (2012)
2. Day, G., Gasparri, E., Aitchison, M.: Knowledge-based design in industrialised house building: a case-study for prefabricated timber walls. In: Digital Wood Design, pp. 989–1016. Springer (2019)
3. ARUP (2019) Rethinking Timber Buildings
4. Himes, A., Busby, G.: Wood buildings as a climate solution. Dev. Built Environ. **4**(100), 030 (2020)
5. Svilans, T., Tamke, M., Thomsen, M.R., et al.: New workflows for digital timber. In: Digital Wood Design, pp. 93–134. Springer, Berlin (2019)
6. Duro-Royo, J., Oxman, N.: Towards fabrication information modeling (fim): four case models to derive designs informed by multi-scale trans-disciplinary data. In: MRS Online Proceedings Library (OPL), p. 1800 (2015)
7. Poli, C.: Design for Manufacturing: A Structured Approach. Butterworth-Heinemann (2001)
8. Svilans, T., Poinet, P., Tamke, M., et al.: A multi-scalar approach for the modelling and fabrication of free-form glue-laminated timber structures. In: Humanizing Digital Reality, pp. 247–257. Springer, Berlin (2018)
9. Willmann, J., Gramazio, F., Kohler, M.: New paradigms of the automatic: robotic timber construction in architecture. In: Advancing Wood Architecture, pp. 13–28. Routledge (2016)
10. Apolinarska, A.A., Knauss, M., Gramazio, F., et al.: The sequential roof. In: Advancing Wood Architecture, pp. 45–59. Routledge (2016)
11. Helm, V., Knauss, M., Kohlhammer, T., et al.: Additive robotic fabrication of complex timber structures. In: Advancing Wood Architecture, pp. 29–44. Routledge (2016)
12. Thoma, A., Adel, A., Helmreich, M., et al.: Robotic fabrication of bespoke timber frame modules. In: Robotic Fabrication in Architecture, Art and Design, pp. 447–458. Springer, Berlin (2018)
13. Adriaenssens, S., Barnes, M., Harris, R., et al.: Dynamic relaxation: Design of a strained timber grid shell. In: Shell Structures for Architecture, pp. 103–116. Routledge (2014)
14. Linkwitz, K.: Force density method. In: Adriaenssens, S., Block, P., Veenendaal, D., et al. (eds.) Shell Structures For Architecture: Form Finding and Optimization, chap 6, pp. 59–69. Routledge (2014)
15. Sullivan, B., Epp, L., Epp, G.: Long-span timber grid shell design and analysis: the Taiyuan domes. In: Proceedings of IASS Annual Symposia, International Association for Shell and Spatial Structures (IASS), pp. 1–10 (2020)
16. Bhooshan, S., Bhooshan, V., Dell'Endice, A., et al.: The striatus bridge. In: Architecture, Structures and Construction, pp. 1–23 (2022)
17. Block, P., Van Mele, T., Rippmann, M., et al.: Redefining structural art: strategies, necessities and opportunities. Struct. Eng. **98**(1), 66–72 (2020)
18. Heisel, F., Schlesier, K., Lee, J., et al.: Design of a load-bearing mycelium structure through informed structural engineering. In: Proceeding of the World Congress on Sustainable Technologies (WCST) (2017)
19. Lu, Y., Seyedahmadian, A., Chhadeh, P.A., et al.: Funicular glass bridge prototype: design optimization, fabrication, and assembly challenges. Glas. Struct. Eng. 1–12 (2022)

20. Bhooshan, S.: Collaborative design: combining computer-aided geometry design and building information modelling. Archit. Des. **87**(3), 82–89 (2017)
21. Saint, A.: Architect and Engineer: A Study in Sibling Rivalry. Yale University Press New Haven, CT (2007)
22. Scheurer, F.: Digital craftsmanship: from thinking to modeling to building. In: Digital Workflows in Architecture, pp. 110–131. Birkhäuser (2012)
23. Lee, J., Van Mele, T., Block, P.: Disjointed force polyhedra. Comput. Aided Des. **99**, 11–28 (2018)
24. Kilian, M., Flöry, S., Chen, Z., et al.: Curved folding. In: ACM Transactions on Graphics (TOG), p 75. ACM (2008)
25. Liu, Y., Pottmann, H., Wallner, J., et al.: Geometric modeling with conical meshes and developable surfaces. ACM Trans. Graph. (TOG) **25**, 681–689 (2006)
26. Poranne, R., Ovreiu, E., Gotsman, C.: Interactive planarization and optimization of 3d meshes. Comput. Graph. Forum **32** (2013a). https://doi.org/10.1111/cgf.12005
27. Schüller, C., Poranne, R., Sorkine-Hornung, O.: Shape representation by zippables. ACM Trans. Graph. **37**(4) (2018)
28. Bhooshan, V., Reeves, D., Bhooshan, S., et al.: Mayavault—a mesh modelling environment for discrete funicular structure. Nexus Netw. J. **20**(3), 567–582 (2018)
29. Jiang, C., Tang, C., Tomičí, M., et al.: Interactive Modeling of Architectural Freeform Structures: Combining Geometry with Fabrication and Statics. In: Advances in Architectural Geometry 2014, pp. 95–108. Springer, Berlin (2015)
30. Prévost, R., Whiting, E., Lefebvre, S., et al.: Make it stand: balancing shapes for 3D fabrication. ACM Trans. Graph. (TOG) **32**(4), 81 (2013)
31. Tang, C., Sun, X., Gomes, A., et al.: Form-finding with polyhedral meshes made simple. ACM Trans. Graph. **33**(4), 70–71 (2014)
32. Akbarzadeh, M., Van Mele, T., Block, P.: On the equilibrium of funicular polyhedral frames and convex polyhedral force diagrams. Comput. Aided Des. **63**, 118–128 (2015)
33. Block, P., Ochsendorf, J.: Thrust network analysis: A new methodology for three-dimensional equilibrium. J.-Int. Assoc. Shell Spat. Struct. **155**, 167 (2007)
34. Rippmann, M.: Funicular Shell Design: Geometric Approaches to Form Finding and Fabrication of Discrete Funicular Structures (2016)
35. Rippmann, M., Tv, M., Popescu, M., et al.: The armadillo vault: Computational design and digital fabrication of a freeform stone shell. Adv. Arch. Geom. **2016**, 344–363 (2016)
36. Vouga, E., Mathias, H., Wallner, J., et al.: Design of Self-supporting surfaces. ACM Trans. Graph. **31**(4) (2012)
37. Bhooshan, V., Louth, H.D., Bhooshan, S., et al.: Design workflow for additive manufacturing: a comparative study. Int. J. Rapid Manuf. **7**(2–3), 240–276 (2018)
38. Louth, H., Reeves, D., Koren, B., et al.: A prefabricated dining pavilion: Using structural skeletons, developable offset meshes, kerf-cut and bent sheet materials. In: Menges, A., Sheil, B., Glynn, R., et al. (eds.) Fabricate 2017, pp. 58–67. UCL Press (2017)
39. Bhooshan, V., Fuchs, M., Bhooshan, S.: 3d printing, topology optimization and statistical learning: A case study. In: Proceedings of the 2017 Symposium on Simulation for Architecture and Urban Design, Society for Computer Simulation International (2017)
40. Pottmann, H., Brell-Cokcan, S., Wallner, J.: Discrete surfaces for architectural design. In: Curves and Surfaces, pp 213–234. Avignon (2006)
41. Bhooshan, S., El Sayed, M.: Use of sub-division surfaces in architectural form-finding and procedural modelling. In: Proceedings of the 2011 Symposium on Simulation for Architecture and Urban Design, Society for Computer Simulation International, pp. 60–67 (2011)
42. Bhooshan, S., Veenendaal, D., Block, P.: Particle-spring systems—Design of a cantilevering concrete canopy. In: Adriaenssens, S., Block, P., Veenendaal, D., et al. (eds.).: Shell Structures for Architecture: Form Finding and Optimization, p. 103. Routledge (2014)
43. Bhooshan, S., Bhooshan, V., ElSayed, M., et al.: Applying dynamic relaxation techniques to form-find and manufacture curve-crease folded panels. SIMULATION **91**(9), 773–786 (2015)

44. Bhooshan, S., Bhooshan, V., Shah, A., et al.: Curve-folded formwork for cast, compressive skeletons. In: Proceedings of the Symposium on Simulation for Architecture and Urban Design. Society for Computer Simulation International, pp, 221–228. San Diego, CA, USA, SimAUD '15 (2015b)
45. Bhooshan, S., Ladinig, J., Van Mele, T., et al.: Function representation for robotic 3D printed concrete. In: Robotic Fabrication in Architecture, Art and Design, pp 98–109. Springer, Berlin (2018a)
46. Panozzo, D., Block, P., Sorkine-Hornung, O.: Designing unreinforced masonry models. ACM Trans. Graph. (TOG) **32**(4), 91 (2013)
47. Schwartzburg, Y., Pauly, M.: Fabrication-aware design with intersecting planar pieces. In: Computer Graphics Forum, pp. 317–326. Wiley Online Library (2013)
48. Maxwell, J.C.: Xlv. on reciprocal figures and diagrams of forces. Lond., Edinb., Dublin Philos. Mag. J. Sci. **27**(182), 250–261 (1864)
49. Rankine, W.: Xvii. principle of the equilibrium of polyhedral frames. Lond., Edinb., Dublin Philos. Mag. J. Sci. **27**(180), 92–92 (1864)
50. Akbarzadeh, M.: 3d graphic statics using reciprocal polyhedral diagrams Ph.D. Ph.D. thesis, thesis Zurich, Switzerland: ETH Zurich (2016)
51. Hablicsek, M., Akbarzadeh, M., Guo, Y.: Algebraic 3d graphic statics: Reciprocal constructions. Comput. Aided Des. **108**, 30–41 (2019)
52. Lee, J., Meled, T.V., Block, P.: Form-finding explorations through geometric transformations and modifications of force polyhedrons. In: Proceedings of IASS Annual Symposia, International Association for Shell and Spatial Structures (IASS), pp. 1–10 (2016)
53. Williams, C., McRobie, A.: Graphic statics using discontinuous airy stress functions. Int. J. Space Struct. **31**(2–4), 121–134 (2016)
54. Akbarzadeh, M., Nejur, A.: PolyFrame Manual. Polyhedral Structures Laboratory, Penn Design. University of Pennsylvania (2018)
55. Mele, T.V., Liew, A., Echenagucia, T.M., et al.: Compas: A Framework for Computational Research in Architecture and Structures (2017). www.compas-dev.github.io/compas/
56. Bolhassani, M., Akbarzadeh, M., Mahnia, M., et al.: On structural behavior of a funicular concrete polyhedral frame designed by 3d graphic statics. Structures **14** (2018)
57. Bhooshan, V., Louth, H., Bieling, L., et al.: Spatial developable meshes. In: Design Modelling Symposium, pp. 45–58. Springer, Berlin (2019)
58. Liu, Y., Lu, Y., Akbarzadeh, M.: Kerf bending and zipper in spatial timber tectonics: A polyhedral timber space frame system manufacturable by 3-axis cnc milling machine. In: Proceedings of the Association for Computer-Aided Design in Architecture (ACADIA) (2021)
59. Reeves, D., Bhooshan, V., Bhooshan, S.: Freeform developable spatial structures. In: Proceedings of IASS Annual Symposia, International Association for Shell and Spatial Structures (IASS), pp. 1–10 (2016)
60. Wang, Z., Akbarzadeh, M.: A polyhedral approach for the design of a compression-dominant, double-layered, reciprocal frame, multi-species timber shell. In: Proceedings of IASS Symposium and Spatial Structures Conference 2022, Innovation Sustainability Legacy. Beijing, China (2022)
61. Adney, E.T., Chapelle, H.I.: Bark Canoes and Skin Boats of North America. Skyhorse Publishing Inc. (2007)
62. Estep, H.C.: How Wooden Ships are Built a Practical Treatise on Modern American Wooden Ship Construction, with a Supplement on Laying Off Wooden Vessels (1918). http://books.google.com/books?id=wwowAAAAYAAJ
63. Wright, R.S., Bond, B.H., Chen, Z.: Steam bending of wood; embellishments to an ancient technique. BioResources **8**(4), 4793–4796 (2013)
64. Ursula, F.: Beyond the Truss [Lecture]. University College of London (2022)
65. Krieg, O., Menges, A.: Potentials of robotic fabrication in wood construction: elastically bent timber sheets with robotically fabricated finger joints. In: Proceedings of the 33rd Annual Conference of the Association for Computer Aided Design in Architecture, pp. 253–260 (2013)

66. Menges, A.: Integrative Design Computation: integrating material behaviour and robotic manufacturing processes in computational design for performative wood constructions. In: ACADIA 2011 Proceedings: Integration Through Computation, pp. 72–81 (2011)
67. Naboni, R., Marino, S.D.: Wedged kerfing. In: Design and Fabrication Experiments in Programmed Wood Bending, pp. 1283–1294 (2022). https://doi.org/10.5151/sigradi2021-85
68. Satterfield, B., Preiss, A., Mavis, D., et al.: Bending the line zippered wood creating non-orthogonal architectural assemblies using the most common linear building component (the 2x4). In: Burry, J., Sabin, J., Sheil, B., et al. (eds.) Fabricate: Making Resilient Architecture, pp. 58–65 (2020)
69. Self, M., Bretnall, C., Dodd, S., et al.: Timber Seasoning Shelter (2014). www.designandmake.aaschool.ac.uk/project/timber-seasoning-shelter/
70. Drexler, A.: Charles Eames Furniture From the Design Collection the Museum of Modern Art. The Museum of Modern Art (1973)
71. Lienhard, J., Alpermann, H., Gengnagel, C., et al.: Active bending, a review on structures where bending is used as a self-formation process. Int. J. Space Struct. **28**(3–4), 187–196 (2013)
72. Neuhaeuser, S., Rippmann, M., Mielert, F., et al.: Architectural and structural investigation of complex grid systems. In: Proceedings of the International Association for Shell and Spatial Structures (IASS) Symposium (2010). https://doi.org/10.1145/2461912.2461958
73. Bhooshan, V., Bhooshan, S., et al.: zspace: A Simple C++ Header-Only Collection of Geometry Data-Structures, Algorithms, and City Data Visualization Framework (2018b). https://github.com/gitzhcode/zspacetoolsets
74. Json.org (1999) JSON. https://www.json.org/json-en.html
75. Schek, H.J.: The force density method for form finding and computation of general networks. Comput. Methods Appl. Mech. Eng. **3**(1), 115–134 (1974)
76. Akbarzadeh, M., Mele, T.V., Block, P.: Three-dimensional compression form finding through subdivision. In: Proceedings of IASS Annual Symposia, International Association for Shell and Spatial Structures (IASS), pp. 1–7 (2015a)
77. Akbarzadeh, M., Van Mele, T., Block, P.: Compression-only form finding through finite subdivision of the external force polygon. In: Proceedings of the IASS-SLTE 2014 Symposium. Brasilia, Brazil (2014)
78. Kremer, M., Bommes, D., Kobbelt, L.: Open volume mesh–a versatile index-based data structure for 3d polytopal complexes. In: Proceedings of the 21st International Meshing Roundtable, pp. 531–548. Springer, Berlin (2013)
79. McRobie, A.: Rankine Reciprocals with Zero Bars. Preprint (2017)
80. Minkowski, H.: Allgemeine lehrs¨atze u¨ber die convexen polyeder. Nachrichten von der Gesellschaft der Wissenschaften zu G¨ottingen. Mathematisch-Physikalische Klasse **1897**, 198–220 (1897)
81. Little, J.J.: An iterative method for reconstructing convex polyhedra from extended gaussian images. In: Proceedings of the Third AAAI Conference on Artificial Intelligence, pp. 247–250 (1983)
82. Graham, R.L., Yao, F.F.: Finding the convex hull of a simple polygon. J. Algorithms **4**(4), 324–331 (1983)
83. Poranne, R., Ovreiu, E., Gotsman, C.: Interactive planarization and optimization of 3D meshes. Comput. Graph. Forum **32**(1), 152–163 (2013b). https://doi.org/10.1111/cgf.12005
84. Bouaziz, S., Deuss, M., Schwartzburg, Y., et al.: Shape-up: Shaping discrete geometry with projections. Comput. Graph. Forum **31**(5), 1657–1667 (2012). https://doi.org/10.1111/j.1467-8659.2012.03171.x
85. Lachauer, L., Block, P.: Interactive equilibrium modelling. J. Int. Assoc. Shell Spat. Struct. **29**(1) (2014)
86. Coxeter, H.S.M., Greitzer, S.L.: Geometry Revisited, vol. 19 (1967). Maa Day, G., Gasparri, E., Aitchison, M.: Knowledge-based design in industrialised house building: A case-study for prefabricated timber walls. In: Digital Wood Design, pp. 989–1016. Springer, Berlin (2019)
87. Catmull, E., Clark, J.: Recursively generated b-spline surfaces on arbitrary topological meshes. Comput. Aided Des. **10**(6), 350–355 (1978)

88. Mcgee, W., Feringa, J., Søndergaard, A.: Processes for an Architecture of Volume, pp. 62–71 (2013). https://doi.org/10.1007/978-3-7091-1465-05
89. Hart, G.W.: Conway notation for polyhedral (2006). http://www.gergehartco/virtual-polyhedra/conwaynotation.html
90. Tolomatic (2019). https://www.tolomatic.com/products/product-details/rsx-extreme-force-electric-linear-actuators. Accessed 01 Aug 2022
91. Mellis, D., Banzi, M., Cuartielles, D., et al.: Arduino: An open electronic prototyping platform. Proc. CHI **2007**, 1–11 (2007)

Unlocking Spaces for Everyone

Mattia Donato, Vincenzo Sessa, Steven Daniels, Paul Tarand, Mingzhe He, and Alessandro Margnelli

Abstract Unlocking spaces for everyone has never been more important for public and private sectors within the Architectural, Engineering and Construction (AEC) industry. Knowing how shape the invisible forces that surround us brings a new dimension to designing buildings. Clients, architects, planners, landscape designers, and engineers, not only influence the lives of the residents of their buildings but touch the lives of everyone else viewing, passing by, and sitting next to them. Buildings have the responsibility to care for and protect the context they sit within. Forward-thinking legislation and the possibilities offered by the Industry 4.0 such as advances in computational capabilities and interoperable open-source tools, unlocked microclimate assessments as never before, in the UK and internationally. Bioclimatic design, a branch of urban climatology, is currently used to transform spaces into destinations, for people, flora, and fauna. By covering aspects of wind engineering, natural and artificial lighting, outdoor thermal comfort, and air quality, bioclimatic designers help shape inclusive, safe, resilient, and attractive spaces, buildings, and infrastructure. This creates more sustainable cities and communities, lowers disparities, heat vulnerability, lack of daylight availability, and poor air quality. This chapter describes the 'why', 'how' and 'what' of bioclimatic design at the city- and building-scale, followed by how it is now fully embedded in the pre-planning stages of medium-to-large buildings across the UK and in many other countries globally.

Keywords Bioclimatic design · Urban climates · Wind microclimate · Daylight · Glare pollution

United Nations' Sustainable Development Goals 11. Sustainable Cities and Communities

---

M. Donato (✉) · V. Sessa · S. Daniels · P. Tarand · M. He · A. Margnelli
AKT II, Bioclimatic Design, London EC1Y 8AF, London, United Kingdom
e-mail: mattia.donato@akt-uk.com

1 Introduction

Merging technologies of Industry 4.0 enabled revolutions in many fields, such as, architecture [1], computing-based disciplines like cloud computing [2], or even forestry [3].

Without advances in computational capabilities, interoperable workflows, and forward-thinking legislation, bioclimatic design as described in this chapter wouldn't exist.

The following paragraphs describe the 'why', 'how' and 'what' of bioclimatic design at the city- and building-scale, followed by how it is now fully embedded in the pre-planning stages of medium-to-large buildings across the UK and in many other countries globally.

1.1 Bioclimatic Design

A commuter who walks to work leaves the comfort of their home and embarks on a journey that may take them through areas that are: shaded or sunlit; next to walls that are warm; across grass that is cool; under the trees that offer shelter from the rain; around a building's corner where the wind suddenly accelerates; along roads where polluted air is inhaled; and so on. As the individual encounters these differing microclimates, the body responds by becoming warmer/cooler, perhaps sweating/shivering, and reacts by changing pace, adjusting clothes, leaning into the wind, or avoiding roads that are busy and polluted [4].

Even if the different features at the street level, the surface covering, the buildings' shapes, and the choices of the landscape around the cities that the commuter may encounter, have been designed as the outcome of countless decisions, all of the different microclimates can be predicted and managed through bioclimatic design.

The term 'bioclimatic design' was pioneered by the Olgyay brothers in 1963, when they published 'Design with climate: bioclimatic approach to architectural regionalism'. This is the first textbook to support the design of 'climate-responsive and sustainable buildings' [5], which is still relevant today.

Almost six decades later, in 2017 Oke et al., published the first design textbook to support the emergent predictive science of 'Urban Climate', once urban climatology had become a recognised subfield of meteorology and climatology [4].

These textbooks pioneered two different and intertwined fields, but they both helped designers (practitioners and researchers) to shape better buildings and cities. We can name the field as 'bioclimatic design' with an additional meaning concerning urban climatology.

In the meantime, COVID-19 has led to the first rise in extreme poverty in a generation [6], but it has reminded us that healthy and livable cities require open public spaces [7], and the importance of investing in a climate-resilient future [7].

Thus, bioclimatic design has become even more complex, filled with the expectation to unlock spaces for everyone and everything, and interconnected with how cities and societies are shaped.

Cities define the urban climate, they set the scene. Buildings, however, alter the urban microclimate around them:

- They change the wind microclimate when they are taller than the average height of the surroundings, to an extent where they change how people use outdoor amenity spaces, or they become life-threatening [8]
- They affect sunlight and daylight availability to nearby buildings and public spaces; they can generate glare for drivers during the day, and light pollution to the surroundings during the night
- Through absorbing and releasing heat, buildings impact the flowering of plants to occur earlier in urban areas; they create lower space-heating costs but higher space-cooling requirements in cities; they cause increased heat stress on human residents in the summer, in less dense fogs, and an increased rate of chemical reactions leading to smog [4].

Even if this is an emergent field, bioclimatic design presents the challenge that the entire Architectural Engineering and Construction (AEC) sector is experiencing: change.

Bioclimatic design, in particular when it comes to wind microclimate and sunlight and daylight, used to rely on physical models that were tested in wind tunnels [9, 10] and heliodons [11] respectively.

With advances in computational capabilities and open-source software tools, wind microclimate and sunlight and daylight studies have been digitalised, bringing a new series of opportunities for leveraging common CAD models and data to assess and interactively present building stack effect, accelerations, outdoor thermal comfort, air quality, disability glare, and light pollution studies.

Practitioners involved during the early design stages must keep the pace of the design teams, iterating through options, and informing about mitigation measures at the first opportunity.

Making good use of the common data, CAD geometry, and computational resources is key to a successful design in the so-called 'Architecture 4.0', post-BIM. It's about computing minimally and strategically, while using experience and expertise to add value to the projects. Nevertheless, the underlying secret is communication. Complex assessments require simple and intuitive means of expression. Software developers help practitioners to seamlessly present gigabytes of information in real-time.

Researchers studying the impact of buildings and cities on their microclimates have been finding strong correlations between a citizen's wealth and their exposure to the adverse effects that urbanization causes: heat vulnerability, reduced daylight, and low air quality.

These studies [7, 12] can guide planners and practitioners to design better and more equal cities in any part of the world: access to green open fields/woods, well-lit spaces, and areas with low air and light pollution are key to make poor areas of cities more resilient to climate change, less dependent from air conditioning, and socially less deprived.

1.2 Case Study

AKT II's office building—the White Collar Factory (WCF), in central London—provides a perfect opportunity to show how the various aspects of bioclimatic design are linked by data.

Bioclimatic designers can harness value through the effective use of data, shared 3D CAD models and interoperability. In particular:

- Wind-induced stack effect, accelerations, and air quality rely on wind microclimate outputs
- Outdoor thermal comfort is linked to wind microclimate and sunlight/daylight assessments
- Light pollution and disability glare, as an extension to the sunlight/daylight studies, uses the same geometry of the other fields.

AKT II's offices are within the fourth and fifth floors of the WCF building (Fig. 1): a redevelopment of a prominent corner site at the Old Street 'Silicon Roundabout' in the heart of London's 'tech city'. The construction was completed in 2017. The scheme provides a new, 16-storey, 237,000-square-foot office tower, together with a low-rise campus of further office, retail and residential provisions, and a new public square [13].

Fig. 1 White collar factory—© Halo

The building fits the ideal case study as it presents:

- Large openable windows, where the building stack effect may reduce their fruition to frail or disabled people and the air quality may compromise their use.
- A corner entrance which is exposed to downwash, funnelling and corner-separation effects.
- A running track and rooftop terraces that are used predominantly in the spring and summer but are both exposed to strong wind and sunlight.
- A large courtyard, where people spend their lunch breaks, and restaurants use as outdoor dining spaces.
- Southern/western facades that are fully exposed to the free stream from the dominant winds in London, which may cause distress at the street and roof levels, and perceived accelerations.
- A congested roundabout to the north, which makes air quality a key assessment considering the use of the outdoor and indoor spaces.
- Nearby residential buildings, with possible adverse impacts on their sunlight levels during the day together with artificial light pollution during the night.
- A busy road network in the proximity of the building, where glare, due to the glossy elements on the facades, may hinder the capability of drivers.

At the time of the building's design and construction, the most up-to-date guidelines and assessments were carried out. However, since its construction, new microclimate guidelines have been published or have become more established across the UK; these will be followed in this chapter:

- Wind Microclimate Guidelines of the City of London—published in 2019 [14]
- Outdoor Thermal Comfort Guidelines of the City of London – published in 2020 [15]
- Daylight and Sunlight Guidelines of the British Research Establishment (BRE)—published in 2011 [16]
- Light Pollution Guidance of the Institution of Lighting Professionals (ILP)—published in 2021 [17]
- Air Quality Management of the Department of Food & Rural Affairs (DEFRA)—published in 2018 [18].

The following pages present assessments of the WCF project that are now carried out in accordance with the most recent best-practice guidelines:

- Wind Microclimate:
  - Wind Comfort and Safety.
  - Wind Loading on Internal Pressure.
  - Wind-Induced Vibrations (Accelerations).

- Natural and Artificial Lighting:
  - Sunlight and Daylight.
  - Disability Glare.

– Light Pollution.
- Outdoor Thermal Comfort.
- Air Quality.

For each assessment, when pertinent, a comparison between two scenarios is shown, to stimulate the reader to think about several aspects that may influence the final result.

2 Wind Microclimate

Wind is perhaps the most common yet mysterious thing in the world. Wind is everywhere, and perhaps its existence is so normal in everyday life that we won't even notice it until it becomes disturbing. Indeed, wind is usually gentle in urban areas where dense built environments around public amenities exhibit a sheltering effect.

However, the growing trend in buildings being higher and slenderer, due to urbanisation [19], has drawn much attention on the potential wind impact from the skyscrapers onto the local environment and vice versa. Wind-microclimate assessment is therefore becoming more and more in demand in today's planning applications and building designs.

The Flatiron Building was perhaps the first skyscraper not only famous for its unique triangular design but also for its northern corner being deemed as the windiest in New York in the early 1900s [20]. The number of tall buildings has since been growing exponentially, and especially in recent years. As the complaints about strong winds around high-rise buildings began to accumulate, the authorities also start to require a consideration of wind comfort when designing new developments. The tragic incident in Leeds in 2011 [8], whereby a pedestrian was killed by a lorry being knocked over by what were believed to be strong winds introduced by a tall building, shows the importance of good wind microclimate, not just for the sake of the public's comfort, but also for their safety.

Wind intensity is not just important within wind comfort; it is also a driving factor within thermal comfort (which will be discussed in detail in the forthcoming chapter). Despite the UK-wide requirement of having a suitable wind microclimate on- and off-site the proposed development for planning approval, there is a lack of official guidelines to regulate the assessment methodology. It was not until late 2019 when the first wind microclimate guidelines [14] were published by the City of London (CoL) Corporation, followed by the first thermal comfort guidelines being released one year later [15].

The next section presents how wind comfort and safety are assessed with particular reference to the state-of-the-art technology context of Industry 4.0.

2.1 Wind Comfort and Safety

Wind by its nature is stochastic and full of randomness, while pedestrian comfort is very subjective. Pedestrian wind comfort is therefore challenging. Fortunately, a number of pioneers, including the giants in wind engineering such as T. V. Lawson and A. G. Davenport, have laid the foundation for assessing pedestrian wind comfort and safety. The method in general relies on the statistics of historical wind data, which are usually collected from weather stations, so that comfort and safety ratings can be obtained based on the probabilities of any wind speeds that exceed the threshold wind velocities at the pedestrian height (at 1.5 m) at the location of interest. Tables 1, 2, and 4, show typical wind comfort and safety criteria using the Lawson London Docklands Development Corporation (Lawson LDDC) [21] and City of London (CoL) Lawson Criteria [21] as an example, of which the former is perhaps the mostly used criteria in the UK and especially within the Great London Authority (GLA).

Case study. On the WCF, both the LDDC and City of London (CoL) criteria have been used to show the difference between the differing wind comfort and distress criteria.

The fundamental input of the entire workflow is the CAD model of the building in question, i.e., the WCF and its immediate surroundings (for this case study, the effect of vegetation has been omitted). The 3D models are then passed to a pre-processing tool, which automatically generates the computational fluid dynamics (CFD) simulation case files based on a number of key parameters as determined by

Table 1 Lawson LDDC wind comfort criteria

Grade	Category	Wind speed with 5% exceedance (m/s)	Description
1	Sitting	4	Sitting use, outdoor dining
2	Standing	6	Standing use, entrances, bus stops
3	Strolling	8	Leisure walking, window shopping
4	Business walking	10	Main objective of the pedestrian is to walk
5	Uncomfortable	>10	Unacceptable for most of the pedestrian activities

Table 2 Lawson LDDC wind safety criteria

Category	Wind Speed with 0.025% exceedance (m/s)	Description
Frail group	15	Unacceptable for vulnerable user groups (kids, elders, frails, disabled)
Others	20	Unacceptable for the general public (not used in the CoL)

Table 3 Lawson city of London wind comfort criteria

Grade	Category	Wind speed with 5% exceedance (m/s)	Description
1	Frequent sitting	2.5	Sitting use, outdoor dining
2	Occasional sitting	4	Standing use, entrances, bus stops
3	Standing	6	Leisure walking, window shopping
4	Walking	8	Main objective of the pedestrian is to walk
5	Uncomfortable	>8	Unacceptable for most of the pedestrian activities

Table 4 Lawson city of London wind safety criteria

Category	Wind speed with 0.022% (m/s)	Description
Unacceptable	15	Unacceptable for all

the user, such as terrain condition (for wind profile), case resolution (whether it is for an initial desk study or a formal planning submission), wind directions (as per requested for each project), etc. Next, the case files can be directly executed via local or cloud computing resources. The simulation results are then post-processed and visualised back within a CAD environment.

Figure 2 illustrates the typical wind microclimate assessment results in compliance accordingly with the Lawson LDDC and the CoL Wind Microclimate Guidelines. Summer comfort is usually the focus as pedestrians are more likely to make use of the public realm in the summer. Results indicate that the wind comfort under the newly published CoL criteria is mostly rated as Standing/Walking Grade around the WCF at ground level, which is similar to the LDDC criteria.

However, the LDDC criteria also reports large areas as being suitable for sitting whereas these turn out to be mostly only suitable for 'occasional sitting' according to the CoL criteria.

Thanks to the flexibility that a CFD-based approach provides, it is possible to identify the critical wind directions and unveil the underlying wind issues by visualising the wind's streamlines. Figure 3 shows the streamlines of the wind from the critical direction around the northwestern corner of WCF.

It clearly indicates that a downdraught is formed after the freestream hitting the western façade of WCF, and is accelerated by the northwestern corner, resulting in unfavourable wind conditions at the pedestrian level. However, the majority of the uncomfortable area is along the main road, where no pedestrian activities are expected to take place. The small uncomfortable area locally along the pedestrian pavement should be mitigated by the in-place landscaping (which is excluded in the current demonstration).

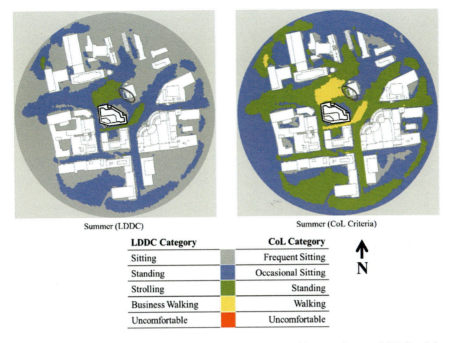

Fig. 2 Comparison of wind comfort in summer at the street level between Lawson LDDC and the CoL criteria—© AKT II

Conclusions. The above demonstrates how powerful CFD is for wind microclimate assessment. It provides the full flow information of the entire simulated domain, instead of the scattered locations with limited information that are obtained through conventional wind tunnel tests. Moreover, streamlines can be plotted using the rich data obtained from CFD simulations, which gives informative indication of the underlying wind issues that cannot be gained within the wind tunnel. Of course, the current industry-standard CFD approach uses Reynolds-Averaged Navier–Stokes (RANS), which provides industry-acceptable turnaround timescales but is limited to the accuracy of the underlying turbulence model. However, in terms of the wind comfort assessment, it is generally deemed to be accurate enough [22]. With the power of Industry 4.0, more powerful computational resources will be available in the near future which may push the industry-standard approach to migrate to a Large-Eddy Simulation (LES) approach, the accuracy of which has been well established through a large number of research projects and case studies [22, 23].

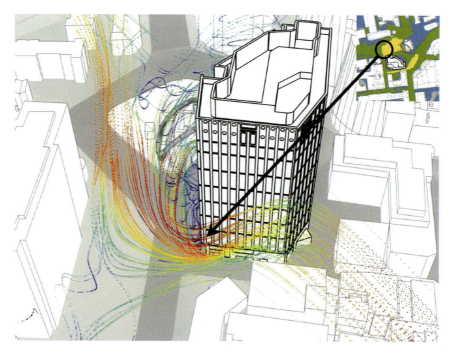

Fig. 3 Streamlines of the wind from the critical wind direction around the northwest corner of the WCF—© AKT II

2.2 Wind Loading on Internal Pressure

In addition to structural implications, the wind loading on a building also has a significant impact on the occupants. This can be due to thermal comfort and buoyancy (stack effect), air quality, or pressure that's exerted on doors and windows. To ensure consistently comfortable spaces, and within the context of sustainable development, many innovative components are often added ad-hoc within the late design stages within modern building designs. These may for example include adaptive façades with motorised building envelope vents or operable windows, however these innovative systems present challenges for the engineer, as they replace conventional cladding [24, 25] while the motorised openings are subject to any power failure [26]. It is desirable to consider this topic and incorporate (or pre-empt) measures within the early design stages instead.

Methodologies. For the past few decades, zonal methods have been the most popular approach when computational resources are limited. Of these, Airflow Network Modelling (ANM) treats the building and systems as a collection of nodes representing rooms, parts of rooms and system components; connections (or elements) between each node represent the distributed flow paths from windows, cracks (leaks), doors, ducts, and Heating, Ventilation and Air-Conditioning (HVAC) systems. These and external frictional forces are explicitly introduced through empirical expressions. Solutions to the internal pressure distribution are achieved very quickly, combined with local weather data, and an annual assessment (hour-by-hour) can be conducted within a reasonable timeframe. The ANM formulation by Walton [27] is used in software such as Energyplus [28] and CONTAM [29], which are widely used in Architectural Engineering & Construction (AEC) industry.

With increasing advances in computational resources, CFD has become more commonplace in industry, providing solutions to fluid-based problems which cannot be easily evaluated analytically. Nevertheless, CFD undoubtably represents the more computationally expensive end of engineering simulation. In the context of internal flow, even with today's computational resources, an annual (hour-by-hour) assessment within a suitable timeframe would be unfeasible.

Case study. We demonstrate these two approaches using the test case from the previous sections. The fifth floor of the building was chosen for analysis. A schematic of the floor layout with the airflow network model setup is shown in Fig. 4. The main focus of this analysis is on the four internal doors, as indicated in red in Fig. 5.

For simplicity, it is assumed that the doors and windows on this floor are fully open, the HVAC system is off (i.e., natural convection is only considered), and the floor is sealed off from the other floors. External window pressures were obtained from the pedestrian wind comfort assessment described above for the prevailing wind direction for London. The resulting pressure distribution for the two methods is shown in Figs. 6 and 7.

It can be seen in Figs. 6 and 7 that the trend of the pressure results between the two approaches is consistent with a small discrepancy of quantitate values. According to BS 8300–2:2018, to allow for access to vulnerable members of the public, the acceptable pressure-driven force on a door (at 30 to 60° open) is 22.5 Newtons, to which all doors in this example are acceptable for both methodologies.

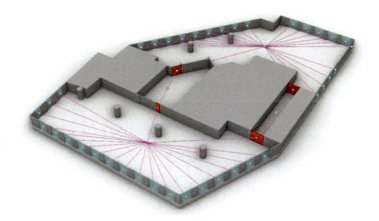

Fig. 4 Network layout for ANM—© AKT II

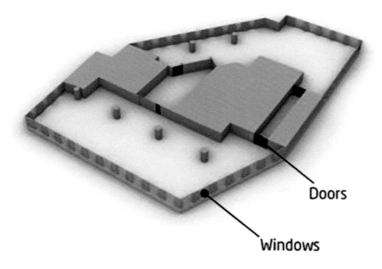

Fig. 5 5th floor layout for WCF and key elements of the analysis—© AKT II

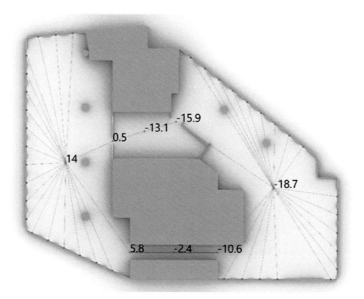

Fig. 6 Internal Pressure calculated using ANM—© AKT II

For the pressure field, this example test case demonstrates strengths and weaknesses between ANM and CFD. On the one hand, the ANM formulation follows the assumptions that form the basis of Bernoulli's principle, notably that the flow is streamlined, i.e. the flow travels between two nodes without deviating from the path.

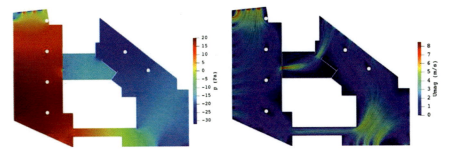

Fig. 7 CFD assessment for the internal flowfield: (left) pressure (right) velocity—© AKT II

The potentially important effects of the room's geometry on the flowfield (incurring recirculation or stagnant regions) are therefore omitted, to which CFD elucidates. On the other hand, as seen from the streamlines in the CFD in Fig. 3, these flow characteristics are insignificant for this example test case, and so the ANM provides the pressure distribution at a fractional computational cost of the CFD. However, the CFD provides additional information for other potential occupant-comfort issues such as the wind gusts around the room. On top of this, topics such as thermal comfort, stack-effect, and air quality are better modelled by CFD, thus making it the more powerful tool for internal airflow.

Conclusions. Advances in CPU technology have allowed designers to consider CFD as a building design tool in industry, as demonstrated in other sections of this chapter. This section demonstrates that the more traditional method of Airflow Network Modelling (ANM) provides an adequate approximation of the pressure field for an internal flow simulation. Quick in its calculation, ANM can compute the annual (hour-by-hour) performance within a reasonable timeframe and accuracy. However, CFD can provide information of local flowfield (e.g. gusts from windows) which occupants may find uncomfortable. It may be more prudent therefore to combine these two approaches: using ANM to conduct and scope the annual performance, and CFD to perform a comprehensive assessment of a few individual cases.

2.3 Wind-Induced Acceleration

In recent years, there has been an increase in demand for buildings that are fast to construct, have large uninterrupted floor areas and are flexible in their intended final use. Modern design and construction techniques enable the sector to satisfy such demands and produce buildings which are competitive in their overall cost. The architectural trend for taller structures, which embodies lightweight floor systems of longer spans, with a tendency to lower natural frequencies with reduced natural damping, has created a greater awareness of the horizontal dynamic performance of the building when subjected to environmental loading, such as wind.

An occupant's perception of wind-induced vibrations of buildings—whether this be physiological or psychological—is largely uncharted. The current industry guidelines and standards provide 'acceptable thresholds' for peak or root square mean (RMS), and accelerations of an occupied building floor (which are typically reported in milli-g, i.e. 1/1,000th of Earth's gravitational acceleration). These include the National Building Code of Canada (NBCC), International Standards Organisation (ISO 6897 (1984), which was more recently updated to ISO 10137 (2007)), and the Architectural Institute of Japan (AIJ) (1991). Academic research surrounding human perception on vibration and tolerance thresholds is very active [30]. From a survey of an office building in New Zealand, Lamb et al. [31] report that 41.7% of respondents were able to perceive the wind-induced swaying of the building, with a similar percentage finding that it was 'difficult to concentrate', with symptoms attributed to the lesser-understood 'sopite syndrome' (drowsiness), which affected their work. Occupants overcame these effects by taking multiple breaks and consuming motion-sickness tablets.

Although vibration may induce a sense of insecurity for occupants, it must be stressed that perception of horizontal vibration does not necessarily imply a lack of structural safety. To determine the wind loading on a building, during the early design stages a structural engineer would conduct a 'desk study' following the calculation procedure outlined in the Eurocode BS EN 1991-1 and NBCC. The Eurocode can be sufficient for standard buildings (e.g. prismatic massing) under 200 m tall. However, if the building geometry is more complex than those outlined in the Eurocode, the calculation becomes very conservative and more of an approximation. The Eurocode also provides limited insight when the building has a low resonant frequency, high slenderness (height-to-width ratio), and susceptibility to 'interference effects' from neighbouring buildings. Conventionally, for several decades, a wind tunnel assessment is conducted to provide this information. Wind-induced acceleration can also be derived from the results. At present however, there are no standards (on wind loading) that describe the use of CFD in this application.

Over the last couple of decades, with encouraging software and hardware advances, there have been an increasing number of publications demonstrating the feasibility (and efficacy) of LES as a design tool for wind engineering applications. In early publications, LES was largely used to provide additional insights into wind tunnel investigations [32] or to investigate the sensitivities associated with the wind tunnel setup—most notably the pressure results to the inflow profiles [33]. In more recent publications, the focus of LES is to act as a design tool, with a focus on determining peak pressure and incurring structural responses [34–37] or wind-induced accelerations [38]; this latter application is particularly promising given LES' ability to resolve large wind gusts.

Case study. We demonstrate the results of LES in determining the wind-induced floor acceleration for the WCF test case. For simplicity we assume that each floor consists of a concrete slab and the (usually estimated) critical structural damping of 1%.

In this example case, the building is 74 m tall with a low slenderness ratio. Therefore, a wind loading assessment would be considered cautious, and the values for

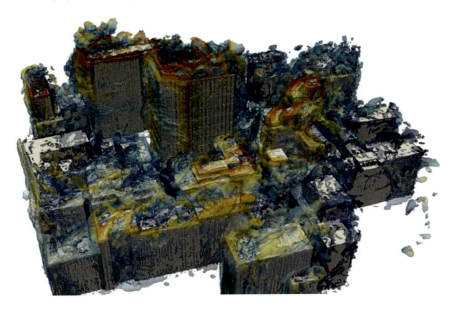

Fig. 8 Wind structures around the WCF and context—© AKT II

the wind-induced acceleration are expected to be small and unlikely to exceed the thresholds specified by ISO 10137:2007 for an office building. Figure 8 shows the simulated wind flow structures around the target building and surrounding context buildings using LES. These structures, and the incurring fluctuating wind pressures on the target building were sampled to obtain the characteristic wind loading on the structure. The wind-induced accelerations are subsequently derived from these results.

The resulting accelerations, assessed without proximity context, for the along-wind (direction of wind), across-wind (perpendicular to wind direction), and torsional motions of the structure for a range of wind velocities are plotted in Fig. 9. We also compare these results to those provided in the NBCC guidelines. As expected, the guidelines provide a more conservative estimation than the LES assessment. An additional benefit is the turnaround time relative to the equivalent wind tunnel assessment, which allows for the engineer to finalise the design before the final assessment.

Conclusions. This section demonstrates the use of CFD, specifically LES, in determining the wind loads and the floor-by-floor wind-induced accelerations of a building. While the Eurocode conventionally requires that the wind loading assessment is determined through wind tunnel testing, quicker solutions provided by LES may be used in the 'desk study' stages to highlight any potential issues with the design that could then be mitigated.

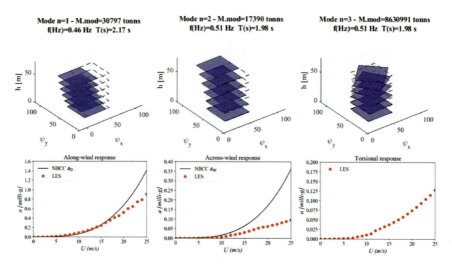

Fig. 9 Calculated wind-induced accelerations for the WCF from LES and the NBCC guidelines—© AKT II

3 Natural and Artificial Lighting

3.1 Natural Lighting

Digital technologies have allowed advanced solar design to become mainstream practice in recent years. Because lighting is such an integral part of how space is experienced, the convergence of these digital technologies has high potential in improving the quality of how we perceive buildings and the built environment. Because sunlight brings both heat and light, solar design has always been a balancing act between the provision of appropriate light and comfort and the avoidance of overheating. Modern workflows allow an integration of many solar indicators into holistic decisions, which allows an effective optimisation of the overall energy balance.

Open-source software such as Radiance [39] and its GPU-based implementation Accelerad [40] have provided a flexible backbone for building simulations. This has allowed designers to move away from graphical rule-of-thumb methods, and for highly accurate and computational physics-based assessments to become a standard practice.

For bioclimatic design, such workflows include energy, daylighting, light pollution, thermal comfort, and reflected glare. The computational cost of these workflows however may be high. When working on large masterplans for planning, or when developing in densely built-up urban centres with large digital models, they could potentially take hours to run on standard desktop PCs.

Cloud computing unlocks these large models, but presently still demands specialist knowledge to set up, and may incur unacceptable financial costs. Designing workflows for small workstations remains a specialist area. On the other hand, the accessibility of solar data and digital models for most parts of the world has removed a barrier in applying the technology to virtually any project.

Another exciting avenue, when it comes to solar design, is the visualisation of 'feels like' lighting conditions using augmented reality (AR) or virtual reality (VR) equipment. Designers are able to experience the design with its true lighting conditions, and also with physical measurements of light levels within and around the building, as it is being designed. If this modelling is achieved with an imperceptible latency (equating to real-time), on appropriately powerful hardware, it can transform the designer's experience of the proposed space.

The following sections focus on a selection of three indicators that are used to assess comfort, safety and inclusivity within the design of daylight conditions for the built environment. These are 'vertical sky component', 'sun hours', and 'reflected glare'.

Vertical sky component. New constructions affect the daylight availability within the surroundings, by overshadowing adjacent areas. Vertical sky component (VSC) is used to measure the amount of diffuse light from the sky that is available on a vertical façade. VSC is expressed as a fraction of the hemispherical light from the sky that is visible from a specific point of view. It is measured from 0 to 40% (on a vertical façade). Table 5 shows the categorisation of the ranges of VSC, and their meaning for design [16]. Digital 3D models have been used as a standard practice for determining the daylight availability through visibility checks, within the model, of the sky in all directions. Using this information, a further optimisation would be to compute the maximum height and width of a building's massing, which could be proposed for a small site in a built-up area, while complying with the necessary right-to-light and other similar regulations for daylight access. Taking this a step further, the layout of an entire masterplan could be informed by continually assessing VSC, together with other planning targets, while iterating the design. Detailed digital models of cities furthermore allow the study of VSC to be scaled up into a quick assessment of all buildings that could be impacted by a proposed construction project.

Sun hours. This describes the amount of time that a location receives direct sunlight. For a location in the open field, this will amount to most of the year, whereas for a courtyard in a built-up area it could be less than a quarter of the year. Knowing

Table 5 BRE, Vertical Sky Component (VSC) thresholds

Grade	Category	VSC range	Description
1	Unobstructed	>27%	Conventional window design will usually give reasonable results
2	Slightly obstructed	15% < VSC < 27%	Provide larger window than usual
3	Obstructed	5% < VSC < 15%	Difficult to provide adequate daylight unless fully glazed
5	Heavily obstructed	VSC < 5%	Often impossible to provide reasonable daylight, even when fully glazed

the daily and annual sunlight patterns for spaces that receive a lot of shadowing is essential in order to provide suitable conditions for outdoor activities. Landscape design, and the selection of appropriate plant species for the daylight conditions, is a common use for such data, which itself can be obtained at a very low cost using the 3D geometry from a project's early stages. Sun hours and radiation data are also the basis for many other analyses that are required for planning, including the outdoor thermal comfort studies.

Disability glare. In addition to measuring solar access using sun hours and the vertical sky component, in a dense urban setting it will also be necessary to check for any adverse impact that could be caused by the sun within the inter-building space. Sunlight reflected from highly glazed facades may cause disturbances nearby. This phenomenon is known as reflected or disability glare. Glare impact is usually visual, such as a bright reflection near a traffic signal, crosswalk or airport runway that could cause an inability to see and a high risk of accident. In rare instances, the reflection can also be so intense that it also causes significant thermal effects, such as overheating in adjacent buildings, damage to property like parked cars, or burning landscape features. High profile cases of thermal impact include the curved facades of the 'Walkie Talkie' tower at 20 Fenchurch Street in London [41], and the Walt Disney Concert Hall in Los Angeles [42].

Such impacts make glare a material concern for space users. However, as opposed to the study of VSC and sun hours, there has been a lack of standards and validated digital tools for the assessment of reflected glare within most design practices, which has resulted in these studies historically only being applied on special projects. Even then, in many cases, it has been an afterthought, and only if mitigation has been required for completed projects. Digital project models, and the emergence of robust analysis workflows in the public domain in recent years, have made it easier to create spaces where reflected glare impacts are known and are controlled within the design stage.

Opportunities and challenges. The goal of daylight and glare studies is to optimise the design for local conditions and to minimise any adverse impacts at sensitive locations around the site. As mentioned above, while daylight access methodologies are well developed, this is not the case for reflected glare. Due to a lack of glare standards, a range of digital approaches to the same problem have been published [43–48]. Before digital tools, the main methodology was geometry-based desk study using solar protractors [49]. This was problematic because only spot checks could be made around the site, plus the work was manual, prone to errors and time-consuming, and also it was not possible to assess reflection intensity.

Modern digital workflows model the environment in 3D and then use rendering engines to fully map the distribution, intensity, and duration of solar reflections. Particular focus has been on visibility within road and rail networks. For air traffic for example, this could be reflections from solar photovoltaic (PV) installations next to the runway which could blind pilots. The Hassall methodology is the most widely used for such cases [45]. A single HDR render of the driver's view is computed and used to calculate the precise timings and intensity of glare throughout the year. While this could also be done using a digital camera at the site, this latter technique only works where the construction already exists.

With detailed 3D models already available from the daylight analyses that are done on a project, no further specialist information is required for the glare assessment. By integrating the already available project data, and with the sensitive locations known, the study can be automated, which makes it faster, more accurate and cost-effective when compared with desk-based methods. This also allows for rapid iteration of the geometry should any significant impacts be found. Often the required adjustments to the glazing's orientation or its reflectance properties are so small that they remain invisible to the naked eye.

Some glare regulations and guidelines have been issued in recent years around the world, which helps to fill the relative gap in comparison with the well-developed daylight access regulation. This includes the City of London [50], Singapore [51] and US Federal Aviation Administration technical guidance [52]. Because glare regulation is mostly qualitative and vague, there is no standard metric or approach. Standards now need to catch up with the technology, to allow the practice to become standardised and accessible for the wider AEC industry.

Finally, one area where digital tools can help to make a large difference is in the improvement of the human connection to the experiencing and understanding of technical data. Integrating workflows with video gaming engines, such as Unity, can allow the conditions on a specific street to be experienced, either locally or remotely, using AR/VR goggles. With such goggles, the planners, designers and other stakeholders could then be able to intuitively explore how the new construction will impact the daylight access and glare for the surrounding areas. This is a matter of data integration. Making such tools accessible to a wider audience will be a step towards creating shared value in the built environment, and towards making safe and accessible places for everyone.

Case study. Further to the wind assessments, the 3D model of the WCF was then used to quantify the natural lighting metrics that are mentioned in this sub-chapter, which are namely VSC, Sun Hours and Reflected Glare. The VSC was investigated for key locations, and shows the adverse impact of vegetation on sunlight availability.

Figure 10 shows the current VSC on a representative location around the WCF versus Fig. 11, which shows the negative impact that vegetation may have on sunlight levels, even if it may have a beneficial impact in reducing strong downdrafts.

The public courtyard to the South of the WCF requires checks concerning its solar exposure. Modelling annual sun hours or lux levels is an efficient way of determining which courtyard areas could be fit for which activities [16]. Sun hours (and radiation) results can inform the landscaping design and the locations of seating areas or outdoor dining. The results are best visualised using contour maps for seasons or for various periods of the day (Fig. 12—Sun hours). As for other solar studies, an important benefit of digital methods here, as opposed to traditional desk study graphical techniques, is the fine granularity of the results, both temporally and spatially. These insights can be achieved cost-efficiently, in a matter of minutes. In this case, the direct sunlight is restricted in the courtyard, but outdoor space making can benefit from leveraging these patterns.

Fig. 10 VSC on nearby buildings—© AKT II

Fig. 11 VSC on nearby buildings with evergreen trees—© AKT II

Unlocking Spaces for Everyone

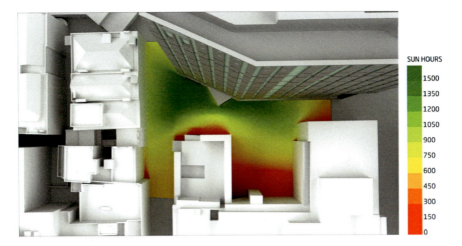

Fig. 12 Sun hours—© AKT II

Tall buildings such as WCF have a greater likelihood of causing glare as the reflections can be seen from a larger area. Sensitive locations such as crosswalks or traffic intersections can readily be mapped digitally by integrating GIS data from sources such as Google Street View. Glare intensity and the sources of reflection were assessed at a location by creating a glare intensity render using a 3D model of the proposal and its surroundings. The render is showing the sensitive part of a driver's field of view (covering 30° around the area of focus), looking toward a traffic light. The false coloured glare intensity field in the render is calculated with Hassall methodology (Fig. 13—Reflected glare intensity caused by the White Collar Factory calculated at a nearby traffic intersection).

Fig. 13 Reflected glare intensity caused by the White Collar Factory calculated at a nearby traffic intersection—© AKT II [left], © Google [right]

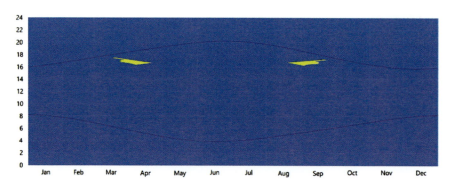

Fig. 14 Annual times that the viewpoint at the crosswalk receives solar reflections are shown in yellow—© AKT II

Intensity is mainly a function of the distance from the centre of the field of view, the glass reflectance, and the sensitivity of the eye. In this case, there may be a glare hazard because the driver must see the traffic light or pedestrians but may be unable to do so if the glare intensity is above 500 cd/m^2. Thankfully, visors help mitigate the issue in this specific scenario. Times and dates when this situation might occur with clear weather can be visualised on an annual map to an accuracy of one minute or fewer (Fig. 14).

Glare becomes problematic if it is occurring for more than 50 h per year. It is time-consuming to derive this data with traditional graphical methods, and especially if the design goes through multiple iterations within its development. The digital approach moreover allows glare studies of large areas to become accurate, quick and cost-efficient, so that any adverse impact on the surrounding space can be spotted and mitigated.

Conclusions. Solar design for space making has in recent years benefitted from the convergence of a number of digital technologies and trends. These include freely available GIS data, open-source analysis software, affordable cloud computing power, visual programming tools to set up workflows such as Grasshopper, and the popularisation of the use of video gaming engines such as Unity within the AEC sector. By leveraging these technologies, the practice of solar design will continue to improve the quality of climate-based building design as well as the user's experience of the built environment. The mainstreaming of modern workflows has the high potential to create shared value between a building's designers and users, and to provide comfortable and safe spaces for everyone to enjoy.

3.2 Artificial Lighting

Artificial lighting improves the safety of roadways and pathways, it enables outdoor and night work and evening sports activities, and it provides decorative effects to all built developments and in particular to monuments and iconic buildings.

However, artificial light can also disturb people by causing distraction if it is visible beyond the area that is supposed to be lit. The combined effects of building exterior and interior lighting, lit advertising, car park lighting, office lighting, streetlights, and the lighting from outdoor sporting venues together cause the so-called 'light pollution' that is one of the by-products of industrial civilisation.

Light pollution leads to adverse health effects (i.e. sleep deprivation), disrupts ecosystems, and interferes with starlight and astronomical observation. The most severe impacts affect highly industrialised and densely populated areas of North America, Europe and Asia. More than half of the population of the UK is unable to see the Milky Way in the night sky, and this problem is spreading into rural areas as the country's artificial lighting increases [53].

This obtrusive light can disturb people and ecosystems in three different ways:

- Sky glow (or upward light pollution), which is caused by the light propagation (or diffuse luminance) into the atmosphere from upward-directed (non-shielded) light sources and/or by reflection from the ground and other surfaces.
- Light trespass (or spill light), which is caused by unwanted light falling beyond the areas intended to be lit and into one's property.
- Glare, which is the discomfort or inability to see when the source of artificial light is much brighter than the surrounding area.

Whether obtrusive or not, the artificial increase in light can also have significant ecological effects on wildlife. Some nocturnal bird species for example may be attracted to or disoriented by bright lights, with induced hormonal, physiological or behavioural changes. Night-flying insects can either cease flying or fly in spirals when exposed to high levels of illumination. Some plants may stop flowering if the night is too short, while others may be flowering prematurely.

Light pollution must be addressed by changing the habits of society so that lighting is used more efficiently with less waste and less unwanted illumination. The Clean Neighborhoods and Environmental Act 2005 introduced artificial light into the list of statutory nuisances. Since then, planning conditions for all new developments have required that specific lighting conditions are applied according to the Institution of Lighting Professionals (ILP) guidelines [17].

The focus of current guidelines is mainly on external lighting, however unwanted spill light from interior lighting into nearby residential premises is also a concern. Because the interior lighting is often not yet finalised during the planning stage, a typical lighting installation is assumed, as the basis for calculating the surrounding illuminance levels.

Some indications of the illuminance level (which is measured in lux) may be obtained by looking at:

- The distance between the new development and the surrounding existing buildings.
- The dimensions and orientation of the new development.
- The glazing fractions of the new development's facades.
- The properties of the glazing.
- The interior lighting design (depending on the intended use).
- Light mitigation measures (such as shading devices).

Guidance for a preliminary design of interior lighting is given by 'Light and lighting—Lighting of workplaces' (BS EN 12,464-1:2021). Depending on the intended use (e.g. office space), the guidance provides this information:

- Average lux value and uniformity on the working plane.
- Height, dimensions and grid resolution of the working plane.
- Reflectance of all surfaces (ceiling, floor and walls).
- Average lux value near to hallways, elevators, stairs and entrances.

Case study. In the pictures below, light pollution from the WCF (in a dark sky) was assessed on the surrounding buildings Fig. 15. Office spaces were assumed on each floor, with an average lux value of 500 and a uniformity of 0.52 on the working plane. A 0.5 m band was removed from the working plane close to the wall, with a grid resolution of one metre at a height of 0.8 m from the floor. The lighting simulations were carried out by assuming a reflectance value of 0.6 for the walls, 0.8 for the ceilings and 0.3 for the floors.

The IPL guidelines [17] provide illuminance limits for pre- and post-curfew (which is typically at 11.00 PM) depending on the location. The limits refer to the surrounding 'brightness' level for four different environmental zones. For high district brightness areas, the average lux limits on the centre of a surrounding residential window should be not higher than 25 lx pre-curfew and 5 lx post-curfew (environmental zone E4).

Figure 16 shows that the average illuminance level on the windows of the residential building on the south-east side of the WCF is around 20 lx, which is below the pre-curfew limit but above the post-curfew limit, and that mitigation measures (i.e. shading devices, on/off switching sensors, dimmable lights, automatic curtains, etc.) should be integrated.

4 Outdoor Thermal Comfort

Forward thinking local regulations [14, 15] fuel the development of innovative workflows to capture the complexity of the urban climate at the building planning stage. It is clear from the City of London Corporation's research that the most important

Fig. 15 Greyscale representation of light trespass

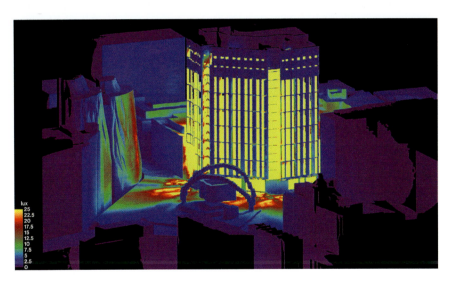

Fig. 16 Lux level on any surface caused by light trespass

factor related to the quality of a public space is the outdoor thermal comfort, which is a combination of solar exposure, wind, air and radiant temperatures and humidity.

This is the 'feels like' quality of the microclimate, which the City of London Corporation terms 'thermal comfort'. For example, a sunny open space in February might appear to be an appealing and comfortable place to dwell, but if the air temperature is low with high humidity and there is a strong northerly wind, it is likely to feel significantly colder and uncomfortable, even in the sun. This is the perception of thermal comfort as experienced by those using the space [15].

Thermal comfort is an important, recognised aspect of the urban climate. It is what we perceive.

However, because of its complexity and subjectiveness, over the years several criteria have been established to identify the thermal state of individuals outdoors. Below are some of the most adopted indices, alongside many others that are available in literature and research:

- Physiological Equivalent Temperature (PET): a rational index which is defined as the air temperature at which, in a typical indoor setting (without wind and solar radiation), the heat budget of the human body is balanced with the same core and skin temperature as under the complex outdoor conditions to be assessed [54].
- Standard Effective Temperature (SET): a rational index which is defined as the dry-bulb air temperature of an environment at 50 percent relative humidity for occupants wearing clothing that would be standard for the given activity in the real environment [55].
- Predicted Thermal Sensation (PTS), which is obtained from the SET, as a modified version with an adaptive approach [56].
- Universal Thermal Comfort Index (UTCI): a rational index that is defined as the air temperature of the reference environment. It produces the same response index value as the dynamic physiological response that is predicted by the model of human thermoregulation [57].
- Thermal Sensation Vote (TSV): a questionnaire, with values ranging from -3 (cold) to $+3$ (hot), with a neutral value at 0. There is also Thermal Preference Vote (TPV), with values going from -3 (a lot colder) to $+3$ (a lot warmer), with a neutral value at 0 [58].
- Wind Chill Index (WCI): a direct index and comfort criterion for establishing the impact of wind when the dry-bulb temperature is below 10 degrees centigrade [59].
- Humidex: a direct index and comfort criterion for identifying the impact of humidity on warm temperature [60].

In the UK, UTCI is the preferred index for establishing outdoor thermal sensation. The reason behind this choice is that the Universal Thermal Comfort Index (UTCI), which was developed by the COST Action 730 [61] between 2005 and 2009, satisfies these requirements:

- Thermophysiologically significant throughout the whole range of heat exchange
- Valid in all climates, seasons, and scales
- Useful for key applications in human biometeorology (e.g. daily forecasts, warnings, regional and global bioclimatic mapping, epidemiological studies, and climate impact research)

- Independent of a person's characteristics (i.e. age, gender, activities, clothing, etc.).

Moreover, the development of an operational procedure via a polynomial regression function [57] to calculate the UTCI without solving the Fiala Model of human heat transfer and thermal comfort [62] lowered the computational cost and unlocked the opportunity to adopt this index for building design and planning purposes.

Since 2020, any new development in the City of London requires an outdoor thermal comfort assessment, based on real meteorological conditions recorded in the Square Mile, and a simplified methodology to perform an hourly assessment around the site and on any balcony/terrace for five years [15].

Outputs are similar to the wind microclimate; alongside the percentage of the time within a comfortable UTCI range (between 0 and 32°) per season (Table 6), an additional annual usage category is also represented (Table 7).

Case study. This methodology was applied to the WCF because of its vicinity to the City of London together with the absence of a local thermal comfort policy (Tables 6 and 7).

Figure 17 shows the seasonal percentage of time within comfortable range for spring. For planning purposes, this plot is repeated for each season at the street level and on any terraces. Results for spring show that close to the WCF the comfortable

Table 6 UTCI—city of London seasonal comfortable range

Range (%)	Colour schemes	Description
0–25	Blue	Typically leads to transient annual destination uses
25–49	Blue	Typically leads to transient annual destination uses
50–69	Light-Blue/Green	Typically leads to short-term seasonal annual destination uses
70–89	Yellow/Orange	Typically leads to short-term/seasonal annual destination uses
90–100	Red	Typically leads to all season annual destination uses

Table 7 UTCI—city of London destination use

Grade	Category	UTCI within comfortable range	Description
1 (Green)	All Season	$\geq 90\%$ in each season	Year-round use (e.g., park)
2 (Purple)	Seasonal	$\geq 90\%$ spring–autumn AND $\geq 70\%$ winter	Suitable for use during most of the year (e.g., outdoor dining)
3 (Cyan)	Short-term	$\geq 50\%$ in each season	Appropriate for short duration and infrequent use year-round
4 (Orange)	Short-term seasonal	$\geq 50\%$ spring–autumn AND $\geq 25\%$ winter	Short duration activities during most of the year
5 (Red)	Transient	$< 25\%$ winter OR $< 50\%$ any other season	Suitable for public spaces where people are not expected to linger for extended periods

Fig. 17 Seasonal % of time within comfortable UTCI range (Spring)

periods vary between 70 and 100%. For the assessment, considering that this is a commercial area, the considered hours have been trimmed from 8 am to 8 pm, as recommended in the thermal comfort guidelines [15].

Figures 18 and 19 show the summary of the seasonal results with an annual recommendation of programmatic uses. At street level, this corresponds to suitability for short-term and seasonal activities within the courtyard south of the WCF. The several internal courtyards around the site, which are less permeable, show comfort grades between seasonal and all season—which translates to a year-round use of these spaces.

At the rooftop level, conditions are mainly suitable for short-term activities year-round.

Conclusions. Outdoor thermal comfort is what we perceive. It is a combination of several factors and it is subjective. The recent guidelines [15] have unlocked repeatable and fast assessments of the UTCI (a complex rational index). Results show that UTCI has a strong correlation with the hourly wind speed, and therefore if we want to design spaces for year-round activities, the protected areas (even without direct radiation) will be more comfortable than those fully exposed to sun and wind intensity. This is valid for a London climate. Given the novelty of the approach,

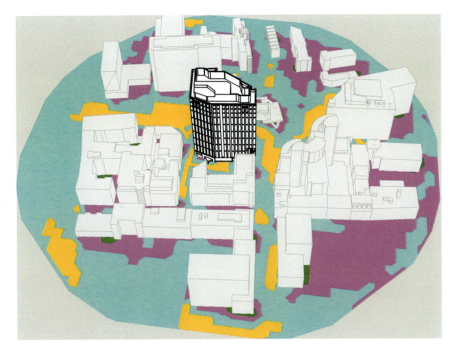

Fig. 18 Suitable destination use according to the thermal comfort results at street level

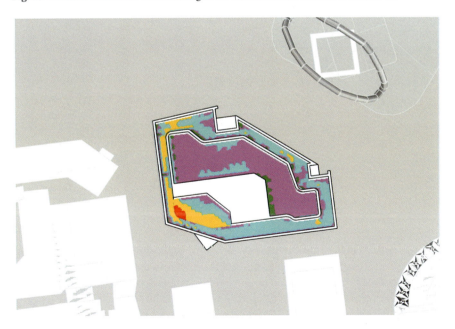

Fig. 19 Suitable destination use according to the thermal comfort results at the accessible rooftop

further work is required to capture the various boundary conditions and building designs.

5 Air Quality

Air pollution is the greatest environmental danger to human health, which is exacerbated by the global population growth, increased urbanisation, and the impact of climate change on atmospheric conditions and weather variability.

Fuel combustion in motor vehicles, coal/oil/gas-fired power stations, and industrial boilers are the major emission sources of air pollutants in urban areas. Sufficient concentrations in the atmosphere are potentially harmful to humans, animals, and plants. For example, exposure to fine particles reduces lung functions in the short term and increases cardiovascular and respiratory diseases in the long term.

In 1984, a catastrophic accident caused the unintentional release of methyl isocyanate from a storage tank in India. Driven by the prevailing wind, the toxic cloud moved close to the ground and enveloped buildings. The accident caused the immediate death of around 5,000 people and more than 20,000 afterwards [63].

The understanding of atmospheric dispersion of pollutants in urban areas is essential in order to predict the potential threat of hazards and accidental events and to identify areas that are at minor or major risk.

More generally, the accurate analysis and prediction of pollution dispersion offers a broad range of beneficial implications:

- Improve air quality at the street, city, and regional scale.
- Decrease human exposure to harmful concentrations of air pollutants in high sensitivity areas (e.g. bus shelter design).
- Design of outdoor activities (e.g. play areas, dining areas).
- Design of green infrastructure as a passive control system for air pollution (e.g. vegetation barriers).
- Lessen the impacts of climate change.
- Improve biodiversity and ecological resilience.

Air quality assessment. Air quality is influenced by the type of development (buildings and/or infrastructure), location of the site, designated uses, and whether a suitable transport strategy is in place. Air quality issues are to be addressed at the planning stage to ensure a strategic approach through the design.

Under the Environment Act 1995, the local authorities of England, Scotland, Wales and Northern Ireland are charged with air quality duties to ensure compliance with EU/English legislation [64] of the air quality objectives (i.e., Nitrogen Dioxide NO_2, Particulate Matter $PM10$) and to draw up an action plan of remedial measures for any areas of relevant public exposure.

To provide a statutory process for local authorities, a technical guidance [18] is also designed to support secondary users (such as transport and planning consultants, air quality experts, etc.) with guidelines on how to monitor, assess and take action to improve local air quality (air quality assessment).

A typical air quality assessment for planning applications includes:

- Determination of the baseline scenario
- Assessment of potential air quality impacts during the construction phase
- Assessment of potential air quality impacts during the operational phase
- Assessment of site suitability
- Assessment of air quality neutrality
- Identification of required mitigation measures

The outcome of an air quality assessment is mainly driven by the potential air quality impacts during the operational phase, where a 3D analysis of pollution dispersion from combustion plants, gas-fired boilers, industrial sources and road traffic is performed.

The 3D analysis is usually conducted using simulation software that is based on the Gaussian plume model [65], which is considered the simplest mathematical model for high-level predictions of pollutant concentrations. This model is frequently applied to continuous, buoyant air-pollution plumes or non-continuous air-pollution puff and is used in combination with empirical correlations to evaluate the growth of the plume dimensions with the distance from the source location.

Concentrations of pollutants are predicted at existing nearby receptors, which are classified into high, medium, and low sensitivity (i.e., hospitals, religious buildings, outdoor dining areas). Pollutants' concentrations are also reported at future receptors of the proposed building to assess the site suitability.

The modelling results of pollutant concentrations are then calibrated to the measured values from nearby monitoring stations, to check compliance with local and national regulations. Any area of non-compliance requires mitigation measures to be integrated within an action plan to improve local air quality.

Atmospheric dispersion modelling. In real urban scenarios, the dispersion of pollutants is a 3D-complex phenomenon that is affected by several environmental conditions (i.e., wind direction, thermal stratification, source location, type of source, buildings' configuration, building area density, etc.). However, pollution dispersion is mainly driven by wind flow with two separate mechanisms: advection and diffusion. Due to the first mechanism, the pollution plume is transported along the average wind direction at the speed of the average wind magnitude. Due to the second mechanism, the pollution plume 'grows' in three directions depending on the wind turbulence in the atmosphere. On a separate note, the wind turbulence (and so the pollutant diffusion) is affected by the atmospheric thermal stability, which can enhance or dampen vertical wind accelerations depending on the geographical location, time of day, and human activities in urban or rural environments. For example, atmospheric urban areas are frequently characterised by the so-called unstable thermal stratification, which is due to the 'injection' of heat in the atmosphere from greater human activities in comparison to those of rural environments. Furthermore, the amount of solar heat in urban areas, which is reflected in the atmosphere, is far greater than in rural areas, due to the high-built density area.

Based on the above mechanism description, interestingly, strong wind speeds, which may adversely affect pedestrian comfort and safety, contribute to the local dilution of pollution concentrations with beneficial effects on public exposure to toxic/dangerous contaminants in the air. Similarly, frequent changes in wind direction lead to beneficial dilution of pollutant concentrations.

Regarding the atmospheric thermal stability, surprisingly, the atmospheric conditions over urban areas are much more favorable to pollution dilution than in rural environments. However, pollutant emissions in urban areas (from transportation, construction, human activities, etc.) are typically higher than in rural areas, therefore, pollutant concentrations in urban areas may be always significantly higher than in the surrounding regions.

Case study. CFD has a long history of use in computing fluid flow around obstacles [66] and pollutant dispersion in urban environments [38, 67]. CFD has been applied to a wide range of problems, and most recently as an official tool to compute wind and thermal comfort. By using CFD, the urban geometries are reproduced in computer simulations with reasonable accuracy, including complex shapes and narrow canyons. Topography, vegetation, and mechanical features such as stacks, and air intakes, are easily included. With the increased computational power, realistic meteorological conditions such as wind direction and magnitude, temperatures, and atmospheric stabilities are simulated in CFD within a reasonable timeframe.

Traffic pollution dispersion around the WCF was simulated for several wind directions using 3D CFD. Emissions of NO_x and PM_{10} were calculated from the daily average number of vehicles on each segment of the surrounding road network. Background pollutant concentrations were included to predict compliance with UK regulations [69].

Figure 20 shows the normalized 3D Pollution contours around the WCF driven by southerly (left) and northerly (right) winds. 3D contours provide useful information on the dispersion mechanisms (e.g. meandering, 'chimney effects', accumulation zones, etc.). For example, the left picture shows a plume rising on the northern façade of the WCF, which is due to the strong wind recirculation in a region of low wind pressure that is induced by southerly winds.

Fig. 20 Normalized 3D Pollution contours around the WCF driven by southerly (left) and northerly (right) winds

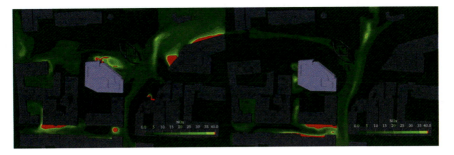

Fig. 21 Plan view of nitrogen oxides (NO_x) around the WCF for southerly (left) and northerly (right) winds

Non-negligible values of pollutant concentrations may reach the second or third floor of the WCF, and will thus affect the indoor air quality at those levels.

For both scenarios (Fig. 20), the narrow street on the south side of the WCF (Old Street Yard) and the adjoined courtyard are characterised by relatively clean air, due to frequent strong winds that localised at the WCF's south-east corner.

Figure 21 shows a plan view of nitrogen oxides (NO_x) around the WCF for southerly (left) and northerly (right) winds.

As mentioned above, a very low concentration of NO_x is predicted for both of the scenarios within the narrow street on the southern side of the WCF and within the adjoined courtyard. Differently, high concentrations are predicted on the northern side of the WCF for southerly winds, again due to the wind recirculation in the area of low wind pressure. It is worth noting that medium-to-high values of NO_x are predicted within the roundabout for both scenarios, because this location is surrounded by pollution sources on each side.

6 Conclusions

As described in the chapter, without the rise of the Industry 4.0, including advances in computational capabilities and interoperable workflows, and forward-thinking legislation, bioclimatic design as described in this chapter wouldn't exist.

Knowing how to shape the invisible forces that surround us brings a new dimension to designing buildings. Clients, architects, planners, landscape designers, and engineers, not only influence the lives of the residents of their buildings but touch the lives of everyone else viewing, passing by, and sitting next to them.

Buildings have long been designed for themselves: striving for the best structural and energy performance, while achieving the right aesthetic. However, buildings have the responsibility to care for and protect the context they sit within.

Bioclimatic designers address all the microclimate variables that make a space into a destination. We ensure that these spaces are attractive, comfortable, safe and inclusive, as part of unlocking them for everyone. This is the recipe for a successful development with a long-lasting legacy.

At the urban level, bioclimatic designers solve a variety of issues that are ultimately different aspects of the same discipline:

- Wind microclimate (comfort and safety)
- Wind-driven rain and snow (not described in this chapter)
- Wind-induced structural loading and cladding pressure (not described in this chapter)
- Wind-induced acceleration
- Wind-induced internal pressure (or building stack effect)
- Sunlight, daylight and overshadowing
- Disability glare
- Landscape strategy (not described in this chapter)
- Light pollution
- Outdoor thermal comfort
- Walkability (not described in this chapter)
- Natural ventilation
- Air quality assessment
- Pollutant dispersion
- Heat and smoke dispersion (not described in this chapter)
- Noise and vibration (not described in this chapter).

And most of the aspects mentioned above are required for planning, which is an early design stage that, until recently, was not so demanding in its deliverables. This prior trend has now changed. There is now an awareness that making the 'right' choices early can generate real value for the design, and the pre-planning stage is no longer a simple tick-box exercise. Industry 4.0 unlocked the potential of bioclimatic design, which has now become a crucial part of the design process.

Planning authorities are aware of this. They are increasingly demanding more and more microclimate assessments during the pre-planning stage, and they want to be involved in the design to an unprecedented extent.

This is the right approach. The public interest needs to be protected, while clients want their buildings to be successful within a harmonious 'urban climate'.

As discussed so far in this chapter, the challenge that the discipline is facing is 'change':

- Change of mentality. Clients have to be convinced to invest in the early design stages (before planning), just as they invest into e.g. RIBA Stages 3 and 4.
- Change of tools. Thanks to open-source software, and readily available cloud services, the complex analyses can be performed and visualised faster than every before. Clients and public authorities however, in the same way that they accepted Radiance to replace the Heliodons, also have to accept software such as OpenFOAM to replace the wind tunnel testing for wind microclimate assessments.
- Change of skillset. As the complexity increases, the challenges of staying up-to-date with the codes of practices, guidelines, standards and tools also increases.

The challenge is to maintain a holistic approach, and to not silo the aspects of bioclimatic design, so as to maintain a truly interdisciplinary branch of building design.

- Change of national policy. There is not a prescriptive requirement at the national level in the UK. The Tall Building policy [70] is too vague when it comes to microclimate assessments, and it is related to tall and large buildings. In London, each Borough has its own policy, which most of the time refers to the city-wide London Plan and is not locally prescriptive. In 2014, the Supplement to the 2011 London Plan [71] identified the Lawson Criteria as the criteria for establishing whether the windiness of the area is suitable for a specific designated use, but this new instruction didn't include a methodology for generating the results. The London Dock Development Corporation have their own LDDC criteria [21]. The borough of Tower Hamlets does not accept CFD as part of the planning submission [72] but the City of London does [14]. There must be more statutory regulation, and more awareness that the clarity on what to deliver must come from the national government. Clients will otherwise never understand why they should invest in these assessments.

Since the publication of the first textbook regarding Urban Climate in 2017 [4], much has changed. Although they are not enough and they are not harmonised, the new regulations, which are driven by greener and more inclusive agendas, have been shaping a requirement for better buildings and cities.

The urgency of building cities that are more equal, resilient and friendly has never been stronger. It is the responsibility of clients, and the broader design teams, to make this happen.

Acknowledgements The authors want to thank Luke Fox, AKT II enthusiastic and highly professional Technical Writer, for his assistance in finalising and improving the manuscript.

References

1. Figliola, A., Battisti, A.: Post-industrial Robotics. Springer, Exploring Informed Architecture (2021)
2. Lee, I.: Cloud Computing in the Age of the Fourth Industrial Revolution: New Services and Economic Acquisition Decision. Western Illinois University, Macomb (2018)
3. Rubí, J.N., de Carvalho, P.H., Gondim, P.R.: Forestry 4.0 and Industry 4.0: Use case on wildfire behaviors. Comput. Electr. Eng. **102** (2022)
4. Oke, T., Mills, G., Christen, A., Voogt, J.A.: Urban Climates. University of British Columbia, Vancouver (2017)
5. Olgyay, V., Olgyay, A.: Design with cliimate: bioclimatic approach to architectural regionalism. Princeton University Press, Princeton (1963)
6. United Nations, [Online]. Available: https://unstats.un.org/sdgs/report/2021/goal-01/. Accessed 18th April 2022
7. United Nations, [Online]. Available: https://unstats.un.org/sdgs/report/2021/goal-11/. Accessed 18th April 2022.

8. BBC, [Online]. Available: https://www.bbc.co.uk/news/uk-england-leeds-12717762. Accessed 18 April 2022
9. Irvwin, H.: A simple omnidirectional sensor for wind-tunnel studies of pedestrian-level winds. J. Wind Eng. Ind. Aerodyn. **7**(3), 219–239 (1981)
10. Durgin, F.H.: Pedestrian level wind studies at the Wright brothers facility. J. Wind Eng. Ind. Aerodyn. **44**(1–3), 2253–2264 (1992)
11. Rhyner, R., Roecker, C.: An automated heliodon for daylighting building design. In: ISES Solar World Congress. Denver (1991)
12. City of Philadelphia , [Online]. Available: https://www.phila.gov/2019-07-16-heat-vulnerability-index-highlights-city-hot-spots/. Accessed 18 April 2022
13. Derwent London, [Online]. Available: https://www.derwentlondon.com/properties/white-collar-factory. Accessed 18 April 2022
14. CIty, of London Corporation: Wind Microclimate Guidelines for Development in the City of London. City of London Corporation, London (2019)
15. City of London Corporation: Thermal Comfort Guidelines for Developments in the City of London. City of London Corporation, London (2020)
16. Littefair, P.: Site Layout Planning for Daylight and Sunlight, A Guide to Good Practice. BRE Press, Gardston (2011)
17. Institution of Lighting Professional, "The Reduction of Obstructive Light," Institution of Lighting Professional, Rugby (2021)
18. Department for Environment Food and Rural Affairs (DEFRA), "Local Air Quality Management, Technical Guidance (TG16)," DEFRA. London (2018)
19. Rohrmann, G.: [Online]. Available: https://www.smartcitiesworld.net/opinions/opinions/mini-cities-the-rise-of-tall-buildings. Accessed 16 May 2022
20. American Mutoscope and Biograph Co, [Online]. Available: https://www.youtube.com/watch?v=hhadBlokSAA&ab_channel=LibraryofCongress. Accessed 16 May 2022
21. Lawson, T.: Building Aerodynamics. Imperial College Press, London (2001)
22. Blocken, B.: Computational Fluid Dynamics for urban physics: importance, scales, possibilities, limitations and ten tips and tricks towards accurate and reliable simulations. Build. Environ. **91**, 219–245 (2015)
23. Blocken, B.: 50 years of computational wind engineering: past, present and future. J. Wind Eng. Ind. Aerodyn. **129**, 69–102 (2014)
24. Karava, P., Stathopoulos, T., Athlenltis, A.: Wind-induced natural ventilation analysis. Sol. Energy **81**(1), 20–30 (2007)
25. Karava, P.: Airflow prediction in buildings for natural ventilation design—wind Tunnel measurements and simulation—Ph.D. thesis., Montreal: Dept. of Building, Civil, and Environmental Engineering (2008)
26. Cochran, L., Peterka, L.: On breached building envelopes and increased internal pressure. In: ICBEST 2001. Ottawa (2001)
27. Walton, G.: Airflow network models for element-based building airflow modelling. AHRAE Trans. **95**(2), 611–620 (1989)
28. U.S. Department of Energy's Building Technologies Office, [Online]. Available: https://energyplus.net/. Accessed 17 May 2022.
29. U.S. Department of Commerce, [Online]. Available: https://www.nist.gov/services-resources/software/contam. Accessed 17 May 2022
30. Lamb, S., Kwok, K.: The fundamental human response to wind-induced building motion. J. Wind Eng. Ind. Aerodyn. **165**, 79–85 (2017)
31. Lamb, S., Kwok K., Walton, D.: Occupant comfort in wind-excited tall buildings: Motion sickness, compensatory behaviours and complaint. J. Wind. Eng. Ind. Aerodyn. **119**, 1–12 (2013)
32. Daniels, S., Castro, I., Xie, Z.-T.: Peak loading and surface pressure fluctuations of a tall model building. J. Wind Eng. Ind. Aerodyn. **120**, 19–28 (2013)
33. Huang, S., Li, Q.S., Xu, S.: Numerical evaluation of wind effects on a tall steel building by CFD. J. Constr. Steel Res. **63**, 612–627 (2007)

34. Daniels, S.J., Castro, I.P., Xie, Z.-T.: Numerical analysis of freestream turbulence effects on the vortex-induced vibrations of a rectangular cylinder. J. Wind Eng. Ind. Aerodyn. **153**, 13–25 (2016)
35. Thordal, M., Bennetsen, J.C., Capra, S., Holger, H., Koss, H.: Towards a standard CFD setup for wind load assessment of high-rise buildings: Part 1–Benchmark of the CAARC building. J. Wind. Eng. Ind. Aerodyn. **205** (2020)
36. Thordal, M., Bennetsen, J.C., Capra, S., Holger, H., Koss, H.: Towards a standard CFD setup for wind load assessment of high-rise buildings: Part 2–Blind test of chamfered and rounded corner high-rise buildings. J. Wind. Eng. Ind. Aerodyn. **205** (2020)
37. Daniels, S., Xie, Z.-T.: An overview of large-Eddy simulation for wind loading on slender structures. In Proceedings of the Institution of Civil Engineers—Engineering and Computational Mechanics (2022)
38. Elshaer, A., Aboshosha, G., Bitsuamlak, G., El Damatty, A., Dagnew, A.: LES evaluation of wind-induced responses for an isolated and a surrounded tall building. Eng. Struct. **115**, 179–195 (2016)
39. Daniel, F., McNeil, A.: [Online]. Available: https://www.radiance-online.org/. Accessed 19 May 2022
40. Jones, N., Reinhart, C.: [Online]. Available: http://web.mit.edu/sustainabledesignlab/projects/Accelerad/. Accessed 19 May 2022
41. BBC, 'Walkie-Talkie' skyscraper melts Jaguar car parts, [Online]. Available: https://www.bbc.co.uk/news/uk-england-london-23930675. Accessed 26 May 2022
42. Schiler, M., Valmont, E.: Microclimatic Impact: Glare Around the Walt Disney Concert Hall. University of Southern California, Los Angeles (2005)
43. Suk, J.Y., Schiler, M., Kensek, K.: Reflectivity and specularity of building envelopes: how materiality in architecture affects human visual comfort. Arch. Sci. Rev. **60**(4), 256–265 (2017)
44. Jakubiec, J.A., Reinhart, C.F.: Assessing disability glare potential of reflecrtions from new construction. **2449**, 114–122 (2014)
45. Hassall, D.N.H.: Reflectivity: Dealing with Rogue Solar Reflections. University of New South Wales, Faculty of Architecture (1991)
46. Gil, V.L.-R.: Evaluation of solar glare from reflective facades: A general method. Light. Res. Technol. **48**, 512–538 (2015)
47. Ho, C.K., Ghanbari, C.M., Diver, R.B.: Methodology to assess potential glint and glare hazards from concentrating solar power plants: analytical models and experimental validation. J. Solar Energy Eng. **133** (2011)
48. Barker, D.: Immersive experiences of building physics analysis to improve human connection to technical data. Energy Procedia **78**, 507–512
49. Littlefair, P.: Solar Dazzle Reflected from Sloping Glazed Facades. Establishment, Building Research, London (1987)
50. City of London, Solar Glare: Guidelines for best practice for assessing solar glare in the City of London. City of London (2017)
51. Singapore Building and Construction Authority, Regulation on Daylight Reflectance of Materials Used on Exterior of Buildings. Building Plan and Management Group (2016)
52. Federal Aviation Administration, Technical Guidance for Evaluating Selected Solar Technologies on Airports. Washington (2018)
53. Campaign to Protect Rural England: Night Blight: Mapping England's Light Pollution and Dark Skies. Campaign to Protect Rural England, London (2016)
54. Höppe, P.: The physiological equivalent temperature—a universal index for the biometeorological assessment of the thermal environment. Int. J. Biometeorol. **43**, 71–75 (1999)
55. Gonzalez, R., Nishi, Y., Gagge, A.: Experimental evaluation of standard effective temperature a new biometeorological index of man's thermal discomfort. Int. J. Biometeorol. **18**, 1–15 (1974)
56. Sheng, Z., Zhang, L.: Standard effective temperature based adaptive-rational thermal comfort model. Appl. Energy **264** (2020)
57. Brode, O., Fiala, D., Blazejczyk, F.I.H., Jendritzky, G., Kampmann, B., Tinz B., Havenith G.: Deriving the operational procedure for the universal thermal comfort index UTCI. Int. J. Biometeorol. 1–92 (2011)

58. Teli, D., Jentsch, M., James P., Bahaj, A.: Field study on thermal comfort in a UK primary school. In: 7th Windsor Conference: The Changing Context of Comfort in An Unpredictable World. Windsor (2012)
59. Government of Canada, "Glossary," [Online]. Available: https://climate.weather.gc.ca/glossary_e.html#w. [ccessed 25 May 2022.
60. Government of Canada, "Glossary," [Online]. Available: https://climate.weather.gc.ca/glossary_e.html#h. Accessed 25 May 2022
61. Cost Action 730, "730—Towards a Universal Thermal Climate Index UTCI for Assessing the Thermal Environment of the Human Being," [Online]. Available: https://www.cost.eu/actions/730/. Accessed 25 May 2022
62. Fiala, D.: A computer model of human thermoregulation for a widerange of environmental conditions: the passive system. Model. Physiol. 87(5), 72–199 (1957)
63. Mandavilli, A.: The World's Worst Industrial Disaster Is Still Unfolding, [Online]. Available: https://www.theatlantic.com/science/archive/2018/07/the-worlds-worst-industrial-disaster-is-still-unfolding/560726/. Accessed 26 May 2022.
64. UK Government, "The Air Quality Standards Regulations 2010," [Online]. Available: https://www.legislation.gov.uk/uksi/2010/1001/contents/made. Accessed 25 May 2022.
65. Cambridge Environmental Research Consultants, "ADMS 5," [Online]. Available: http://www.cerc.co.uk/environmental-software/ADMS-model.html. Accessed 26 May 2022
66. Xie, Z., Voke, P., Hayden, P., Robins, A.: Large-Eddy simulation of turbulent flow over a rough surface. Bound.-Layer Meteorol. **111**, 417–440 (2004)
67. Sessa, V., Xie, X., Herring, S.: Turbulence and dispersion below and above the interface of the internal and the external boundary layers. J. Wind Eng. Ind. Aerodyn. **182**, 189–201 (2018)
68. Sessa, V., Xie, Z., Herring, S.: Thermal stratification effects on turbulence and dispersion in internal and external boundary layers. Bound.-Layer Meteorol. **176**, 61–83 (2020)
69. DEFRA, "Road traffic statistics," [Online]. Available: https://roadtraffic.dft.gov.uk/#6/55.254/-6.053/basemap-regions-countpoints. Accessed 26 May 2022
70. Council, D.: Tall Buildings, Advice on Plan-Making, Submitting, Assessing and Deciding. Design Council, London (2014)
71. Greater London Authority: Sustainable Design and Construction, Supplementary Planning Guidance. Greater London Authority, London (2014)
72. Tower Hamlets, "Wind Impact Assessment," [Online]. Available: https://www.towerhamlets.gov.uk/lgnl/planning_and_building_control/planning_applications/Making_a_planning_application/Local_validation_list/Wind_Impact_Assessment.aspx. Accessed 22 May 2022
73. Holmes, J., Tse, K.: International high-frequency base balance benchmark study. Wind Struct. 18 (2014)
74. Roland Schregle, C.R.S.W.: Spatio-temporal visualisation of reflections from building integrated photovoltaics. Buildings **8**(8), 101 (2018)
75. Ryan Danks, J.G.R.S.: Assessing reflected sunlight from building facades: A literature review and proposed criteria. Build. Environ. **103**, 193–202 (2016)
76. City of London Corporation.: Solar Glare: Guidelines for best practice for assessing solar glare in the City of London. City of London (2017)

Lotus Aeroad—Pushing the Scale of Tensegrity Structures

Matthew Church and Stephen Melville

Abstract The Lotus Aeroad sculpture was a focal point for the annual Goodwood Festival of Speed. Research shows it is the longest tensegrity cantilever built to date. The concept for the structure was for a scheme which echoed the philosophy of Lotus Cars, that of lightweight, pared-down design. Although the structure is simple in that it is composed of pure axially loaded struts and ties, the structural design was highly complex. Format worked exclusively within parametric environments to progress the scheme from initial form finding to the automatic production of final fabrication information. This allowed for flexibility in the design until the point when the material was ordered. We worked in conjunction with the Artists to form-find a stable tensegrity which matched the artistic intent. The global system was analysed using K2E—an parametric plug-in for Rhino-Grasshopper, developed in-house. This analysis was performed iteratively to optimise the sections used for the struts and ties to reduce the structural weight to an absolute minimum. Any unnecessary mass would require further strengthening and stiffening, adding even more material. The resulting structure was as lean as possible. The parametric workflow included a connection design script which designed the joints whilst the overall structural form was still being finalised, helping the design keep pace with the strict project deadline. Finally, this project was a test of using augmented reality at building scale. Using augmented reality software it was possible to compare the calculated positions of the elements in the 3D model to their positions on site.

Keywords Tensegrity · Steel structure · Lean design · Augmented reality

United Nations' Sustainable Development Goals 9. Build resilient infrastructure, promote inclusive and sustainable industrialization and foster innovation · 12. Ensure sustainable consumption and production patterns

---

M. Church · S. Melville (✉)
Format Engineers Ltd, Second Floor, Riverside North Building, Avon Wharf, Walcot Yard, Bath BA1 5BG, Bath, England
e-mail: sm@formatengineers.com

1 Introduction

The Lotus Aeroad sculpture is the latest centrepiece to be developed for the annual motor sport and classic car festival, Goodwood Festival of Speed, in Chichester, United Kingdom. Unlike previous years where artist vision was paramount, the original concept for the structure was to develop a scheme which was overtly lightweight, visually logical and 'readable' but also an illustration of high technology lead design. It was to be an embodiment of the ethos of the founder of Lotus Cars, Colin Chapman. Lotus were the sponsor and ultimate client for the piece. They have a long history of efficient and lightweight engineering design, which transcends motor sport. Colin Chapman's rationale was very pertinent to the discipline of Structural Engineering and to the Aeroad in particular, namely; 'Simplify, then add lightness' [1] (Fig. 1).

The sponsors appointed production artists Unit 9 to design the piece. Format Engineers were then appointed directly to Unit 9 who then quickly arrived at the choice of a tensegrity structure as that was a very visual example of lean design made real through high technology and one which was felt to be under explored as a frame typology. The Client, Lotus Cars and the organisers of the festival were keen that the structure cantilevered from existing foundations at the centre of an elliptical lawn and over a perimeter pathway used for public circulation. They also wished to push the size of the cantilever as far as possible within a set budget, on the way exceeding the previous record for a tensegrity cantilever.

Fig. 1 The lotus aeroad sculpture outside goodwood house. *Photo Credit* Crate47

Although the structure is, in many ways, simplistic in that it is made from an arrangement of pure struts and ties, the structural design process was on the opposite side of the complexity scale. As with many festival structures the timescale for design exploration, detailed analysis and production of fabrication information was extremely tight. In this instance engineering appointment was in late March 2021 whilst the piece was to be in-place mid June. Because of the very short time scale and their previous experience of working on this particular site, the steel fabricators Littlehampton Welding Limited were appointed early on to work alongside Format Engineers and Unit 9.

From the outset parametric design was used so that structural exploration and form exploration could remain fluid. This meant that, like sculpting in clay, the design could be moulded and tweaked into the artist's vision. Many different forms and iterations could be quickly modelled and presented to the Artist for their assent or to provoke debate on the direction of the design. Another key factor in the use of parametric design was the speed at which the design process could progress. Several different processes could be performed concurrently. For example, whilst the global geometry was still to be finalised, the scripts which designed and checked the connections could be written so the final geometry could be "plugged-in" once it was completed. Format Engineers also scripted the automatic generation of steel fabrication drawings from the final model, another task which could be undertaken in advance of the Artist 'signing off' the final form. Another benefit of the parametric workflow was that many different iterations of foundation support locations could be tested and the support reactions tested against the capacity of the existing foundations. This was important as it avoided the construction of new footings with the associated cost, time and environmental impact.

Although creating parametric scripts takes time up front compared to performing the same process manually, the geometry can then be assessed and altered and reassessed in a matter of minutes once it has been coded. As previously noted multiple variations of form and material volume can be explored and presented to the client.

The structural design of the superstructure was broken down into two distinct parts: the 'Global Structural Design', and the 'Connection Design'. The global structural design focused on determining the overall form which primarily consisted of the node locations and required strut and cable sections as outputs. The connection design consisted of accurate determination of the geometry of the joints plus the detailed finite element analysis of the mild steel plates and tubes to which they were welded. These processes are described in more detail later in this chapter.

The extreme efficiency of the Aeroad tenesgrity structure has been made possible by low cost but powerful computational design techniques. Technology is driving the refinement of low carbon design. It allows designers to explore form in a holistic way. The trade offs between competing factors of shape, material strength and stiffness no longer need to be approximated, they can be balanced accurately. The promise of the fourth industrial revolution of democratisation of machines and machine techniques has led to a rapid take up parametric design tools within the AEC industry and most importantly the ability of practitioners to develop those tools for themselves and not be reliant upon software houses or large scale technology companies. Our workflow for the Lotus Aeroad is a great example of that.

As will be seen later in this paper, the gap between a digital model and the real world was bridged with augmented reality, another example of the premise of Industry 4.0. The power of virtual reality and augmented reality tools lies in the mainstream adoption by design practices who are then able to explore the limits of form and construction of form. This democratisation can only serve to benefit further design practice.

2 Concept Design Aspirations

Our brief at the outset of the project was succinct:

- To design a structure which embodied the ethos of Lotus Cars, i.e. 'Simplify, then add lightness'
- Unit 9 wanted the structure to be 'made more from air than any other material'
- That structure should be as lean as possible with regard to material use as this is a responsible aspiration for any design but also one which would very visibly lead the way in inspiring other designers to do the same.
- To design a tensegrity cantilever.
- The cantilever should exceed the span of any previously built tensegrity.
- The structure should be elevated above the ground, both to prevent the public from climbing it and to emphasise the 'floating' cantilever.
- Whenever possible re-use the foundations from previous years Festival of Speed. New foundations would absorb too much of the budget, take too long to build and would add embodied energy for no discernible benefit.
- In addition to this our role was to ensure that the structure did move or deflect in an uncomfortable looking way, it was safe to build and it could withstand storm force winds.

Materiality

Many different materials were considered during the initial design phase. Timber compression elements were considered as were stainless steel, aluminum and carbon composite tubes. Alternative tie members under consideration included nylon ropes or solid bars. The final choice of stock structural mild steel tubes and stainless steel wire with stainless steel fork ends was made for a number of reasons. The materials were easily available in the limited time available for the design and fabrication process, they had a high strength and stiffness and they could be cut, welded and paint finished by the steel fabricator, Littlehampton Welding Ltd in one location thus eliminating supply chain risks as far as possible.

Given more design research time and if building at a smaller scale we would definitely consider material options such as bamboo or roundwood timber. We would also consider re-used or reclaimed steel. Tube sections especially are becoming more widely accepted and available in the construction industry, therefore, structures of this nature are primed to take advantage of this to achieve a significantly reduced environmental impact. Re-used stock was not used at the time due to the uncertainty of material strength and the limited time in which to organise and report material strength testing. Although it was not a key factor in the design the client has intimated that the sculpture could be re-erected as a permanent feature. With the limited time available we were not able to complete research into the provenance of the right size of reclaimed tubes and hence their fatigue resistance.

New foundations were avoided hence no new concrete was needed in the ground, only high strength grout under the new steel baseplates.

Historical Precedents

Tensegrity structures have a well-documented history with the major instigators in the field being published and quoted in some detail. Russian/Latvian artist Karlis Johanson [2] is credited with being a pioneer in this field. His 'self-tensile constructions' of the 1920s were small scale experiments in pure tension/compression frames. Johanson's work was popularised in the 1960' by Buckminster Fuller [3], who coined the phrase 'tensegrity as a amalgam of 'tensional integrity'.

Buckinster Fuller's student, the artist Kenneth Snelson, was responsible for realising a number of large scale tensegrity works and these were the reference points at the outset of the Lotus Aeroad project. Snelson's Needle [4] and Needle 2 [5] towers are 26 and 30 m high respectively and were structural span lengths that the Lotus team were determined to exceed.

The Schlaich Bergermann and Partner Tower at the Fair of Rostock [6] is 62.3 m tall but this is not a tensegrity structure in the classic sense. Compression members are joined end to end rather than suspended via tension cables and hence this structure, whilst undeniably striking, is discounted from the cannon of tensegrity forms.

Form

The shape of the structure changed considerably from the artist's original intent of a long, linear tensegrity elevated above ground and supported by rows of columns, to the final solution of the dramatic long span cantilever. That development came about following an intensive period of design investigation and dialogue between Artist and Engineer. As previously discussed, the aim from the outset was to push a long-supported span hence the supports were moved to the rear of the structure. Several different iterations of supports were discussed before the adoption of a tensegrity 'cradle' which held the main body of the work. A small back span was included, not necessarily to balance the structure but for aesthetic purposes.

The Artists were keen that the structure was not perceived as a regular, uniform tube but had a degree of visual 'randomness' or jitter along the length. This was accomplished within the parametric script with the addition of a function which randomly pushed frame nodes away from the perimeter of the tensegrity. In addition to an element of irregularity the both Engineers and Artists were keen that the structure be perceived as tapering in cross section towards the tip. This helped further reduce material and hence mass where it was least needed.

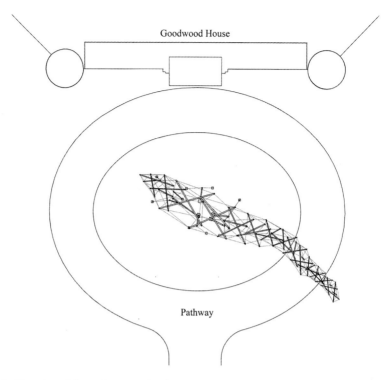

Fig. 2 Plan view of the sculpture on the site

By entirely generating the geometry within a parametric script multiple iterations of form, size and support location could be tested for visual appearance and material volume within a very constrained timescale. With these tools the project would not have been realised.

Size

The structure is 21 m tall from ground to tip of cantilever and 45.5 m long from tip to tip. It is 8.2 m wide at the root of the cantilever above the support cradle and 2.6 m wide at the tip. Plans and elevations are shown below (Figs. 2 and 3).

3 Global Structural Design

The global structural form was generated through an iterative development with two sub-operations. In the first operation the desired form was sculpted and stabilised through form finding, then in the second operation the form and the individual members were checked to the code of practice to increase its structural performance, namely by decreasing the movements under self-weight and wind loading

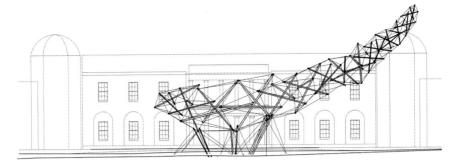

Fig. 3 Elevation view of the sculpture in front of goodwood house

and providing sufficient strength to prevent collapse. The output of this was then exported to the next stages where the connection design and fabrication information was created.

Form Finding

The first step in creating the 3D geometry was to use a form finding procedure which could produce a stable tensegrity form which could also be sculpted through the parametric inputs. This allowed the sketches and ideas from the design team meetings to be turned into a 3D "stick" model which could then be shown, discussed, and then modified to slowly mould the form.

This sculpture-like process was scripted into code within Rhino + Grasshopper. This script was made of several steps which gradually built up to create the final topology. The steps are outlined in the list below and in the Fig. 4:

1. Tensegrities can be created from a huge range of different topologies—Marcelo Pars has explored many different forms and topologies [7]. Early in the design process it was decided that Aeroad would be created by combining multiple "tripod" units together to create the cantilever.
2. The tripod units are stacked on top of each other to assemble a chain of prismatic cells.
3. The full length of the tensegrity was created by stacking 15 units together. To add additional stiffness, more cables are added in each cell connecting one end of each strut to another in the next cell along; this was key as it added robustness to the design. Without these the structure would collapse if a single cable was de tensioned or removed.
4. Once the topology was set, the form was then relaxed using Kangaroo 2 [8] a physics based solving algorithm within Rhino + Grasshopper. This allowed for a stable form to be produced by adding a gravitational force to the element vertices and forcing the cable lengths to shorten—the process of this is very similar to the form finding of cable net structures but with more flexibility in the resulting geometry.

Fig. 4 Development process of the sculptural form

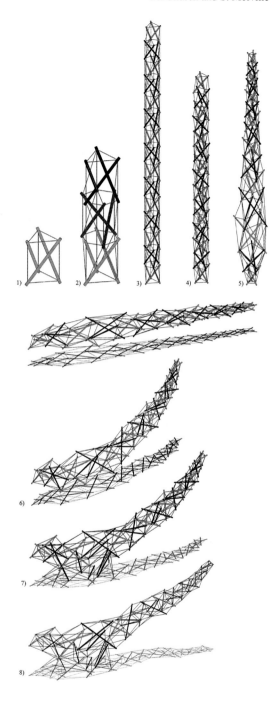

5. The form was then modified to increase the spatial volume around where the structure would be supported and create a taper towards the tip as well as adding some "noise" to create a more unpredictable or 'random' feeling in the sculpture in line with the Artists instructions. This was achieved by using the same Kangaroo 2 algorithm as step 4 but increasing or decreasing the strut and cable lengths in specified regions.
6. The final step in creating the sculpted form was to add the desired curvature of the cantilever. The Kangaroo 2 algorithm was again utilised with additional forces to push and pull the structure into the desired shape.

Initial Global Structural Analysis

The next stage in the development of the global form development took the stable idealised form from the previous stage and altered the cable lengths to create a stiffer structure which also had sufficient prestress to prevent the cables going slack during regular everyday wind events. The structural elements could also be sized so that they would resist the forces which would flow through them during storm events. To speed up this step, only the wind directions which caused the largest deformations were considered as the other directions resulted in an insignificant impact on the overall impression of the structure.

As previously explained the artists were keen that the structure appeared to 'float' off the ground and that the public were deterred from climbing it. A base structure was therefore included in the design which was also to be of a tensegrity form but one which was visibly distinct from the main body of the cantilever—see step 7 on Fig. 4. The base structure is added into the model before the analysis is performed. The base consists of another tensegrity producing a cradle around the four lowest strut ends. These struts and cables are pin supported where they touch the ground meaning that the struts and cables carry only tension and compression forces and ensuring that the entire structure is a pure tensegrity.

To achieve this the structure was analysed using K2Engineering [9, 10]—a parametric plug-in for Rhino-Grasshopper which is an additional set of tools for the Kangaroo 2 solver that accommodates an easy parametric input and output of engineering data such as section sizes and forces. This enables a smooth analysis process for structures which exhibit non-linear behaviour and structures with relatively large movements. These two aspects are essential to capture for tensegrity structures as the cables are non-linear elements because they can only resist tension and exhibit large displacements due to a relatively high flexibility.

Initially the model was analysed with an estimated section for each strut and cable so the dead load and wind load onto each member could be automatically calculated and then fed back into the analysis. The maximum force through each strut and cable could be determined. These forces were then used to calculate the smallest sections that met the strength requirements for each member. Using these new sections the analysis could then be repeated. The process was repeated three times to find the optimal cross section assignment for structural strength—the final output of this process is shown in step 8 of Fig. 4.

Using the results of this the deflected form of the structure under self-weight was then outputted and used for discussions with the design team to assess how it could be changed to better fit to the original aspiration for the form. The reason for doing this was that due to the prestress in the structure and its self-weight it had a different feel to the output from the form finding stage. The shape was then altered by changing the strut and cable lengths in the form finding analysis. This process produced a stable and structural efficient form with a minimum use of material that met the artistic vision.

Complete Global Structural Analysis

In the previous stage the most critical aspects of the structural analysis were performed which included the deflection and element stress under self-weight and the wind loads causing the largest lateral movements at the tip of the cantilever. In the final global design sub-process the remaining load combinations were assessed. The reason for excluding these from the second sub-process was to decrease the analysis duration as these simulations had a long running time due to their non-linearity; doing this meant that the design iterations could be performed quicker. This process only required some small increases to the strut and cable sizes due to higher stresses experienced in the new loading conditions and had only a minor impact on the overall form.

Also, in this process the structure was assessed for the high wind strategy. The purpose of the high wind strategy was to reduce the loads the structure would experience in the event of an unusually high wind by attaching two ties at a point roughly halfway along the cantilever when it was predicted that the wind speed would exceed a determined limit. This meant that the sculpture only needed to resist the peak design wind speed while it had additional supports and therefore kept the structure as lightweight as possible.

4 Connection Design

The purpose of the connection design script was to deal with all the complexity inherent in the form as the elemental simplicity of the structure required each connection to be bespoke. The core idea behind the connection design was to create the nodes from modular pieces which could be welded together in a repeatable process. To do this every tie type had an associated fin plate design. Therefore, the nine cable types plus the two bar types (which were needed as two elements were shorter than the minimum fabricable cable length) needed 11 fin plate geometry types. The fin plates and their associated welds were designed for the maximum tensile capacity in every feasible direction; this meant that any individual connection likely used more material than was necessary, but it led to a much easier fabrication process and reduced the risk of any confusion between connection geometries. In any case, the fin plates remained partially optimised as cables with lower forces would have a small cable diameter which would in-turn result in a smaller fin plate. The resulting fin plates are shown in Fig. 5.

Although the design of the 502 fin plates could be collected into 11 groups which needed designing, the same could not be done for the 102 strut ends. This is because every strut end had a unique combination of fin plate sizes, forces, and inclination angles of the cables. To overcome this every connection needed to be assessed individually by creating a finite element model for every node.

The connections were designed following the process below and in Fig. 6:

1. The first analysis stage was to analyse the connection with the fin plates attached to the strut section with the forces from global analysis. The cap plate which was already required for architectural and finishing reasons was also included in the model to restrain the end of the tube from warping. If the material was shown

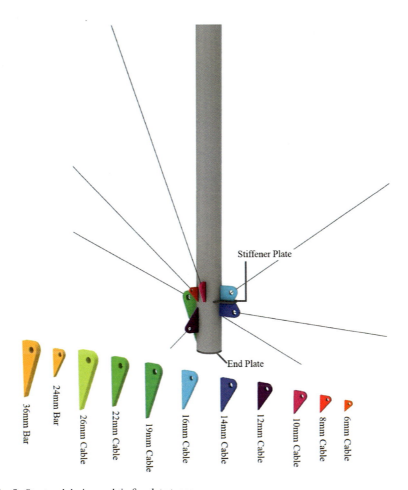

Fig. 5 Strut end design and tie fin plate types

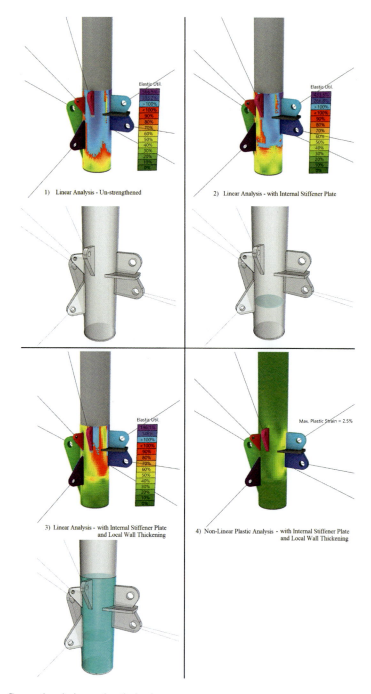

Fig. 6 Connection design and optimisation process

to be below the yield limit, then the connection was determined to be sufficient, if this was not the case then the design went to the next stage.
2. The second stage was to perform the analysis again but including an internal stiffening disc which the contractor confirmed could be welded within the tube by a distance equal to the inner tube diameter. If the material was still yielding, then the connection went onto the third stage.
3. In this stage the same analysis was performed a third time, but the strut element was locally thickened at the end which could be fabricated by welding a thicker CHS section in the region of the connection. If the material was still yielding in this stage, it went onto the final step.
4. The final analysis used the same geometry as step 3 but analysed it using a non-linear solver which could capture the plastic yielding of the material. Therefore, the areas which saw high local stresses would instead yield distributing the force to the area nearby. To check that the connection did not fail from excessive plastic deformation, the von mises strain was checked against the steel plastic strain limit.

The process allowed for connections to be designed in a way in which the material was only added when and where it was needed. For example, if the connection only needed a small amount of stiffening then the internal disc stiffener would be added, if it needed to be a bit stronger then the tube was thickened but only in the region of the connection. This meant that the structure used as little material as was possible for the fabrication techniques employed. Although it may at first appear that optimising the connection design would have had little effect on the total steel weight in the structure, this is not the case in reality. As the sculpture is only loaded by self-weight and wind (which is highly dependent on the cable and strut diameter) any increase in the connection weight may result in higher force through one or more elements which then also increases the self-weight and wind load. This detrimental cycle occurs as the global structural analysis is heavily optimised for structural mass meaning any changes in the loading change the output results. This is a key challenge to acknowledge when optimising a structure to reach the upper limit of material utilisation, as there is no spare capacity for any increases.

5 Fabrication

The previous steps in the design process came together to produce a 3D fabrication model which captured every strut, cable, fin plate, and stiffening element. The final step was to translate this into information the fabricators could use with their typical fabrication techniques to produce each element.

The stainless-steel cables were fabricated by Jakob Rope Systems who produced the cables to the exact lengths required for the structure. The purpose for doing this was to avoid having to adjust the cable lengths in situ as this would have been practically impossible and had been likened to tuning a piano but where tensioning any one string affected the tension in every other string. Obtaining the correct lengths may seem like a simple task at first, however the fabrication model was based on the final geometry once it had been loaded with its self-weight. This meant that the cable lengths in the fabrication model already accounted for some elastic extension. Therefore, so that the cables were the correct length when they arrived on site, Jakob Rope Systems used the loaded cable lengths and load under self-weight to calculate the resting length of the cables.

The struts were fabricated by Littlehampton Welding who ordered the laser cut tube sections and plates and welded them together to form the completed strut elements. The specification of the tube and plate parts were directly outputted from the 3D fabrication model and sent to the cutters. The instructions for how the parts should be assembled to create each strut had to be given via 2D drawings which the fabricators could read from. To achieve this Format Engineers created an automated script which could take the final geometry of each strut and automatically populate it with 2D views that showed every dimension and component tag that made up the struts. This script was created in Grasshopper making use of the EleFront plugin [11] which could access the Rhino 2D drawing tools within grasshopper. Using automation in this way removed hours of tedious CAD drawing work and enabled the complete set of fabrication information to be issued in a much shorter time frame.

6 On-Site

The sculpture arrived on site as a complex assembly of 302 unique strut and cable pieces which needed to not only be connected correctly but this also had to be done while the structure was partially unstable and up in the air (Fig. 8). This is akin to an unstable 3D jigsaw puzzle that needs many of the elements to be in place before it stands under its own weight. Prior to the contractor arriving on site, we had already discussed with them an erection sequence which required least temporary support whilst the structure was unstable. That erection sequence was modelled using segments from the global 3D model and illustrated in a 2D drawing format as a series of cantilever assembly steps as well as the 3D model being issued to the Contractor.

Even with a highly detailed set of drawings and a 3D model the erection proved to be a challenge for the contractor. Because cables run in front of and behind tubes at different angles with empty 3D space the structural pattern is difficult to read from a computer screen and even more so from a flat 2D drawing. In order to help with the process, staff from Format Engineers attended site with copies of the 3D model within the augmented reality (AR) software Fologram [12]. This enabled the 3D model to be viewed on the tablet with the real world as a background (see Fig. 7). The correct setting out the elements could be easily compared with the as built arrangement and corrections made before the error had propagated too far. This was an opportunity to show a typical steel fabricator and contractor how AR could be used to streamline and smooth the process of the construction, which they were fascinated by.

Fig. 7 Comparing the Real and Calculated Position in AR

It's important to note that even with many steps in place errors still crept in and a significant impact of this was that a number of cables remained slack when it had been fully assembled, even though the structure was designed so that every member would be under stress in its final position. Its still unclear what the exact causes of this were but some hypotheses are given below:

- There may have been some slight deviations between the designed base plate locations and the exact locations on site.
- One of the fin plate types was specified with an oversized bolt hole by mistake, but also for tolerance reasons every bolt hole was at least 2 mm larger in diameter than the bolt which was not accounted for in the structural analysis.
- Some of the cables could have been overstressed during assembly which caused them to "bed-in" more than predicted.
- The exact strut lengths, positions of the fin plates, or cable lengths may have been off due to manufacturing tolerance or error.

None the less, the structure was re-analysed including for the slack cables, and it was deemed that it still met the structural criteria as there were extra cables added for robustness. If a tensegrity of this scale and complexity is constructed again in the future special care should be taken to avoid these possible problems, or this issue should be mitigated by allowing the cables to be adjusted on site.

Fig. 8 The final part of the sculpture being lifted into place. *Photo Credit* The Goodwood Estate Company Limited

7 Conclusions

The Lotus Aeroad is a prominent example of a tensegrity structure, its huge length and height belies the amount of material required to span that distance. All the parts of the structure when packed together only take up one half of a standard shipping container. Whilst the elements are simple, being only circular hollow section struts and stainless-steel cables in tension plus small amounts of steel plate at the connection all of which are simple to fabricate the lessons learnt on the Lotus Aerod project are that the structural form finding and the analysis are extremely complex and that the construction of the piece is challenging.

Parametric workflows are crucial to the form finding and the optioneering. Whilst there have many examples of small scale maquettes [7] and demonstration models [13] which have been used to finalise forms a computer based generative environment can cycle through thousands of options, allowing the designer to take the role of 'puppeteer', changing input parameters which may influence the form and getting instant feedback on the appearance. It also enables feedback on structural performance to take a prominent role in the shape making process. This is an important consideration for tensegrity structures where every kg of unwanted material has an over-amplified feedback on the performance. This process within the parametric environment is greatly enriched by the ability to include physics-based solvers within the analysis. The Kangaroo 2 physics solver and the Format Engineers in-house structural analysis toolset built on top of that were critical tools in the development of stable forms and instant feedback on structural performance. A physics based 'kernel' within the design and analysis toolkit was essential as it allowed for the accurate simulation of pre-stress and deflected shape.

Without the parametric based toolset the time scale would not have allowed for any exploration, rigorous design and detailing for practical and safe erection. The latter was particularly important as it was found very early on in the build that the sensible sequence of erection once past the base was not obvious. The 3D model could be modified to reflect different stages of build and check the stability. This and the use of the Fologram augmented reality plug-ins meant that after several false starts the erection was able to proceed at pace and meet the deadline imposed by the festival.

A comparison of different cantilevering structural typologies and their length to weight ratio compared to Lotus Aeroad is outside of the scope of this paper (but would be a useful exercise). However we believe that the tensegrity structure could be demonstrated to be the least volume of material for the span compared to all steel section trusses or other similar more 'conventional' solutions. In this respect the Aeroad was an important step in determining the viability of tensegrity forms in real world scenarios. We now appreciate that although the form can be extremely light and material efficient the analysis of the structure is extremely complex and great care and careful planning is needed in its construction. This type of construction could be suitable for tall and long structures such as communication towers, lightweight enclosures, roofs and bridges, and should be considered more often at the early optioneering stages of design.

References

1. Colin Chapman. https://media.lotuscars.com/en/heritage-people/lotus-heritage-colin-chapman.html
2. Karlis Johansons. Wikipedia. https://en.wikipedia.org/wiki/Karlis_Johansons
3. Buckminster Fuller. https://patents.google.com/patent/US3063521
4. Needle Tower. Wikipedia. https://en.wikipedia.org/wiki/Needle_Tower
5. Snelson, K.: Needle Tower II. Kennethsnelson. [Online] http://kennethsnelson.net/sculptures/outdoor-works/needle-tower-ii/
6. Tower at the Fair of Rostock. sbp. [Online] https://www.sbp.de/en/project/tower-at-the-fair-of-rostock/
7. Marcelo Pars. TENSEGRITY. http://www.tensegriteit.nl/index.html
8. Piker, D.: KANGAROO PHYSICS. food4Rhino. [Online] https://www.food4rhino.com/en/app/kangaroo-physics
9. K2 Engineering. Format. https://formatengineers.com/research/k2engineering.html
10. Brandt, C.: K2Engineering. Github. [Online] https://github.com/CecilieBrandt/K2Engineering
11. Rahimzadeh, K., van der Heijden, R., Tai, A.: ELEFRONT. food4Rhino. [Online] https://www.food4rhino.com/en/app/elefront
12. Fologram. https://fologram.com/
13. Snelson, K.: http://kennethsnelson.net/

Data-Driven Performance-Based Generative Design and Digital Fabrication for Industry 4.0: Precedent Work, Current Progress, and Future Prospects

Ding Wen Bao and Xin Yan

Abstract With the development of computing technology, architectural design has been impacted and changed significantly over the past decade. It led us to rethink the new design methodology and application, such as the data-driven performance-based design method and its relevant digital fabrication for Industry 4.0. The paper explores the theories and practices of "Overall Structure Performance Data-Driven Design" and "Swarm Intelligence-Based Architectural Design" by collecting, reviewing, and analyzing cutting-edge design methodologies and proposes a new algorithm framework that combines performance data with agent-based modelling for design. The paper demonstrates the original process and iterative argument affiliated with the "Multi-Agent-Based Topology Optimization" (MATO) method, as proposed by Bao and Yan in 2021, which has the potential to provide a new path for the future computational design of buildings. Finally, the paper concludes with an analytical study and future expectations for complex bionic morphology digital fabrication generated by the related methodology above.

Keywords Generative design · Swarm intelligence · Topological optimization · Robotic fabrication · Data-driven performance · Additive manufacturing

United Nations' Sustainable Development Goals 9. Industry, Innovation and Infrastructure · 11. Sustainable Cities and Communities · 12. Responsible Consumption and Production

---

D. W. Bao (✉)
School of Architecture and Urban Design, RMIT University, Melbourne, VIC 3000, Australia
e-mail: nic.bao@rmit.edu.au

X. Yan
The Future Laboratory, Tsinghua University, Beijing 100084, China
e-mail: yanxin2022@tsinghua.edu.cn

1 Introduction

In modern engineering, computer-aided design techniques such as Finite Element Method (FEM) and Computational Fluid Dynamics (CFD) have become essential tools for analyzing and evaluating architectural designs with accurate and rapid data feedback. Through quantitative analysis using simulation calculation technology, changes to each design in the architectural design process caused by structural, sound, light, heat, wind, and other performance factors can be timely and accurately assessed. Many software programs based on these computational techniques play significant roles in designing and implementing complex architectural projects. However, most of these programs are used for analyzing and correcting completed building layouts, and their performance data is primarily utilized for analyzing and processing existing models.

On the other hand, with the development of digital architecture theory, biomorphic model-based form-finding technology is gradually moving from pioneering design to practice. Swarm Intelligence algorithms are increasingly being used by architects and researchers in architectural design due to their simple logical architecture, excellent portability, and wide application prospects. Swarm Intelligence is a formal simulation logic of complexity that originates from flocks of birds, swarms of bugs, human social grids, and urban operational systems [1]. By applying the behavioural logic of organisms in nature to buildings, architects can enable intelligent buildings to exhibit biological and environmental response strategies. However, the application of swarm intelligence algorithms to architectural design is still in the experimental stage, and this dynamic design logic often fails to converge in a final design due to the lack of data feedback from the building itself and its environment.

In this context, the research of applying the performance data of the building on the multi-agents to realize the performance data-driven bionic computational architectural design will be helpful to break through the above dilemma and provide a new way of thinking for future architectural design. Bionic computational design based on performance data is becoming increasingly popular, and emerging architects are introducing multiple computer numerical control (CNC) or robotic construction tools to realize these designs and using building information modelling systems to manage and manufacture building prefabrication. Computational design methods and advanced manufacturing processes will gradually update the paradigm of architectural design and more accurately use data to enhance the performance of future buildings.

2 Structural Performance Data-Driven Design

2.1 Architectural Computational Modeling and Performance Data Feedback

Since the 1960s, computer-aided simulation methods such as finite element methods and computational fluid dynamics have gradually been applied in many engineering fields. We can establish an interrelated path between complex building geometry models and building performance analysis to achieve quantitative feedback on building performance through computer graphics and related scientific theories. The overall logic of this workflow can be divided into three steps: model discretization, setting input parameters, and computational simulation analysis.

First, after the architect obtains a conceptual geometric model based on the design intent, computer graphics technology is used to discretize the complex form into more regular elements to obtain a computational analysis model; this computational model shares the same nodes between neighbour elements to transfer performance data. For different analysis objects, different types of elements can be selected for model discretization: one-dimensional truss elements subject to axial forces only (corresponding to a truss structural system) and beam units considering bending moments (corresponding to the beam-column structural system), while more complex 3D models, there are shell elements overlaid on the surface and solid elements filled inside the continuum. Then, only the model property coefficients (material properties) and environmental condition parameters (boundary conditions) need to be set according to the actual situation to enter the computational analysis simulation process. Through discretization, the analytical difficulties caused by the complex geometric model can be replaced by a large number of relatively simple theoretical numerical calculations, and the architect can then modify the solution through performance data feedback until the requirements are met.

This workflow, shown in Fig. 1, essentially follows the paradigm of "post-rationalized geometries" [2], which involves optimizing an existing free-form design into a reasonable alternative model. The advantage of this approach is that each profession is less dependent on the others and can work independently. However, this independence can result in a lot of repetitive cycles of revision during the architectural design process. At the same time, the linear workflow inevitably overlooks many possibilities of morphological solutions due to the lack of overall control over the performance data of all aspects of the building.

With the development of digital architecture theories and technologies, some approaches to building morphology generation aimed at "pre-rationalized geometries" [2] are gaining the attention of architects and scholars. These methods use performance data as the basis and change the above one-way linear modification mode from "geometric model-performance analysis-manual modification" to "geometric model-performance analysis-computer reconstruction" through relevant theoretical and procedural algorithms. The "geometric model-performance analysis-computer reconstruction" approach realizes efficient linkage between building form

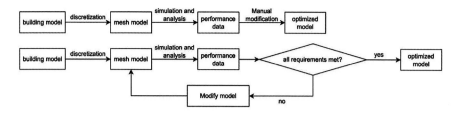

Fig. 1 Comparison between workflows of "Post-rationalized Geometries" and "Pre-rationalized Geometries"

revision and building performance evaluation, making it easier to find the "preferred solution" of building form that can meet various needs simultaneously.

2.2 Performance Data-Driven Architectural Design and Form-Finding

As one of the most significant aspects of architecture that affect morphology, the method of form finding has long been of great interest to architects and structural engineers. In the past two decades, more and more researchers have started to focus on quantitative structural form-finding algorithms based on finite element analysis. Compared to the qualitative methods represented by graphical statics, these methods based on performance data can not only consider the real material resistance and damage characteristics but also facilitate the breakthrough of the existing structural system to find a new and efficient structure for a specific situation.

The Evolutionary Structural Optimization (ESO) [3] and Bi-directional Evolutionary Structural Optimization (BESO) [4], proposed and developed by Mike Xie's team at RMIT University, have been used to develop a new approach to structural optimization. ESO and BESO can determine whether structural elements should be removed or added at each iteration based on the relative magnitude of the sensitivity numbers, such as strain energy density, calculated from the structural performance data of the model elements. The BESO method can be applied to the structural analysis of completed buildings, such as the analysis of the tree structure of Gaudi's Sagrada Familia [5] and the analysis of the shell structure of Palazzetto dello Sport of Rome [6] (Fig. 2).

Through the BESO algorithm and its plug-in Ameba, architects and designers can introduce real mechanical behaviours into the scheme design to generate innovative, efficient and organic architectural forms, such as the new structural designs for traditional Chinese architecture from DigitalFUTURES 2019 workshops supervised by authors (Fig. 3). In practical projects, architects can even apply gravity and wind load on floor slabs, façade systems and central cores to create various types of leaning towers by authors (Fig. 4).

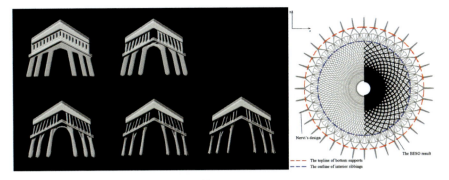

Fig. 2 Analysis of Sagrada Familia Façade [Left] and Dome of Palazzetto dello Sport [Right]

Fig. 3 New structural designs for traditional Chinese architecture

Fig. 4 Diverse structural designs for leaning towers using the BESO method by Nic Bao, Xin Yan and Yulin Xiong

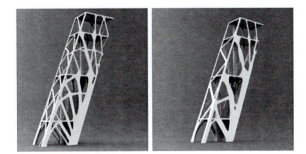

The topology-optimized structure, driven by performance data, meets the requirements of complex working conditions, which is incomparable to qualitative form-finding algorithms. The Akutagawa River Side project in Takatsuki City, completed in 2004, was the first practical project to apply the ESO method for structural optimization of shape finding. Moreover, a renowned Japanese engineer, Mutsuro Sasaki, has collaborated with architects such as Toyo Ito, SANAA, and Arata Isozaki to create innovative buildings using his original Sensitivity Analysis Method and Extended ESO Method. The Sensitivity Analysis Method is suitable for adjusting curved surfaces in architectural designs to meet structural rationality requirements, as seen in the Kitagata Community Centre (Fig. 5), Kakamigahara Crematorium (Fig. 6), and Rolex Learning Center (Fig. 7). The Extended ESO Method can automatically generate reasonable structures under predefined conditions, with a process similar to standard structural topology optimization, as demonstrated in the Qatar National Convention Center (Fig. 8) [7]

In addition to the large-span buildings described above, topology optimization techniques can still be used to drive high-rise building design. Unlike the large-span buildings above, which mainly deal with gravity, in high-rise building design, the main problem that the building structure has to deal with comes from the lateral loads caused by the wind flow at high altitudes. However, for the performance data-driven design approach, wind load changes in the load conditions in the computational

Fig. 5 Kitagata community centre

Fig. 6 Kakamigahara crematorium

Fig. 7 Rolex learning center

Fig. 8 Qatar national convention center

model, which certainly helps to simplify the process of designing the form and structure of high buildings. For example, in the CITIC Financial Center project in Shenzhen, the engineers obtained the optimal arrangement of diagonal intersecting grids of different orders through topology optimization and modified them based on the requirements of building lighting and construction to obtain the final results (Fig. 9) [8]. Zaha Hadid Architects has also conducted a data-based analysis to find the shape of a super-tall building structure, such as the One Thousand Museum condominium in Miami (Fig. 10).

Fig. 9 CITIC financial center Shenzhen

Fig. 10 One Thousand Museum condos Miami

2.3 Performance Data-Driven Optimization of Building Components

Besides providing a holistic morphological concept for buildings, performance data-driven generative design methods continue to play important roles at smaller scale components. In real-world projects, local building components often need to be as material efficient or aesthetically simple as possible while meeting certain structural performance requirements. In the face of practical needs, performance analysis data can be used to meet specific functional requirements and individual choices more precisely. In recent years, beamless floor slabs have been a popular research topic in lightweight structures. Through finite element analysis, researchers can easily obtain the force distribution of the floor slab and determine the reinforcing part of the floor slab through topology optimization or stress line extraction. For example, the Block Research Group (BRG) and Digital Building Technology (DBT) at ETH Zurich used principal stress lines obtained from finite element analysis to determine the location and morphology of the reinforcing ribs of the floor slab [9, 10]. The XtreeE research group has used topology optimization to design the performance of columns [11]. The Smart Node, a collaboration between RMIT University and the Chinese University of Hong Kong, optimizes the connection nodes in the pavilion's wooden structure to meet mechanical requirements while satisfying arbitrary angular connections (Fig. 11) [12].

In addition, RMIT Centre for Innovative Structures and Materials (CISM), led by Prof Mike Xie, conducted pioneering work on the theoretical development and practical application of various structural optimization techniques. CISM also optimized floor slabs, partition walls, columns and bridges using topology optimization techniques [13–15] (Fig. 12, 13, 14 and 15).

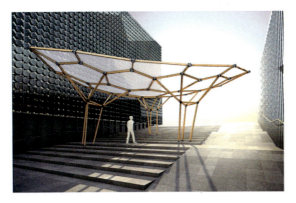

Fig. 11 Smart node pavilion

Fig. 12 Topologically optimized floor slab by Jiaming Ma, Mohamed Gomma, Nic Bao and Mike Xie

3 Swarm Intelligence-Based Architectural Design

3.1 Biomimetic Morphogenesis Design Based on Swarm Intelligence

The emergence of complexity theory in the last 40 years has significantly impacted on our understanding of form generation. The conceptualization of form has shifted from the macro level to a focus on the operation of the underlying complex systems of

Fig. 13 Topologically optimized partition wall by Nic Bao, Xin Yan and Mike Xie

Fig. 14 Tree-like columns and pavilion by Nic Bao, Xin Yan and Mike Xie

Fig. 15 Topologically optimized FootBridge by Mike Xie and Nic Bao

form generation. Swarm Intelligence algorithms, which create multi-agent systems based on rules abstracted from the biological world, offer architects a more dynamic and comprehensive approach to studying the self-organized form generation of architectural structures.

3.2 Swarm Intelligence in Architectural Design

Architecture has long been interested in natural and biomorphic simulation. Within the last decade, architectural firms such as Kokkugia and Biothing have used multi-agent algorithms to explore the morphogenetic potential of swarm logic to generate new architectural forms. Swarm-based design research has emerged at various academic institutions, including Columbia University (GSAPP), the Architectural Association School (AA), University College London (UCL Bartlett), the Southern California Institute of Architecture (Sci-Arc), the University of Pennsylvania (UPenn), and the Royal Melbourne Institute of Technology (RMIT). Pioneering architects such as Alisa Andrasek, Roland Snooks, Robert Stuart-Smith, and Marc Fornes have advanced the computational morphology of swarm intelligence and multi-agent algorithms through their laboratory design practices. Roland Snooks, in particular, has focused on formal research on behavioural design and the use of algorithmic computing through multi-agent algorithms, which closely links algorithmic design, machine learning, robotic fabrication and advanced materials to generate intricate architectural forms (Fig. 16). He has also proposed using finite element structural analysis to extend the structural agent system and establishing a global evaluation system for the whole structure by assigning structural behaviours to multi-agents and getting real-time feedback. These ideas have inspired the author's exploration of swarm intelligent computational design methods driven by structural performance data [16].

Fig. 16 Kazakhstan Symbol using the multi-agent algorithm by Roland Snooks

4 Bioinspired Generative Design Approach Based on Structural Performance

4.1 Framework of New Design Methodology

As an emerging digital architectural design method, performance data-driven design derives from the scientific and rational analysis of architectural design concerns such as structure, sound, light, heat, and wind, giving buildings material rationality with their objective and precise characteristics. Meanwhile, swarm intelligent design simulates the social behaviour of biological groups in nature to produce complex formal functions, adding a perceptual biological perception to architectural design. By combining the advantages of both approaches, we propose a new performance data-driven bionic swarm computational design approach [17]. This approach consists of two abstract data layers: the Finite Elemental Analysis Layer (FEA layer) and the Multi-agent System Layer (MAS layer). As shown in Fig. 17, this algorithmic architecture involves a typical building morphology model located in the FEA layer, presented as a mesh model. Due to its natural discrete cell structure, it can be easily imported into various engineering analysis systems to obtain performance data. Meanwhile, multi-agents are introduced into the MAS layer to generate new forms through interactions based on vector-based behaviour rules. The multi-agent system can be linked to the architectural mesh model using the Material Interpolation Scheme [18] and the Marching cube algorithm [19], and the behaviour vectors of the multi-agents are computed based on performance data from the mesh model and various constraints from multidisciplinary sources. Thus, the closed-loop evolutionary logic of "computing performance data from the mesh model, influencing the behaviour of the agents from the performance data, and mapping the agents to new building forms" is achieved.

By applying the above algorithmic framework, architects can use performance data to drive swarm intelligence to generate and optimize building forms. The benefits are twofold: first, the simple vectorized behaviour rules of the swarm intelligence facilitate different users to develop behaviour rules for the agent based on various functional requirements and constraints in order to avoid other conflicts and defects such as poor ventilation, difficult evacuation, and pipe collisions when optimizing the building form performance; second, the complex role of the swarm intelligence allows the building form to evolve in a variety of directions, thus generating various optimized building forms, and avoiding the problem of forming monotonous unique solutions under the traditional engineering optimization form-finding method.

4.2 Data-Driven Performance-Based Generative Design

Applying the above algorithmic framework, we have developed a multi-agent-based topology optimization (MATO) method that applies the algorithmic framework

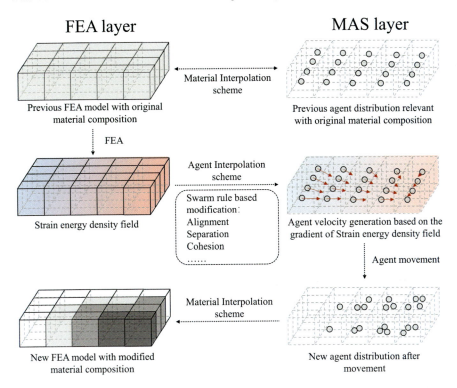

Fig. 17 The diagram of the performance data-driven bionic cluster swarm computational design

described above to exploit the respective advantages of performance data-driven design and swarm intelligence algorithms. With this approach, swarm intelligence is structure-aware and can simultaneously generate diverse topology-optimized forms with high structural performance. We attempt to combine the local self-organizational generation methods of swarm computing with recursive global structure analysis, providing an open framework for architects to program multi-agents based on their intent and behaviour. This enables collaboration between multiple disciplines, integrating local and global awareness in the design process.

The MATO method establishes a correspondence between the mesh model and the agent model through a material interpolation model, allowing the mesh model to change easily with the movement of multi-agents in evolutionary iterations, providing a model basis for the next structural performance calculation. Within this context, an increase in the number of agents within the range of a mesh element enhances the material properties of that element and vice versa. The structural performance data obtained based on the finite element technique is applied to each agent through an interpolation function, generating their gradient vector in the strain energy density field as the initial velocity of the agent. These initial velocities move the multi-agents toward high-strain energy density, enabling material convergence and structural optimization.

Meanwhile, the three most basic behavioural rules of swarm intelligence species, namely separation, alignment, and cohesion, are used to modify the velocity of the agents, as shown in Fig. 18. The Alignment Rule is a local smoothing method that can adjust the movement direction of an agent according to the average movement direction of its neighbours to achieve local coordination. The Cohesion Rule can guide the agent to consider the global optimal position movement in the global scope, avoiding the problem of material dispersion, which often occurs in structural optimization problems. After experiments, the MATO method has been shown to have significant advantages over traditional topological optimization methods in generating diverse optimized structures, as 2D results shown in Fig. 19 and 3D results are shown in Fig. 20.

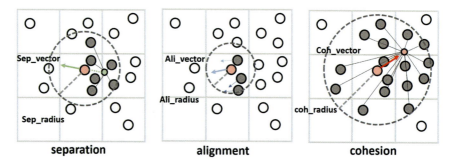

Fig. 18 Three behavioural rules of swarm agents

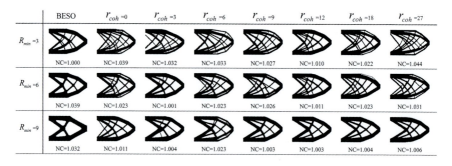

Fig. 19 Diverse optimized layouts by MATO method

Fig. 20 Space grid structure based on data by MATO

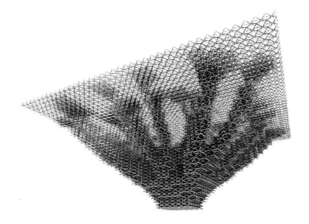

5 Advanced Digital Fabrication

5.1 Large-Scale Robotic 3D Printing Technology

With the arrival of Industry 4.0, robotic construction has become increasingly important in the development of today's construction industry. Factories can manufacture non-monolithic prefabricated building components cost-effectively in mass production, thanks to digital processing simulations based on industrial robots and customized tool designs for different construction materials [20]. Internationally renowned architecture schools and pioneering academic architects, such as the Institute for Computational Design (ICD) led by Achim Menges at the University of Stuttgart, the Gramazio Kohler Research and Robotic Fabrication Lab chaired by Fabio Gramazio and Matthias Kohler at ETH Zurich, Autonomous Manufacturing Lab directed by Robert Stuart-Smith at the Weitzman School of Design, University of Pennsylvania, and the Tectonic Formation Lab founded by Roland Snooks at the School of Architecture and Urban Design at RMIT University in Australia, are all pioneers in this field.

To realize the bionic swarm intelligence designed forms, Roland Snooks has equipped the Tectonic Formation Lab with a variety of robotic fabrication technologies and is currently focusing on 3D printing research in polymers, ceramics, and metals for the additive manufacturing of large-scale custom architectural components (Fig. 21). The lab is dedicated to exploring cutting-edge pilot projects and large-scale prototype development to establish the potential of its research architecture and extend the meaning of its architectural practice. Some examples of their work include the designing and fabricating of an interior partition wall for Monash University's SensiLab and the "Floe" art installation for the NGV Triennial, which were both created using their innovative methods (Fig. 22).

Fig. 21 Robotic fabrications by RMIT Architecture | Tectonic Formation Lab

Fig. 22 "Floe" art installation at National Gallery of Victoria (NGV) by Roland Snooks

5.2 *Augmented Reality Holographic Digital Construction Technology*

Fologram, a technology company founded by Gwyllim Jahn and Cameron Newnham, faculty members from RMIT School of Architecture and Urban Design, has taken the world by storm in the last two years with its augmented reality holographic digital construction technology. This program overlays digital guidance in the workspace to assist in the construction of complex projects requiring a series of measurements, verification, and targeted management, providing step-by-step guidance for masonry work during construction (Fig. 23). Architects and builders have already used Fologram to realize complex structures and art installations, such as the "Steampunk" pavilion at the 5th Tallinn Architecture Biennial in Estonia (Fig. 24), which was also designed with swarm intelligence algorithms, and the live construction of the "Floe" art installation at the National Gallery of Victoria.

Fig. 23 AR-assisted construction using Fologram

Fig. 24 Steampunk Pavilion by Gwyllim Jahn, Cameron Newnham, Soomeen Hahm Design, and Igor Pantic

6 Conclusion

In summary, both performance data-driven design and bionic swarm optimization morphogenesis are digital technologies that have gradually improved with the development of computing technology. Over the past decade, these cutting-edge theories have evolved into project practices. Based on the review and development of these two emerging generative design methodologies, this paper also proposes a performance data-driven computational design method based on bionic swarm optimization, which provides a new path for the future computational design of buildings by deeply combining the two. In this approach, rational performance simulation data is applied to the behaviour of swarm intelligence in the form of vectors. Together with other perceptual human control factors, this data can be derived into various "preferred solutions" of building forms. This process realizes a two-way cycle of "geometric model-performance analysis-computer reconstruction," which is an efficient, open computational design framework that integrates rational analysis and perceptual generation.

At the same time, the complex and variable building forms generated by this method can be easily realized by introducing advanced manufacturing in construction. The rich and comprehensive data information about building performance obtained during the generation process will play a crucial role in the subsequent management of the building information system and the manufacturing process of building prefabricated parts. This data will ensure that robotic construction and augmented reality holographic digital technologies can be successfully implemented in the subsequent construction industry. The complete data chain from computational design to the advanced manufacturing process will gradually update the paradigm of architectural design, thus empowering future buildings to be more precise and rational.

References

1. Reynolds, C.W.: Flocks, herds, and schools: a distributed behavioural model. SIGGRAPH Comput. Graph. **21**(4), 25–34 (1987). https://doi.org/10.1145/37402.37406
2. Ceccato, C., Hesselgren, L., Pauly, M., Pottmann, H., Wallner, J.: Advances Architectural Geometry. Springer, Vienna (2010). https://doi.org/10.1007/978-3-7091-0309-8
3. Xie, Y.M., Steven, G.P.: Evolutionary Structural Optimization. Springer, London (1997). https://doi.org/10.1007/978-1-4471-0985-3
4. Huang, X., Xie, Y.M.: Evolutionary Topology Optimization of Continuum Structures: Methods and Applications. Wiley, Chichester (2010). https://doi.org/10.1002/9780470689486
5. Burry, J., Felicetti, P., Tang, J., Burry, M., Xie, Y.M.: Dynamical structural modeling: a collaborative design exploration. Int. J. Archit. Comput. **3**(1), 27–42 (2005). https://doi.org/10.1260/1478077053739595
6. Yan, X., Bao, D.W., Cai, K., Fang, Y., Xie, Y.M.: A new form-finding method for shell structures based on BESO algorithm. In: Proceedings of IASS Annual Symposia, IASS 2019 Barcelona Symposium: Form-finding and Optimization, pp. 1–8 (2019)
7. Januszkiewicz, K., Banachowicz, M.: Nonlinear shaping architecture designed with using evolutionary structural optimization tools. In: Conference Series 2017: Materials Science and Engineering, vol. 245, p. 082042 (2017). https://doi.org/10.1088/1757-899X/245/8/082042
8. Alessandro, B., Neville, M., Mark, S., Bin, P., Li, S., Cheng, H.: Structural Optimization for An Innovative Structural System: Shenzhen CITIC Financial Center project, vol 37 (2016). https://doi.org/10.14006/j.jzjgxb.2016.S1.023
9. Block, P., Mele, T.V., Rippmann, M., Ranaudo, F., Barentin, C.C., Paulson, N.: Redefining structural art : strategies, necessities and opportunities. Struct. Eng. **98**(1), 66–72 (2020). https://doi.org/10.56330/UJFI2777
10. Meibodi, M.A., Jipa, A., Giesecke, R., Shammas, D., Bernhard, M., Leschok, M., Graser, K., Dillenburger, B.: Smart slab: computational design and digital fabrication of a lightweight concrete slab. In: 38th Annual Conference of the Association for Computer Aided Design in Architecture (ACADIA) on Proceedings, pp. 434–443. Mexico City, Mexico (2018). https://doi.org/10.52842/conf.acadia.2018.434
11. Gaudillière, N., Duballet, R., Bouyssou, C., Mallet, A., Roux, P., Zakeri, M., Dirrenberger, J.: Large-scale additive manufacturing of ultra-high-performance concrete of integrated formwork for truss-shaped pillars In: Willmann, J., Block, P., Hutter, M., Byrne, K., Schork, T. (Eds.) Robotic Fabrication in Architecture, Art and Design 2018. ROBARCH 2018. Springer, Cham. https://doi.org/10.1007/978-3-319-92294-2_35
12. Crolla, K., Williams, N.: Smart nodes: a system for variable structual frames. In: 4th Annual Conference of the Association for Computer Aided Design in Architecture (ACADIA) on Proceedings, pp. 311–316. Los Angeles (2014). https://doi.org/10.52842/conf.acadia.2014.311
13. Ma, J., Gomaa, M., Bao, D.W., Javan, A.R., Xie, Y.M.: PrintNervi: design and construction of a ribbed floor system in the digital era. J. Int. Assoc. Shell Spat. Struct. **63**(4), 122–131 (2022). https://doi.org/10.20898/j.iass.2022.017
14. Bao, D.W., Yan, X., Xie, Y.M.: Fabricating topologically optimized tree-like pavilions using large-scale robotic 3D printing techniques. J. Int. Assoc. Shell Spat. Struct. **63**(2), 241–251 (2022). https://doi.org/10.20898/j.iass.2022.009
15. Bao, D.W.: Performance-Driven Digital Design and Robotic Fabrication Based on Topology Optimisation and Multi-agent System. RMIT University, Australia (2022)

16. Snooks, R.: Behavioural Formation: Multi-agent Algorithm Design Strategies. RMIT University, Australia (2014)
17. Bao, D.W., Yan, X., Xie, Y.M.: Encoding topological optimisation logical structure rules into multi-agent system for architectural design and robotic fabrication. J. Int. Arch. Comput. **20**(1), 7–17 (2022). https://doi.org/10.1177/14780771221082257
18. Bendsøe, M.P., Sigmund, O.: Material interpolation schemes in topology optimization. Arch. Appl. Mech. **69**, 635–654 (1999). https://doi.org/10.1007/s004190050248
19. Lorensen, W., Cline, H.Y.: Marching cubes: a high-resolution 3D surface construction algorithm. ACM SIGGRAPH Comput. Graph. **21**(4), 163–169 (1987). https://doi.org/10.1145/37402.37422
20. Yuan, F., Menges, F., Leach, N.: Robotic Futures. Tongji University Press, Shanghai, China (2015)

Parameterization and Mechanical Behavior of Multi-block Columns

D. Foti, M. Diaferio, V. Vacca, M. F. Sabbà, and A. La Scala

Abstract The research aims at studying the mechanical behavior of a monumental structure to preserve the historical-architectural heritage as recommended by the Sustainable Development Goal 11.4 to "strengthen efforts to protect and safeguard the world's cultural and natural heritage". This research focuses on multi-drums classical columns, a very common typology in the architecture of ancient Mediterranean civilizations. These columns are made of stone drums of considerable size compared to its entirety, thus, their resistance to vertical and horizontal loads is entrusted to simple support and friction between the drums. So, the present paper proposes a parametric study of the geometrical characteristics, mechanical properties and interactions of the blocks with the aim of describing their influence on the dynamical response of columns, highlighting fundamental aspects concerning their vulnerability. The analysis is performed by means of the Distinct Elements Method (DEM). Many numerical analyses have been conducted to investigate the effects of parameters that influence the dynamic behavior of the examined structural elements and identifying their stability domains.

Keywords Parametric structural analysis · Cultural heritage conservation · Multi-drum column · D.E.M. · Rocking

United Nations' Sustainable Development Goals Goal 11 Make cities and human settlements inclusive, safe, resilient and sustainable

---

D. Foti (✉) · M. F. Sabbà · A. La Scala
Department of Architecture, Construction and Design, Polytechnic University of Bari, 70124 Bari, Italy
e-mail: dora.foti@poliba.it

M. Diaferio
Department of Civil, Environmental, Land, Construction and Chemistry, Polytechnic University of Bari, 70124 Bari, Italy

V. Vacca
ANAS SpA, Struttura Territoriale Basilicata, 85100 Potenza, Italy

1 Introduction

1.1 The State of Art

Reality and what surrounds us is often presented to our eyes as something extremely complicated and sometimes difficult to understand; this is especially true in the field of scientific research, but it also undoubtedly extends to the professional field of engineering and architecture.

One of the key objectives to achieve is therefore to provide a simplified, yet accurate, representation of reality, to define a mathematical model which allows us to fully estimate the behavior of the object under study. To do this, the current state of the art envisages the use of the so-called discontinuous element modelling, which consists, after simplifying the geometry as much as possible, in subdividing the object to be investigated into elementary parts (i.e. mesh).

This chapter is intended to provide a general overview of the state of the art on the numerical modelling approaches in the structural field and, in particular, to refer to the Discrete Element modelling Method (DEM). The main aim is to discuss the applicability of such approach to structures with a cultural, historical, and architectural value, and that for cultural, social, and economic reasons must be preserved. The present study wants to highlight some peculiarities of their response to loads, which must be considered by the decision makers. The application of this methodology to a series of real case studies will also be shown, having as object of investigation the stability analysis of some stone columns.

There are two different approaches to numerically model a 3D structure: the Finite Element Method (FEM) and DEM. Both methods aim to provide approximate solutions to problems described by systems of partial differential equations, reducing the latter to simpler systems of algebraic equations that can be solved by means of well-known algorithms.

The FE method can be easy applied to materials that are homogeneous and isotropic or that can be approximated to these behaviors (like concrete), while it encounters some difficulties when representing and solving problems involving more complex media with varying characteristics.

Therefore, if modelling of heterogeneous solids with purely orthotropic behavior is required, such as masonry, the FE method requires the use of appropriate interface elements, i.e. it is necessary to model the mortar joints, so that the masonry blocks can be considered as a continuous and homogeneous medium, surrounded by these joints.

From a design point of view, the FE method is the most spread and used in modern calculation software for the design of structures, as it enables these software to solve the problem of estimating stress and deformation states induced by several load conditions and that, due to the complexity of the problem and the number of unknowns, cannot be solved by means of closed form analytical solutions.

Thanks to its versatility, this approach has been widely used since the early' 70 s in various engineering problems, from fluid dynamics to electromagnetism, as well as structural and geotechnical calculations, and it is still the most widely used method for numerical simulation and problem solving in such fields.

DEM, on the other hand, is a numerical modelling method that can represent systems composed by several distinct bodies maintaining their geometric boundaries, in such a way as to leave the separation between the elements clearly defined and, thus, by modelling their interaction.

In the case of masonry, for example, the DE method allows the masonry texture to be represented as a set of distinct blocks, thus enabling the researcher or designer to precisely analyze the behavior of the individual blocks, modelling the mortar joints as contact surfaces.

The evident geometric and constitutive non-linearity of a medium such as masonry requires a precise description of the possible responses in structural terms, including those related to local mechanisms, such as sliding phenomena along the joints or effective separation of the blocks themselves. To investigate these problems, the DE approach appears particularly reliable, as it operates by considering all the local mechanisms and adapts the analysis to the consequent variation in the initial geometric configuration.

From an application point of view, this method is usually favored in the analysis of structural collapses, by means of pseudo-static or dynamic processes. An example is the use of DEM software in the analysis of historical load-bearing masonry buildings for the compilation of the so-called "Seismic Vulnerability Sheets".

Historically, the DE method developed in parallel with FEM, initially for geotechnical applications; it was used for the analysis of slidings in hard, compact rock beds, where the failure mechanisms are defined by the discontinuities in the rock.

The representation of the object in question as a set of blocks, therefore, represents the fundamental characteristic of the DE method, contrary to what happens in FE models, where the starting point is the representation of a continuous body.

The first to propose the possibility of a numerical approach to the resolution of this problem was Cundall in 1971 [1, 2]; the need to update consequently to the sliding positions of the blocks, led Cundall to develop the method starting from the integration of the equations of motion of the blocks themselves, thus obtaining the possibility of considering large displacements. The further innovation was that this method was able to consider viscous damping coefficients even in static analysis, just as in approximate dynamic methods.

As far as masonry structures are concerned, the DE method is an extension, to complex systems, of the analytical techniques used since the early XIX century for the analysis at the collapse of such constructions [3].

In conclusion, the advantage given by DE method for the analysis of historical buildings in natural stone or masonry blocks, is mainly due to the best approximation of non-linearities and geometric material; secondly, it allows for the modelling and analysis of structures of considerable size, without suffering too much from an increase in computation times, contrary to what happens in the classic FE models.

1.2 DE Method

Many different DE methods can be found in the technical literature [4, 5]; it is, however, possible to identify some common features:

- DE models start basically from the hypothesis that the blocks are rigid, and that the deformability depends on the mortar joints; it should be noted, however, that recent developments in technology have led to deformable block models.
- The blocks interact by means of edge-to-edge contact points or groups, but not through a uniform contact surface; this prevents uniform redistribution of the stresses over the entire block surface.
- Thanks to the possibility of considering a complete separation of the blocks, DEM analysis can be utilized in the field of large displacements.
- "Time-stepping" algorithms are used in DE models to solve quasi-static problems.

Given the importance of the boundary conditions in solving the static and/or dynamic problems for DE modelling methods, it is obvious that a correct representation of the contact surfaces between the individual blocks is of fundamental importance for an accurate calculus.

In many DE models, the representation of the contact surfaces is done by synthetizing the surface into a cloud of points, called 'contact' points, in which the contact forces are applied, the latter ones depend on the displacements of the blocks at the corresponding point.

Although idealized as a series of points, the need to still apply the classical constitutive laws to each joint (i.e. in terms of relative stresses and displacements) has led to the need of defining the area of the contact surface represented by the contact point. Thus, the properties associated to each contact point depend on the area to which it is related.

Regarding the numerical analysis, it is assumed that velocities and accelerations are constant during each time step. DE methodology is based on the concept that this time step is sufficiently small, so that no variation can propagate between the discrete elements during the calculation step. For this purpose, very small-time intervals of about 10^{-6} s are considered, so that in the simulated system, collisions and contacts between bodies can be satisfactorily captured.

In the present paper, DEM was applied by adopting the software 3DEC [6]. 3DEC is a three-dimensional numerical software based on the discrete element method for discontinuous modelling.

The basis of this software is the numerical formulation widely tested and used in the two-dimensional version, UDEC [6].

3DEC is based on a Lagrangian calculus scheme that is particularly suitable for modelling large displacements and deformations in a system consisting of blocks.

The features of 3DEC can be summarized as follows:

- The analyzed structure can be modelled as a 3D assembly of rigid or deformable blocks;

- The discontinuities due to mortar joints are modelled through the contact points located between the boundaries of distinct blocks;
- The continuous or discontinuous behavior of the joints is generated on a statistical basis;
- 3DEC employs a time-explicit solution algorithm that accommodates both large displacements and rotations, and allows calculations in the time domain;
- 3DEC is programmable using the FISH language (short for FLACish, FISH was originally developed for two-dimensional continuous finite difference software, FLAC [6]). With the FISH programming language, it is possible to write your own functions to expand the possibilities provided to the user by the program to meet specific needs.

2   Dynamic Impulsive Analysis of Stone Columns

The monumental patrimony of Italy and other Countries in the South of Europe has a deep-rooted constructive tradition which adopted the stone as its characteristic element of construction; it is in fact since ancient times that it is possible to find the presence of constructions and monuments made of local stone, which differ deeply according to the construction materials available in each area, for example, in the South of Italy the "Trani" marble, the "Lecce carparo", etc. are well known.

The abundance of quarries has therefore allowed the spread of stone structures that have come to the present day, but it should be noted that the characteristics of such structures are quite different from each other and a unique general approach to their study is still a challenge. In fact, masonry buildings (including those made of bricks) are characterized by a greater variety of structural elements differently from the modern structural schemes that adopts three structural elements: beams, pillars and plates.

It is evident that the urge to create majestic works using materials that do not provide any tensile strength, has brought the ingenuity of the master builders of the time to extreme levels, and this is the reason why today we can find in this type of constructions the presence of many structural elements: vaults (at least dozen different types are recognized), arches (characterized by several different geometrical configurations) and columns.

In particular, the present chapter deals with the study of the structural behavior of the latter.

The Countries which stand in the Mediterranean area are, as previously said, traditionally linked to the use of stone as a fundamental material for buildings; already in prehistoric times to which belong religious monuments such as the dolmen.

Moreover, in the subsequent periods, the inclusion of several Countries in the region of Magna Graecia and the consequent Greek and Roman dominations, led to the spread of Hellenistic culture and consequently to constructions that go from the classical characters of ancient Greece to the post-Alexandrian eclecticism and the overlapping of the already consolidated Ionic, Doric and Corinthian orders.

Subsequently, after the fall of the Roman Empire, due to the continuous state of war, there was the stripping and the "recycling" of the built, also favored by the depopulation of the cities; this phenomenon continued until the threshold of the year 1000, when there was a gradual resumption of city activities.

An example of reuse of existing buildings is the crypt of Basilica of San Nicola in Bari (Italy), where the columns belong to completely different eras and styles.

The end of the High Middle Ages saw a flourishing of building activities in the cities of the South of Italy, which led to the construction of monuments that have come down to the present days, and that have given an incredible architectural variety to the territory.

Recurring elements are precisely the columns, which represent the key elements and supports in all structures made before the advent of concrete. In fact, in literature there are many studies regarding the behavior of single and multiple columns structures [7–11]. In the present chapter the results of the analyses carried out by applying the DE method to the case of stone columns are reported and discussed.

The columns were subjected to harmonic pulses of varying amplitude and frequency values with the aim of describing the activation of local/global instabilities induced by seismic events. The objective is acquired by plotting the column rocking curves [12, 13], also known as stability domains, which describe the mechanical behavior of these structures.

The curves were drawn by varying some fundamental parameters that influence the dynamic response of the system: slenderness of the column (which is defined by the ratio between the total height H and the diameter of the base B), number of drums, contact properties and damping.

The range of slenderness has been varied in the range $4 \leq H/B \leq 13$ to cover all the types of columns belonging to the Doric and Ionic orders characteristic of the architectures of ancient Mediterranean civilizations.

2.1 The Geometric Model

The investigated columns, which stand in the Mediterranean area, are characterized by several geometrical parameters depending on the dimensions of the overall structure. Thus, a parametric analysis is requested to consider the high variability of such geometric parameters. This led to the need of preparing a routine that would allow automating the construction of the geometry for several cases, starting from few significant parameters.

The cross-section of the column, identified as an icosagon, has been assigned and the following parameters have been considered:

- Column slenderness (H/B) defined as the ratio between the height and the diameter of the section at the *imoscape* (the lower base section of the column);
- Column diameter at the *imoscape* of the column;
- Number of drums in the column shaft (maximum number allowed 48);

- *Entasis*, the ratio between the diameter of the column at the top and that at the *imoscape*;
- The ratio between the length of the column shaft with a constant cross-section and the overall height of the column;
- Four quantities linked to the geometry of the capital, in detail the height and the width of the echinus and abacus;
- Three different "scaling coefficients" along the three directions x, y and z that allow the column shape to be scaled homogeneously or differently.

This approach allowed defining a great number of possible geometrical configurations which can be considered representative of the existent patrimony of such structures.

2.2 The Mechanical Characteristics

In the mechanical modelling, the blocks were considered infinitely rigid, to obtain greater overall reliability of the results, and because of the need to reduce the time required by the computer for the solution of the equation of motion.

The drums have been considered made in limestone, which is widely used in the construction of these structures because it is easy to find and work. A density of 2,200 kg/m^3 has been assumed for limestone.

The behavior along the joints between the drums is defined by Coulomb's law which, in turn, depends on the cohesion c and internal friction angle Φ.

The stiffness coefficients of the joints in the normal and tangential direction are assumed to be 100.0 MN/m, as for the 3DEC software [6] they are equal to the ratio of the relative elastic modulus divided by the dimension of the drum in the direction of the force.

The interaction between drums is governed by complex phenomena concerning impacts and the consequent energy dissipation. The amount of energy that is lost during the motion is difficult to quantify except through in-depth studies of the materials as shown in [14]. It was therefore considered appropriate to assign an average damping coefficient of 0.5 for the tests carried out, and, to examine how the variation of this parameter could affect the global response of the structure, different values of the damping coefficient have been considered.

2.3 The Load Conditions

Impulsive loads, specifically, harmonic pulses defined by a co-sinusoidal and sinusoidal acceleration laws were applied to the base of the columns (Fig. 1) [4, 5].

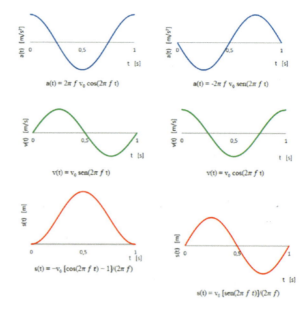

Fig. 1 Harmonic pulses considered in the analysis of columns expressed in terms of acceleration, velocity and displacement:
a co-sinusoidal pulse;
b sinusoidal pulse

The results obtained were then plotted in the plane which presents along the axes the frequency and the peak acceleration of the harmonic pulse according to [17].

2.4 Results

A dynamic analysis of the modelled columns subjected to the harmonic pulses has been conducted and the following four behaviors were observed:

1. the column remains intact despite more or less evident displacements and rotations of the drums (Safe Area);
2. the column collapses entirely after it oscillates at least once around the base, i.e. the collapse occurs after the base of the column exhibited one or more impacts with the supporting surface (mode I);
3. the column collapses entirely without oscillating around the base, i.e. without the base impacting the ground before integral collapse (mode II);
4. the column undergoes a partial collapse (mode III).

By plotting the observed results in a frequency-peak acceleration graph of the harmonic pulse (Fig. 2), the regions of stability/instability of a column can be identified.

The points in red identify the line of overturing without impact (mode II); all points above this line represent regions of total collapse; the points in yellow identify regions of overturing with impact (mode I), while the points in fuchsia identify the region

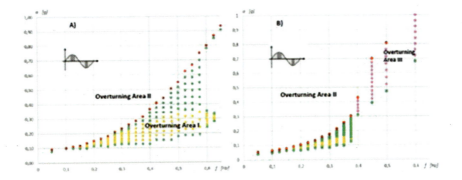

Fig. 2 Frequency peak acceleration of the harmonic pulse diagram of a column, with a slenderness equal to 4, subjected to a sinusoidal impulse in acceleration at the base; in red mode II, in yellow mode I, in green stability, in fuchsia partial collapse: **a** monolithic column, **b** column composed by 4 drums

of partial collapses which is clearly only present in multi-drum columns; all other points are part of the region of stability for the column.

The region of safety in the frequency-peak acceleration plane clearly depends on different parameters that influence the dynamic response, in particular: (a) the dimensions; (b) the column slenderness H/B ratio; (c) the number of drums that compose the column; (d) the damping coefficient; (e) the contact properties between the drums. Thus, to acquire a detailed knowledge of the muti-drums columns, a parametric analysis was performed by varying the aforementioned parameters in wide ranges of values. In the following, the results of such parametric analysis are presented and discussed with a detailed attention to the influence of each parameter for multi-drums columns.

2.4.1 Effects of Column Slenderness on the Stability Domains

As previously mentioned, columns of the classical Doric and Ionic orders have been analyzed, characterized by a slenderness ranging from 4 to 13 [15, 16] composed by a fixed number of drums (in the case here discussed such number is set equal to 8), with a base diameter B equal to 1.0 m. However, to conduct a parametric analysis, the column height was varied from 4 to 13 m. In Fig. 3 a sketch of the cases considered in the parametric analysis is shown.

According to Dimitrakoppoulos and DeJong 2012 [14] the results will be plotted in a dimensionless plane whose axes are defined according to the following relationships:

$$a[adm] = \frac{a_g}{g} \cdot \frac{H}{B} \tag{1}$$

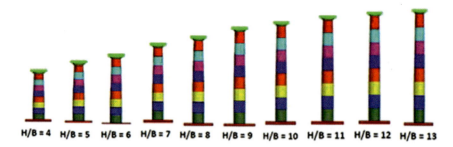

Fig. 3 Cases considered in the parametric analysis with a slenderness varying in the range from 4 to 13

$$f[adm] = \frac{2\pi f}{\sqrt{\frac{3g}{2\sqrt{H^2+B^2}}}} \quad (2)$$

where g is the acceleration of gravity, a_g and f are the amplitude and the frequency of the applied harmonic pulse, respectively.

The left column of Fig. 4 shows the graphs in the frequency-peak acceleration plane; the right column shows the graphs in the same plane but with dimensionless axes according to the relations (1) and (2).

Figure 5 shows the overall results obtained for the 10 columns analyzed, that means the ten cases of slenderness.

The diagrams obtained in the dimensional plane show that the stability region of the multi-drums columns decreases as the H/B ratio increases, especially for low frequency values: for the column with slenderness H/B = 4 at the frequency f = 0.05 Hz the acceleration that induces the collapse is equal to 0.102 g, while for the column with slenderness H/B = 13 the acceleration corresponding to the collapse is equal to 0.037 g (almost three times lower than that for H/B = 4). For increasing frequency values, the green curves tend to get closer to each other, but globally they always confirm the same order: the highest one is related to the stockiest column and the curve appears lower than the first one as the slenderness of the column increases.

The analysis of the yellow curves highlights that the area enclosed by them decreases significantly as the slenderness of column increases and therefore, contrary to what might be expected, a squat column will be more susceptible to global collapse after oscillation than a slimmer column. The red curves occur at lower frequency and acceleration values for increasing slenderness; in detail, for slenderness H/B = 4 the bifurcation occurs at f = 0.298 Hz while for H/B = 13 it occurs at f = 0.150 Hz. Therefore, there is a region in which partial collapses occur and whose area increases as the slenderness of the column increases.

Similar analyses were carried out by stressing the columns with sinusoidal pulses.

Differently from the case with a co-sinusoidal pulse, the columns subject to sinusoidal pulses show the following behavior, as it can be observed in Fig. 6: the areas

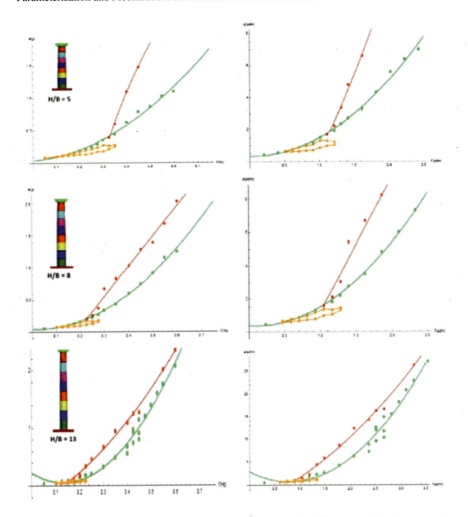

Fig. 4 Frequency-Peak acceleration diagram for a column with H/B ratio variable from 4 to 13 subjected to a co-sinusoidal pulse acceleration at the base: the left column presents the graphs with dimensions of the axes respectively in Hz and in g, the right column shows the dimensionless axes

surrounded by the yellow curves are much smaller than that obtained by the sinusoidal pulses, so that they do not occur in the case of the squat column with H/B = 4. Furthermore, the red curves break away from the green ones at a frequency between 0.10 and 0.15 Hz, while in the previous case this behavior began at frequency values lower than those now considered for higher slenderness ratios and in a wider frequency range (0.15–0.35 Hz).

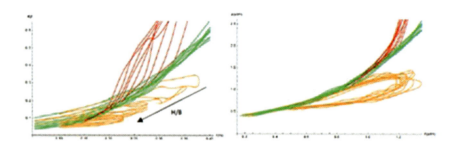

Fig. 5 Frequency-Peak acceleration diagram for the 10 columns with H/B ratio variable: the left column presents the graphs with dimensions of the axes respectively in Hz and in g, the right column shows the dimensionless axes

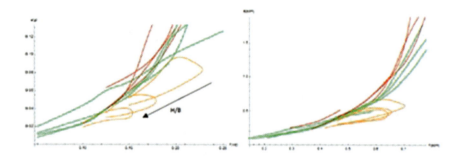

Fig. 6 Frequency-Peak acceleration diagram for columns with H/B ratio ranging between 4 and 13 subjected to sine pulse acceleration at the base, on the right dimensionless graph

2.4.2 Effects of Global Column Dimensions on Stability Domains

It is known that the global dimensions of structures, even if characterized by the same geometrical ratios, significantly influence the dynamic behavior. The study of the variation of the basic dimensions is intended to observe the scale effect on the behavior of the multi-drums columns. Therefore, the slenderness of the columns will be fixed and their base dimension, and consequently their height, will be varied to respect the established H/B ratio.

The analyses were carried out by subjecting many columns to an acceleration that varies in accordance with a co-sinusoidal harmonic pulse. The following values of the parameters have been set:

- column slenderness H/B = 12;
- diameter at the column base B ranging between 0.25 m and 2 m;
- number of drums equal to 8;
- internal friction angle equal to $\Phi = 30°$;
- damping coefficient 0.5.

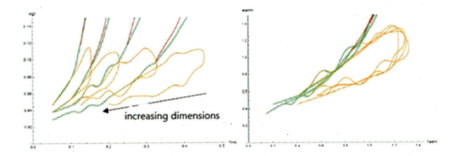

Fig. 7 Frequency-Peak acceleration diagram for columns with H/B = 12 subject to co-sinusoidal pulse acceleration at the base, diameter at the column base B ranging between 0.25 m and 2.00 m, number of drums 8; internal friction angle $\Phi = 30°$; damping coefficient 0.5; on the right dimensionless graph

The results are plotted in Fig. 7.

The results confirm that the larger columns are less vulnerable to collapse as the area of safety increases with the scale of the column. It is possible to notice that the trend of the green curves grows faster as the column scale increases; the yellow curves have the same extension along the ordinate axis (they cover a range of extremes 0.04–0.115 g), while they present a reduction along the frequency axis as the scale increases. The red curves follow the same pattern as the green curves, growing faster and appearing at lower frequencies as the scale increases.

2.4.3 Effects of the Damping Coefficient on the Stability Domains

During the motion, the interaction between the drums, which compose the column, generates energy dissipation phenomena. These latter are considered in the numerical simulation through 3DEC by introducing in the model a damping factor.

In the following analyses, the value of the damping factor has been varied to establish its influence on the overall behavior.

The analyses were carried out by subjecting the columns to accelerations, which are co-sinusoidal harmonic pulse, and setting their characteristic parameters as follows:

- column slenderness H/B = 12;
- diameter at the column base B = 0.50 m;
- number of drums 8;
- internal friction angle $\Phi = 30°$;
- damping coefficient varying in the range $0.3 \div 0.7$.

The results are plotted in Fig. 8.

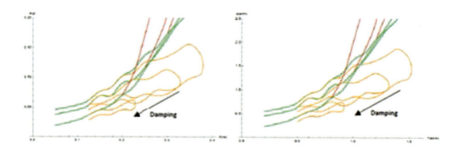

Fig. 8 Frequency-Peak acceleration diagram for columns with H/B = 12 subject to co-sinusoidal pulse acceleration at the base, diameter at the column base B = 0.50 m, number of drums 8; internal friction angle $\Phi = 30°$; damping coefficient ranging between $0.3 \div 0.7$; on the right dimensionless graph

The diagrams plotted in Fig. 8 show the typical monotonous increasing trend for the green curves, obviously, there is a reduction in the value of the intercept as the damping coefficient increases.

It also turns out that, by increasing the damping factor, the curves grow much faster, increasing the stability zone at high frequencies. In fact, at the same frequency f = 0.6 Hz, for a damping factor of 0.3, acceleration a = 0.633 g is recorded, while a factor of 0.7 results for a = 1.033 g.

The yellow curves have the same frequency of origin at f = 0.125 Hz, and corresponding peak accelerations that decrease as the damping increases.

The area enclosed by the curve decreases as the damping coefficient increases.

As the damping factor increases also the red curves increases, and in detail, they become detached from the green curves at lower frequencies, and the area enclosed by them, which indicates partial collapse, decreases progressively due to the growth of the green ones as explained above.

3 Conclusions

The aim of the present work was to study the dynamic behavior of multi-drums stone columns using the method of discrete elements and performing a parametric analysis to highlight the aspects that influence the response.

The analysis has been carried out by starting from the following considerations:

- multi-drums columns do not have a vibration mode (except when it comes to very small shifts), as the adoption of a continuum model is quite far from the effective behavior;
- stability is highly influenced by the variation of the parameters of the structural system;

- the vulnerability of such structures is influenced by the dominant frequency of the earthquake.

Analyzing the obtained results, it is possible to state that a slimmer column at low frequency values is less stable due to the progressive decreasing of the area of the stability zone.

This phenomenon, however, corresponds to a reduction in the collapse zone after oscillation of the base wheel. This evidence allows us to assert that the columns become less and less sensitive to this type of collapse as the H/B ratio increases.

Finally, at high frequency values, as the slenderness increases, there is a widening of the zone in which the structure can be defined as stable.

Through the study of the variation of dimensions in scale, it was possible to conclude that columns with larger dimensions are generally more stable than those of smaller dimensions; the zone of stability appears to increase as the scale increases. This phenomenon is associated with the reduction of the zone in which the collapse is preceded by oscillations, which contributes to the increase of the stability zone in larger columns.

Finally, the influence of the damping coefficient on the overall dynamic response was evaluated. At low frequencies, by increasing this coefficient, the interface of the stable zone tends to be lower, while at high frequencies this tendency is reversed, causing the increase of this zone: a higher damping coefficient makes the structure less stable at low frequencies, while at high frequencies it increases its stability.

Conversely, assigning a lower damping coefficient will make the behavior at low frequencies more stable, and vice versa. The variation of this coefficient also influences the area in which the collapse occurs through oscillations: the area, which represents it, decreases as the damping increases, thus making the column less subjected to this type of instability.

A series of diagrams with zones of stability and collapse have been obtained from the described analyses. The construction of these diagrams allows the prediction of the stable or unstable behavior of the structure, based on the characteristics of the column and the frequency and peak acceleration of the acceleration impulse.

References

1. Cundall, P.A.: A computer model for simulating progressive large scale movements in blocky rock systems. Contribution title. In: Proceedings of the Symposium of the International Society for Rock Mechanics, Society for Rock Mechanics (ISRM), France, II-8 (1971)
2. Cundall, P.A.: The measurement and analysis of accelerations in rock slopes (1971)
3. Heyman, J.: The stone skeleton. Int. J. Solids Struct. **2**(2), 249–279 (1966)
4. Cundall, P.A., Strack, O.D.L.: A discrete numerical model for granular assemblies. Geotecnique **29**, 47–65 (1979)
5. Cundall, P.A., Hart, R.D.: Numerical modeling of discontinua. In Proceedings of the First US Conference on Discrete Element Methods. CSM Press, Colorado (1989)
6. Itasca Consulting Group, Inc. Mill Place 111 Third Avenue South, Suite 450 Minneapolis, Minnesota 55401 USA

7. Mitsopoulou, E., Doudoumis, I.N., Paschalidis, V.: Numerical analysis of the dynamic seismic response of multi-block monumental structures. In: Proceedings of the 11th European Conference on Earthquake Engineering, Paris (1998)
8. Chatzillari, E.T., Tzaros, K.A.: Influence of the friction coefficient in the rocking response of rigid multi-block columns via nonlinear finite element analysis. In: 11th HSTAM International Congress on Mechanics (2016)
9. Sarhosis, V., Asteris, P., Wang, T., et al.: On the stability of colonnade structural systems under static and dynamic loading conditions. Bull. Earthq. Eng. **14**, 1131–1152 (2016). https://doi.org/10.1007/s10518-016-9881-z
10. Dimitri, R., De Lorenzis, L., Zavarise, G.: Numerical study on the dynamic behavior of masonry columns and arches on buttresses with the discrete element method. Eng. Struct. **33**(12), 3172–3188 (2011)
11. Psycharis, I.N., Papastamatiou, D.Y., Alexandris, A.P.: Parametric investigation of the stability of classical columns under harmonic and earthquake excitations. Earthq. Eng. Struct. Dyn. **29**(8), 1093–1109 (2000)
12. Minafò, G., Amato, G., Stella, L.: Rocking behaviour of multi-block columns subjected to pulse-type ground motion accelerations. Open Constr. Build. Technol. J. **10**(1) (2016)
13. Pena, F., Prieto, F., Lourenco, P.B., Campos, C.A., Lemos, J.V.: On the dynamics of rocking motions of single rigid-block structures. Earthq. Eng. Struct. Dyn. **36**(15), 2383–2399 (2007)
14. Dimitrakopulos, E.G., DeJong, M.: Revisiting the rocking block: closed-form solutions and similarity laws. Proc. R. Soc. A **468**, 2294–2318 (2012)
15. Rocco, G.: Guida alla Lettura degli Ordini Architettonici Antichi, Volume I: Il Dorico. Liguori Editore, Napoli, Italy (1994). ISBN 88-207-2256-9 (in Italian)
16. Rocco, G.: Guida alla Lettura degli Ordini Architettonici Antichi, Volume II: Lo Ionico. Liguori Editore, Napoli, Italy (2003). ISBN 88-207-3461-3 (in Italian)
17. Foti, D., Vacca, V.: Rocking of multiblock stone classical columns. WIT Trans. Built Environ. (2017)